Contents

Acknowledgements *6*
Foreword Sir George White Bt *7*
Introduction *8*

1. **Why here? It's all to do with the springs** *11*
1.1. Hotwells: development and decline
1.2. Sion Hill development
1.3. Sion Hill tunnel
1.4. Turkish baths revival
1.5. Revival of a Spa in Clifton
1.6. Plans to build a cliff railway
1.7. George Newnes, MP enters the scene
2. **The People Involved: Investors, Engineers and Master Blasters** *23*
2.1. Sir George Newnes financier
2.2. Sir George Croydon Marks engineer
2.3. Cliff railways built by Newnes and Marks
2.4. Philip Munro architect
2.5. George George gang master and master blaster
2.6 Christopher Albert Hayes
2.7. George White
3. **Construction story Trials and Tribulations** *36*
3.1. Surveying and air shafts
3.2. Rock drills and landmark achievements of tunneling and concrete
3.3. Concrete
3.4. Tunnelling in Clifton area
3.5. Hotwells Bristol Port Railway and Pier 1865-1921 (north of Suspension Bridge)
3.5.1 Clifton Down to Avonmouth
3.5.2 Clifton Rocks Railway 1893-1934 (south of the Suspension Bridge)
3.6 Explosives
3.7 Skewed brick tunnel construction
3.8 Dimensions of the tunnel and brickwork
3.9 Bonding in brickwork
3.9.1 Bricks and brickwork in Clifton Rocks
3.10. Construction starts
3.11. Firing of the first shot, March 1891
3.12. Disaster strikes
3.13. Compensation for the victim's wife
3.14. Job advertisements
3.15. Decision to line the tunnel with brick
3.16. Claim for damages
3.17. Electric signals
3.18. First test journey
4. **Changes in Architecture and Operation** *57*
4.1. Façades
4.2. Interiors
4.3. Employees of the Railway
4.4. Lighting
4.5. The cars
4.6. Operation of the cars
4.7. Water balance principle
4.8. The cables and wheels
4.9. Engines, pumps and workshop equipment
4.10. The brakes
4.10.1 Cable failure
4.10.2 Speed govenor
4.10.3 Deadman's handle
4.11. First journeys
4.12. Opening day announcements
4.13. Signalling and communication
4.14. Plans for opening day and the commemorative medallion
4.15. Opening day reports
4.16. Finances
4.17. Receivership
4.18. Change of façade by BTCC in 1913
4.19. Operating expenditure 1925-27
4.20. Passenger figures and prices
4.21. Other news after the opening
4.21.1. Hotel and railway up for sale
4.21.2. Railway closure and neglect
5. **Other Transport Links** *84*
5.1. Port and Pier Railway
5.2. Trams
5.2.1. Tram operating times and fares
5.3. Paddle steamers
5.4. Rownham Ferry
5.5 Summary
6. **Maintenance** *91*
6.1. Woolley Brothers Engineers
6.2. Woolley's table of railway repairs 1904-1908
6.3. Regular maintenance
6.3.1. Three different specialist engineering insurance companies
6.4. CRR year bankrupt maintenance expenditure
6.5. Maintenance after BTCC takeover
7. **Railway Artefacts** *99*
7.1. Table of Railway artefacts found
7.2. Crossley engines
7.3. Cable fittings
7.4. Brakes
7.5. Top station valve
7.6. Clack valve
7.7. Pulleys
7.8. Telegraphic depth indicator
7.9. Turnstiles
7.10. Ornamental ironwork inside
7.11. Grating
7.12. Folding scissor gates
7.13. Partition
7.14. Pottery
7.15. Beer bottle
7.16. Plaster and marble
7.17. Whistle

7.18.	Ornamental ironwork outside	10.20.	Change of mind – the BBC need to continue to use the tunnel
7.19.	Gas and oil lighting		
7.20.	Electrical fittings	10.21.	Earmarked by the Home Office
7.21.	More things brought back	10.22.	Back to the BBC negotiations
7.22.	What we would like back	**11.**	**Barrage Balloons**
8.	**Railway Memories** *116*		**Operation and Memories** *173*
9.	**Pump Room and Spa**	11.1.	Introduction of the use of barrage balloons during the war
	Development from 1880 *124*		
9.1.	Problems with the plans	11.2.	Balloon Command comes to Clifton and Bristol
9.2.	Opulence, lots of marble, no expense spared	11.3.	Clifton Grand Spa Hotel becomes the main registry for Air Transport Auxiliary
9.3.	Opening preparations		
9.4.	The Pump Room	11.4.	Squadrons embodied
9.5.	Grand opening	11.5.	Balloon locations
9.6.	Use of the Pump Room	11.6.	Wartime telegraph posts
9.7.	Turkish bath and Hydro	11.7.	Permission granted to use Clifton Rocks Railway
9.8.	Quality of Hotwell spring water	11.8.	More Balloon Command activity and the end of the war
9.9.	Proposals		
9.10.	Funding	11.9.	Barrage Balloon artefacts
9.11.	Spa competition from abroad	11.10.	Memories – Barrage Balloons
9.12.	Closure of the Pump Room and progression with the Hydro	11.11.	BOAC memories
		12.	**Air-raid Shelter Operation and Memories** *189*
9.13.	Opening the Hydro	12.1.	Air-raid Precaution operation
9.14.	Licensing issues having opened the Hydro, and other news	12.2.	Clifton Rocks Railway shelter
		12.3.	Shelter artefacts
9.15.	Sale of site and conversion of Pump Room to cinema	12.3.1.	Beverages
		12.3.2.	Cigarettes
9.16.	Failure of the cinema and licensing battles	12.3.3.	Domestic artefacts found
9.17.	Sale of the Hotel and Railway again	12.3.4.	Toys found
9.18.	War breaks out	12.3.5.	Signs
9.19.	Attempts at revitalising the Spa	12.3.6.	Artefact summary
10.	**New Uses for the Tunnel**	12.4.	Memories of people using Clifton Rocks Railway as a shelter
	Wartime Conversion *150*		
10.1.	Ownership of Clifton Rocks Railway	12.4.1.	*Secret Underground Bristol* 2005
10.2.	Air-raid Precautions in Bristol	12.4.2.	Individual memories
10.3.	Ministry of Works	12.5.	Portway tunnel
10.4.	Bombs in Clifton and Hotwells	12.6.	Portway tunnel memories
10.5.	Imperial Airways and ARP seek use of the tunnel. BBC London departments move to Bristol	12.7.	General wartime memories
		13.	**BBC Operation and Memories** *217*
		13.1.	BBC West Region headquarters comes to Bristol
10.6.	Plans for the top station and the tunnel	13.2.	BBC West Region search for a more secure broadcasting station site
10.7.	Plans for the bottom part of the tunnel		
10.8.	Clifton Rocks Railway conversion begins	13.3.	The Home Service and transmitter frequencies
10.9.	The BBC want the use of a tunnel	13.4.	Conversion starts
10.10.	Prolonged correspondence between the BBC, Ministry of Works, BTCC and the town clerk to get a lease and avoid restoring the Railway	13.5.	Generator room
		13.6.	Air-conditioning plant room
		13.7.	Cooking area
		13.8.	Toilet facilities
10.11.	BBC conversion starts	13.9.	Control room (the nerve centre)
10.12.	Still wrangling about the lease	13.10.	Recording facilities
10.13.	Transmitting starts	13.11.	Studio facilities
10.14.	Still more delay with the lease	13.12.	Transmitting room
10.15.	More space needed	13.13.	Aerials
10.16.	Damp wall	13.14.	After the war
10.17.	Drips in BBC section contaminated by sewage?	13.15.	Employment
10.18.	BBC tenancy of the cave	13.16.	Training
10.19.	Closing down the station by the BBC		

The ups and downs of
Clifton Rocks Railway
and the
Clifton Spa

THE DEFINITIVE HISTORY

Maggie Shapland BEM, BSc

Avon Industrial Heritage
Fifty Years of Work 1967-2017

Bristol Industrial Archaeological Society
www.b-i-a-s.org.uk

CLIFTON ROCKS RAILWAY
(Charitable Trust No. 1126999)

First published 2017, by Bristol Industrial Archaeological Society on behalf of the Clifton Rocks Railway Trust
97 Princess Victoria Street, Clifton, Bristol BS8 4DD
www.cliftonrocksrailway.org.uk

ISBN 978-1-908905-05-5
British Library Cataloguing-in-Publication Data
A catalogue record for this book is available from the British Library
All rights reserved. Except for the purpose of review, no part of this book may be reproduced, stored in a retrieval system, or transmitted, in any form or by any means, electronic, mechanical, photocopying, recording or otherwise, without the prior permission of the publishers.

Project production, design and typesetting by Stephen Morris; set in Garamond 11/13
Printed in the Czech Republic via Akcent Media Ltd

13.17.	Play	
13.18.	BBC artefacts	
13.19.	Memories	
13.19.1.	Secret Underground: BBC, What does it feel like now?	
13.19.2.	Individual memories	
14.	**Life after the BBC 1960-2004: Planning Problems and Dereliction** *253*	
14.1.	The stability of the bottom station	
14.2.	Regular break-ins at the bottom station	
14.3.	Deterioration and undergrowth	
14.4.	Contentious planning application	
14.5.	Bristol Junior Chamber feasibility study	
14.6.	Permission to demolish the Railway and Pump Room	
14.7.	Business Plan by Bristol Junior Chamber	
14.8.	Change of ownership and listing	
14.9.	Memories post-war	
15.	**Life from 2004** *266*	
15.1.	My involvement	
15.2.	Restoration highlights	
15.3.	Exciting times in 2005, something new every month	
15.3.1.	Railings	
15.3.2.	Railway track and cable wheels	
15.3.3.	First open day	
15.3.4.	Clearing the rest of the railway lines	
15.3.5.	Retrieving a turnstile	
15.4.	2006: a year of tidying up, media interest and feasibility study	
15.4.1.	Discovering a new tunnel under Sion Hill	
15.5.	Work begins again 2014	
15.5.1.	New railings for the top station	
15.5.2.	New windows for the top station	
15.5.3.	Port and Pier Drinking Fountain	
15.5.4.	Replacement of George Newnes' sign	
15.5.5.	New staircase	
15.5.6.	Replacement of the George White Sign	
15.6.	Number of visitors	
16.	**Conservation vs. Restoration: the Mary Celeste experience** *295*	
16.1.	Introduction to the layers of history	
16.2.	Heritage Value	
16.3.	How to determine future use	
16.3.1.	What does Restoration mean?	
16.3.2.	What does Preservation mean?	
16.3.3.	What does Conservation mean?	
16.4.	Statement of significance	
16.4.1.	Public significance	
16.4.2.	Historical significance	
16.4.3.	Assessment of cultural significance	
16.5.	Survey	
16.6.	Problems of getting a railway running again	

Glossary *303*
Abbreviations *309*
Endnotes *310*
Contributors of memories *312*
Image sources *313*
Index *314*

Acknowledgements

The information in this book has been gathered over twelve years. I would first like to thank all those people who have been so enthusiastic about the Railway over the years to let me use their photographs and family information, suggest leads to more research, tell me their memories and show so much continued interest in the project.

I would like to thank Mike Taylor for being my rock through my ups and downs, and the Railway ups and downs. He is my stalwart soul-mate and is also good at making ironwork and helping me keep my cars going, as well as being a guide. Others have described him as a treasure since he wants to help everyone in their time of need.

I would like to thank my 'boys' Andrew and Michael, who both are Doctors of Archaeology. They have done really well for themselves and I am very proud of them. They have helped me with advice, helped with proof reading, and have encouraged and supported me especially in this last year in my time of need.

Stephen Morris who has laid this book out for me, no mean feat with over 430 images. He has also taken a huge interest in the content and has made many useful suggestions, so we have worked together.

I would like to thank the Rocks Railway refurbishment group for working so well with me. We are an enthusiastic, loyal team who know each others capabilities, support each other, and who have a good skill set between us. Many have worked with me from the beginning. Our success on the project is down to the team, not just one person. No one person could have done it on their own. We have also been able to have a laugh together and learn together. So, I would particularly like to thank them, giving their names in alphabetical order:

Peter Davey for being such a great chairman. He gives so many talks all over the country about the Railway and trams, and has accrued a huge collection of historic photographs. He has taken many groups down the tunnel and can go up the steps faster than me despite being somewhat older. He is always so cheerful and goes down well on open days when he guides people round. He has a wonderful charisma.

Alan Griffiths for helping me on our work days virtually from the beginning, helping on group trips, and helping on open days. He helped me make the car profile that sits on the railway lines so I put his face in one of the windows (along with Peter and mine). He is very good at weeding.

Dave Hewgill (Demolition Dave) who has helped with many of our work days from the beginning and is always the gate controller on open days.

Dominic Hewitt who looks after our website, and books all the people on the individual tours. He also helps on open days and is our trustee publications officer.

June Jeffreys who started painting railings way back in 2005 and has been an open day guide ever since. She has been a blue badge guide for many years. Pete Luckhurst for teaching me about risk assessments, and helping with our work in our first year. Donna Luckhurst for doing all the original design work and creating our first information boards, and organising the work to begin with.

Sasha Lubetkin who is always so cheerful, has helped on our open days, and done an amazing job with proof reading. She is a near neighbour who I have known many years and has been incredibly supportive this last year, and supplied me with many foods containing ginger to keep me going. Michael Nelki who was a GP, has helped on many work days and on open days, and was also a proof reader. He has also been supportive when I needed help.

Jon Picken who has been with us since 2015, and has 15 years' expertise in running festivals managing stewards and volunteer teams at large-scale, outdoor, music festivals. He finds helping on our open days very enjoyable and a total change, and his expertise is useful to us too. He has been quick to learn and willingly did my rock plan for me (plate 3.28). He also loves taking groups down the tunnel.

Marion Reid for helping me on some of our work days, being a tail ender on group trips, helping on open days and being a thorough, inquiring proof reader.

Ed Scammell for helping me on our work days virtually from the beginning, being our diligent treasurer, and helping on open days. He is one of the 'Power Workers'. He was really good at finding treasures under the ledges with his eagle eye. Tom Scammell (brother of Ed) and his wife Steph for helping me on our work days virtually from the beginning, being our secretary, and helping on open days. Tom was our other Power Worker. This year they helped me do some research in the library – going through Wrights directories looking for various names and making useful comments on the text.

Mike and Dave Steadman for setting up the first website and creating a video of Peter talking about the Railway. Sue Stops and Pauline Barnes who have been with us from the beginning and helped 'woman' the back office on all our open days, and providing cake for the volunteers. Dave Strawford who has turned up to all our open days to be a guide. He has also helped on some of our major projects such as removing a turnstile as well as installing the replica Sir George White sign. James Tonkin who is the chairman of the Trust and a skilled joiner. He started off the work at the Railway, and made windows for us, as well as being a negotiator in times of need due to his experience of being a Councillor.

Nicola Williams, one of our volunteers, who is a keen photographer and belongs to the Bristol Photographic Society, took them down the tunnel with me on a photo-shoot. As a result, she created a fantastic exhibition of photographs that we have been able to use in various locations, including the Records Office and in the Avon Gorge Hotel.

Thank you to Mr Robert Peel, the original owner of the Avon Gorge Hotel who let us have full access to the tunnel to see what we could make of it. Later owners have continued to see us as a community project, and continued to let us use their facilities, and show the Railway to visitors. Thank you to Rachel Schofield, Simon Pitcher and Tim Rew who helped me with the Junior Chamber of Commerce information, to Patrick Handscombe for letting us use his letter and photographs from Gerald Daly, the Engineer in Chief of the BBC in Bristol, and Richard Hope-Hawkins for letting us use his research on the Railway history. I would also like to thank Bristol Records Office who have been so helpful to us and have such a wonderful archive – which gets bigger every time I go down to see them, Caversham for letting me look at their BBC archive, and the National Archives at Kew who also have an amazing archive. Bristol Central library have also been very helpful and it is great to be able to see the local newspapers from the 1800s and the directories. Their staff are very helpful too. The Bristol Museum and Art Gallery also has an amazing collection of local paintings and have let me use a couple of their images.

Thank you to Clifton village Co-op for providing some funding from the Co-op Community Fund, to Dorothea Restorations who helped us make the top station signs and Mike Brett of Bristol Foundry who did all our castings. Thank you to the Haematology and Oncology Department at the BRI for the treatment they gave me and their diligence and care. And thank you to my neighbours who help me on open days, and have offered me much personal support, particularly this last year. And thank you for the public for continuing to support us. It was gratifying to see the huge queues at our last Doors Open Day. Without your donations, buying merchandise, and wanting to do trips down the tunnel we would not have had the funding to do what we have done. Your comments on everything have been very useful, and your applause on open days have spurred us on to continue.

Foreword

In 1880 my ancestor, George White, proposed to the Society of Merchant Venturers, that he should build a funicular railway on the rock-face of their part of the Clifton Gorge. Had he done so it would undoubtedly have morphed from being a practical nineteenth-century mode of transport into one of Bristol's best-loved twenty-first century tourist attractions. The views from its cars would have been incomparable, the trip from the river to Brunel's Suspension Bridge, breath-taking. As it was the Merchants turned him down and it fell to George Newnes fourteen years later to take up the challenge and build a railway inside the rock, rather than on it.

This Herculean task was accomplished at great expense and great risk to the workforce. But like so many major engineering projects, the huge initial cost made the original company unviable and George White was able to take it over, on behalf of his Bristol Tramway and Carriage Company. The BTCC ran the railway successfully for many years.

By 1934 road transport had improved access to Clifton to such an extent that the Victorian funicular had ceased to have any useful function. Having served its purpose honourably, the steeply sloping tunnel was effectively locked up and abandoned. After a brief respite during World War Two, when it served as what must have been one of the most remarkable (and secret) broadcasting studios ever constructed, and one of the most awkward air-raid shelters ever contemplated, it was closed again. What remained of this exceptional example of nineteenth-century engineering and its unique mid-twentieth-century broadcasting station, was left to slumber in the cliff, trussed and bound by complications of ownership and restrictive covenants.

It took the trustees, volunteers and friends of the Clifton Rocks Railway Company, a twenty-first-century charity, not only to appreciate the educational potential of this hidden gem, but through immensely hard physical work, to turn the concealed railway tunnel into the tourist attraction the cliff-face railway might have been. This carefully researched book by Maggie Shapland, one of the most tenacious of the Clifton Rocks Railway campaigners, is part of this long and challenging process. It is very much to be hoped that Maggie's scholarship will lead directly to the preservation of the old tunnel in perpetuity, as it most certainly should.

Sir George White Bt

Introduction

Maggie Shapland vice-chairman, and restoration officer of the Clifton Rocks Railway Refurbishment Group and secretary of the Clifton Rocks Railway Trust

I did my degree in Maths and Computing at the Bristol College of Science and Technology (now the University of Bath) between 1965 and 1969. I used to go dancing in the Pump Room of the Grand Spa Hotel. I moved back to Bristol in 1972, and have lived in Clifton since 1978, when we bought a derelict garage to live in and keep the cars in one place. We spent three years turning a site with hardly any floors and an asbestos roof into a home, only with the help of a bricklayer, a plasterer and my father.

I have always been interested in history from when I was small, and how things worked, and was delighted to get involved with the Clifton Rocks Railway refurbishment in 2005 at a time in my life when my two sons (who both have Doctorates in Archaeology) had left home. I lived near the site and had always wondered what was behind the hoardings. A lot of Clifton was rather derelict in the 1970s and 1980s, and the fact that this was derelict too was just a sad fact of life, I did not ever think I could do anything about.

The history of the railway has led to unexpected discoveries and research. It is a story spread over 530 years, which befits such a large site. I will be telling the story interspersed with oral histories, research, and artefacts the volunteers have found during their work since 2005. Visitors love receiving a history lesson, social history, engineering history – all mixed into one. Their enthusiasm has helped our enthusiasm. They also have helped us understand the story. It is not just a tunnel, it is a tunnel that has had two distinct roles to play, courted controversy when it was suggested that a lift should be built in the 1880s and even now courts controversy as to whether it should run as a railway again, despite it going bankrupt twice and having war time conversions built on top of the tracks.

I researched artefacts and tried to understand the jigsaw of what was built and when and how and why. Even now, I am still I am still wondering. I am lucky to have had the internet to help look up old newspapers, and was lucky to have the resources of Bristol Records Office, the National Archives at Kew, and the BBC archives at Caversham. Many experts in their fields suggested more things which needed research, and so the story got bigger and my files grew larger and larger. I retired from the University of Bristol Computer Centre in 2013 (having working in that department for over 40 years) so then had a little more time (but not much since I was involved in several local societies, and still enjoying driving my vintage cars).

It was not until I saw the deeds of my garage last year that I realised that it had been bought by the hotel in 1905, and in 1929 it mentions a mortgage charge of £10,000 had been taken out for Clifton Rocks Railway when the Hotel and Railway was then sold to the Grand Hotel in Broad Street, central Bristol. When the Hotel was requisitioned during the war by the Barrage Balloon section, my garage was requisitioned too, to keep wartime vehicles in, presumably belonging to the Barrage Balloon squadron. Under the layers of paint on the doors, we found the original lettering from when it was a garage (and states the keys are at the 'Spa'). This has all been lovingly restored. So I have a lot of connections with the Railway.

I had threatened to write a book for several years, but it was not until October 2016, when I was unexpectedly diagnosed with advanced pancreatic cancer with only a few months to live, that I started. After the initial shock, I wanted to leave a legacy, so I immediately started to write a definitive book about the railway – a massive story. I was able to carry on doing tunnel trips to keep fit, and proof read sections of the book when in hospital all day every other week. In December, Mike Taylor (who has tirelessly worked by the side of me ever since we started on the project) and I got married, and after the reception at the Avon Gorge Hotel we took visitors a trip down the tunnel. Where else? It was a lovely cheerful wedding and friends and neighbours were so kind and supportive.

In February I was told my cancer was stable, and again in May when I finished my treatment. I was told this was because of my positive attitude, being so fit (by going up and down the Railway steps so often), and being focussed on writing this book. I was amazed I could survive the open days, which are very challenging, without stopping. So I am still here a year later, much to everyone's surprise, mine especially, and I completed the contents in September. The book has enabled me to gather all my research together, reminded me of things I had forgotten, and made me ask myself more questions when describing the huge photographic archive. I have also taken the opportunity to add the kind of information I keep on getting asked when I do the tunnel trips, and what I point out to visitors.

So the work has been a massive voyage of discovery, both for myself as well as for the Railway. I have had my ups and downs, just like Railway, but we have got there in the end.

Do have a look at www.cliftonrocksrailway.org.uk for more information about this amazing, fascinating site. There is a facebook page too. If you have any information about the railway, or your memories, or anything I have left out of this book, then please contact me at 97 Princess Victoria Street, Clifton, Bristol BS8 4DD, or email me at:

Maggie.Shapland@gmail.com.

I intend to be around for many years to come.

Maggie Shapland Bristol, December 2017

Standing by my trusty 1924 Lanchester at the Merchants' Hall on Clifton Down, looking clean and tidy for a change, having just received the British Empire Medal for services to Clifton in 2012. Lord Lieutenant Mary Prior on the right, presented the medal to me. The badges I was wearing were a 'Clifton Rocks Railway' badge and a 'History Matters' badge.

1.1. This aerial photograph shows the steepness of the Avon Gorge. Anderson Noted are the locations of the Railway, Colonnade, Avon Gorge Hotel and other key sites. The location of Clifton Suspension Bridge is indicated. (Bob Steer)

one

Why here?

It's all to do with the springs

This chapter tells the story from 1480 to 1890, to explain why the Railway was built here, and the setting. All the passer-by sees today is the restored top station railings, ornate ironwork, and a bricked-up tunnel at the bottom of Sion Hill, and wonders why here? If they venture down the zig-zag path to the bottom, or drive past (they can't stop), they find a derelict façade with a narrow pavement by the side of the very busy Hotwell Road (the A4 which leads to the Portway), and a derelict landing stage by the river (which they are unlikely to connect with the Railway). The chapter also describes the involvement of the Merchant Venturers, and attempts to build a cliff railway. Minutes of the Merchant Venturers and contemporary newspapers have been quoted where possible.

The Rocks Railway is an important part of the Clifton and Hotwells conservation area which has a unique character influenced by local topography and geology. It is located in the west of Bristol, east of the Hotwell Road and Avon Gorge, north of the Floating Harbour, and west of the city centre.

When the Clifton and Hotwells conservation area was first designated in 1972, the area was classified as being of outstanding interest. The steep escarpments at the bottom of Sion Hill, where the top station of Clifton Rocks Railway is located, provide spectacular views towards the Suspension Bridge and the Avon Gorge. The Avon Gorge Hotel (until the 1980s it was known as the Grand Spa Hotel) terrace has the best views in Bristol for a commercial property. More details about the conservation area and its history can be found in the Character Appraisal[1] published by Bristol City Council and can be viewed at their website.

1.1 Hotwells: development and decline

The story starts at the bottom of the gorge and the development of Hotwells followed by the development of Clifton at the top of the hill from 1746. The earliest mention of the Hotwell is in William Wyrcestre's *Itinerary* in 1480, (William was a fifteenth-century chronicler), but he says nothing about the medicinal qualities of its water. Latimer[2] tells us that the first record of its healing properties was in 1628, as a cure for leprosy and scurvy; but it afterwards grew to be regarded as a sort of panacea for consumption, diabetes and other diseases, so Hotwell water started to attract invalids and the fashionable, including royalty. In 1662, a carriage road was cut out of the rock along the riverside as there was no easy approach by land.

In 1676, the Society of Merchant Venturers purchased the Manor of Clifton, and hence owned the Bristol side of the Avon Gorge. In 1695, the

> Hotwell was leased by the Hall [executive body for the Venturers] for 90 years at £5 p.a. It enjoyed enormous popularity for much of this time, but the Society was not in a position to make money out of it. This must have been very irritating for the Hall.[3] The lessee covenanting to expend £500 in building a pump room.[4]

But St Vincent's spring – 1.5km north of the Hot Well, and first mentioned in 1703[5] – was a failure due to its inaccessibility and expensive rent.

The Grade II* Listed Colonnade in Hotwell Road is all that remains of the once fashionable Spa. It was built in 1786, by Samuel Powell, to meet the need for a sheltered walk. It was designed with shops below and living quarters above. It has a gently curving brick front of 13 bays with shopfronts beneath a deep Tuscan colonnade supported by pillars.

From the 1790s, Hotwell spa declined in popularity because it was expensive, attracted less desirable people, and because the spring was polluted – the Avon was an open sewer (though 32,000 bottles of water from the Hotwell had been sold by 1795). Seaside resorts became the new fashion.

There was then the slump, in 1815, following the Napoleonic Wars. The Society of Merchant Venturers

1.2. The old Hotwell House was built in 1696 and the spa enjoyed success for 100 years. The Colonnade is to the right. (Samuel Jackson, BMAG M923)

1.3. Matthew's map of 1825 showing the road did not go beyond Hotwell House.[6] (BRO)

decided to demolish the Old Hotwell House in 1822 to build a road (creating Bridge Valley Road to wind round the side of the rocks) from Hotwells to Clifton Down and partially plant trees in the extensive commons. There was a free tap in a back yard of the pump room which was stopped in 1831 but reopened in 1837 after a pressure group threatened a law suit.

Initially, Granby Hill provided the primary route between the Hotwells and Clifton. A steep and treacherous flight of steps also offered a link between Clifton and the Spa from the Colonnade to the bottom of Sion Hill, although this route was improved with the construction of the 'zig-zag walk' in 1828 from north of the Hot Well.[7] Projecting rocks had already been removed from the footpath in 1816 and 1817. A succession of bad harvests had led to high food prices, and so much suffering from the poor that a subscription was raised to employ labourers to quarry and break stones.

4. Before the old Hotwell House was demolished, a smaller pump-room (the Hotwell House) was built behind it to avoid disruption to the spring industry. Demolition was well under way by October 1821. This painting by Samuel Jackson shows both houses clearly and the promontory that the old house is standing on. In 1840, it was written 'These waters issue from an aperture in the rock, about ten feet above low water mark; their mean temperature is about 71° fahrenheit; they contain a portion of sulphuric acid, but are peculiarly soft and pleasant to the taste, and free from any offensive smell. A new pump-room, with hot and cold baths, and containing also apartments for the residence of invalids, a neat building of the Tuscan order, has been erected at an expense of £8000, by the Society of Merchants of Bristol, who are lords of the manor of Clifton.'[8]
(BMAG, M969)

1.5. This photograph clearly shows the location of the new Spa in relationship to the Colonnade. Look at plate 1.1 to see where the bottom station of the Railway is in relationship to the Colonnade. The Railway is sited further back than the spa. (BRO)

1.6. The Royal Clifton Spa on the Hotwell Road
In 1851, the Spa was granted royal status and became known as Royal Clifton Spa (even though it was in Hotwells at the bottom of the hill) and James Bolton the proprietor made additions and improvements:[9] 'With the view that the salutary qualities of this water should be more generally known, and that, for the good of the community, its former celebrity should be revived.' It now offered ten warm baths made of marble, porcelain or metal, as well as vapour, shower, medicated baths, douche baths, and a tepid bath. It had a new entrance and refreshment room that could be hired for balls and concerts. But in 1855 the lease was available. On 29 March 1856, an advert: 'Respectable Female Wanted, as BATH ATTENDANT, and to make herself generally useful'. On 29 November 1856, James Bolton states 'The APARTMENTS at the ROYAL CLIFTON SPA are now VACANT. They are particularly suitable for invalids requiring baths of any description or any person wishing for a warm, cheerful and healthy residence during the winter. The Spa water is recommended in consumption, indigestion, and general debility; and in some diseases of the kidneys, particularly diabetes, it is considered a specific.'
In March 1857 it is still advertising for respectable bath attendants over 25 years old. Bolton seems to have been here from 1854 to 1864 by which time he had stopped selling Hotwell water.

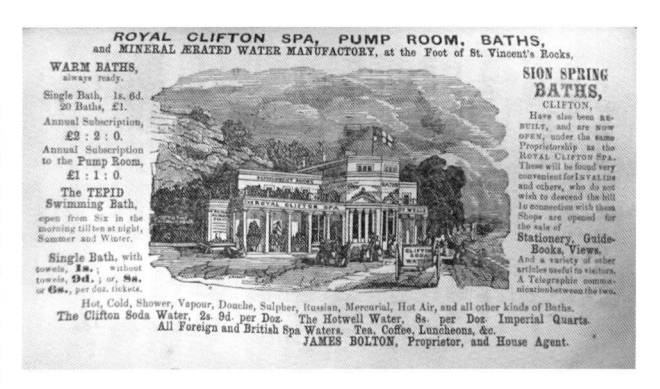

In 1858, a landing stage (constructed by Messrs Stothert & Co) was located near the Hotwell[10] for passengers of the Cardiff and Newport steamers, and in 1859 it was reported that another was needed to cope with more than one boat at a time (one for upstream traffic and one for downstream traffic) – thus lessening the risk of boats running aground while waiting for embarkation. Clearly steamer trade was buoyant.

Bristol was losing shipping trade to Liverpool owing

1.7. 1828 The Old Hotwell House has gone, the new House is shown. There is now a road leading to the Downs on the promontory. Two zig-zag footpaths are shown. The earlier zig-zag leads from the Colonnade. This was little more than a track and difficult in wet weather. The later zig-zag, north of Hotwell House, was used by visitors to the Spa in Hotwells and is still there. There is a stable across the road from Sion House where the Railway would be built. Note the circular mark in the middle of the Sion House garden – perhaps a horse gin used to lift water. (Bristol City Council, knowyourplace)

1.8. Hotwell Road grotto north of bottom station. The grotto is still there, but the fountain was removed to Underfall Yard during WWII, and has subsequently vanished. (Facebook, Bristol Then and Now, posted by Elle Tomlinson)

to the difficulties of navigating the Avon Gorge, leading to many vessels being grounded at this part of the river. The Avon Gorge has the second highest tide range in the world and is very narrow at this point. In 1865, Bristol Corporation, at the insistence of the Docks Committee, decided that the time had come to build a new and larger entrance lock at the City Docks, north of Brunel's short-lived lock (which had replaced Jessop's lock) designed for Bristol Dock Company in 1844. In 1866, the Society of Merchant Venturers sold Hotwell Point, Hotwell House and Hotwell Spring to the Docks Committee with a lot of other property in the Cumberland Basin area. The buildings were then demolished to straighten the approach to the 1865 lock, which was completed in 1873, and is the one in use today.

So, despite all James Bolton's efforts and many public protests, since the Spa had been one of the area's chief sources of early wealth, the Royal Clifton Spa was demolished in 1867. Hotwell Point was then removed in order to make the Gorge wider.

The Merchant Venturers retained the rights to all wells, springs etc. on the land and the right to sell the water, including from the Hotwell. Following public outcry, the Merchant Venturers requested the Docks Committee build a grotto in 1877, which contained an ornate cast iron pump to continue to make the water available until 1913 when it was closed owing to pollution.

On 11 February 1884, there is a sad article about 'Grandeur in decay at the Hotwells' and

> how rank and fashion have long since deserted Hotwells, and taken themselves to the heights above, where noble crescents and stately homes adorn the fashionable suburb of Clifton, The Hotwells Spa and Pump Room have disappeared, and in the spacious chambers of once luxurious mansions, where votaries of fashion have toiled and

dallied their fortunes away, poverty stricken people are now receiving parish relief.[11]

Looking at the area now it is clear that the bottom station of the Railway was built where the Hotwell House had been because it is next to the Colonnade (see figure 1.1). The adjacent grotto is still to be seen, but the pump was removed during WWII (see chapter 13). At low tide, it is said that the spring can be seen still flowing; the reality is it is probably sewage. So, a sad ending to the Spa in Hotwells.

1.2 Sion Hill development

Regarding development at the top of the hill, Clifton in 1746 was a small farming settlement with a dozen houses. Clifton derives its name from:

> its romantic situation on the acclivities and summit of a precipitous cliff, apparently separated by some convulsion of nature from a chain of rocks on the Somersetshire coast. The River Avon, which at spring tides rises to the height of 46 feet, and is then navigable for ships of very large burthen, flows with a rapid current through this natural chasm, forming the southern and western boundaries of the parish, and dividing the counties of Gloucester and Somerset.[1]

The wealthy merchants began to live at the top of the hill to get away from all the smog of the heavy industries in the centre – coal, iron, glass, brass, etc. – a handful of merchants' mansions appeared such as the Goldney estate. Investors turned to Clifton from Hotwells and began a building boom. In 1840, the population had risen to 14,177 inhabitants:

> now of piles of stately edifices of Bath stone, forming, from the beauty of their architecture, a conspicuous and imposing feature in the landscape for many miles.[13]

A shaft was drilled at the bottom of the Sion Hill to tap the Sion Spring in 1796[14] by

> Mr Morgan, an attorney of Bristol, who, being about to build a house [Sion House] on the hill above the Hotwell, determined to obtain water, if possible, on his premises. With this view, the miners dug and blew up the rocks, till they came to the depth of two hundred and forty-six feet, before they accomplished their object, when, all of a sudden, a stream gushed in upon them so impetuously, that they had great difficulty in escaping the inundation. Mr Morgan, on discovering that this water had the same properties as that of the lower house [Hotwell House], erected an engine for raising it daily, built a spacious pump room, and prepared bathing places adjoining. When taken from the pump, it raises the thermometer to seventy three degrees, though drawn from so great a depth. Mr Aitkins' Reading Room and Public Library, at Sion Spring House [adjacent to Sion House], comprehend a valuable collection of modern publications; and new works of merit are immediately added, to suit the various demands of his subscribers; he also sells perfumery of all kinds.

A topographical description of the springs in 1840 states:

1.9. 1833 map: the red lines show the pipework from the Sion Spring, the blue lines the pipework installed by the Merchant Venturers to the mill across the Gorge. (BRO 40662/1)

Sion House and Spring House are across the road from the Railway. During 1811-13, water pipes were laid from Sion Spring to most of the houses in the neighbourhood, which had previously been supplied by water cart. The map shows the location of the baths (which are quite small) on Sion Hill and Hotwell House in Hotwell Road. Hotwell Point was removed later. Other springs such as the Richmond Spring and Buckingham Spring in Gordon Road were drilled in 1820.

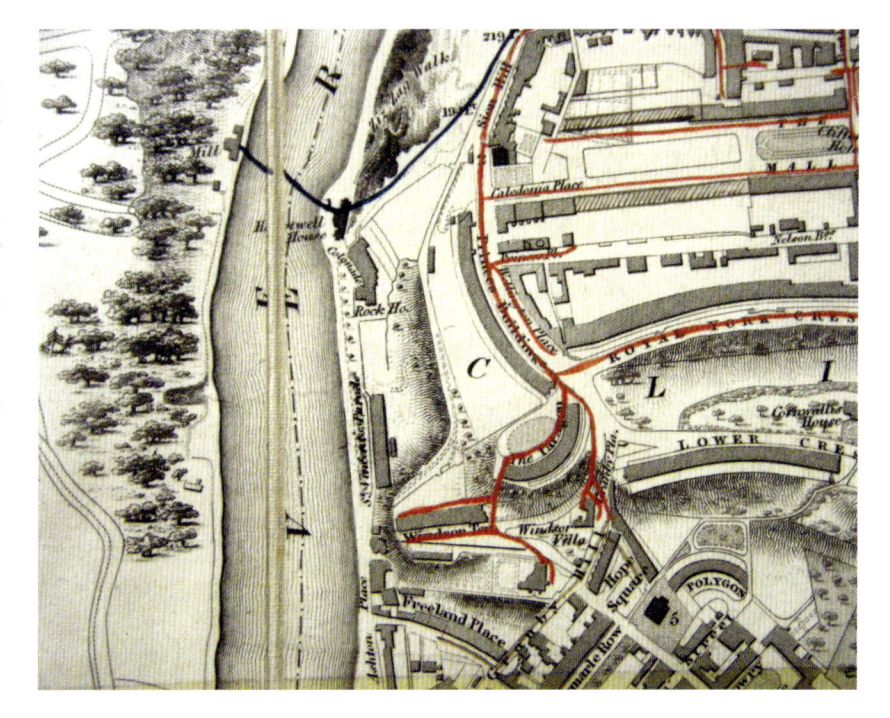

This portion, like the Hot Wells, owes its origin and rapid increase to the efficacy of a similar spring issuing from the rock into a well 320 feet in depth, sunk at an immense expense, in 1772, and from which 30,000 gallons of water are daily raised by a powerful steam-engine, and afterwards propelled to an additional height of 120 feet, and distributed through pipes to most of the respectable houses on the hill. There are some splendid ranges of buildings, and handsome hotels, with every requisite accommodation, and commanding beautiful and extensive views.

On 11 June 1814, Schweppes in Wine Street 'established a manufactory of genuine Soda water at the upper of Sion Spring, Clifton.'[15]

The spring could yield 33,560 gallons per day and served 304 dwellings in 1845. The deeds of Caledonia Place and West Mall explicitly state that owners can not bore their own springs, but must buy their water from the Clifton Water Company for four guineas a year (as well as not keep pigs in the Mall Gardens or keep a house of ill-repute, or a noisy inn, etc.). Residents who did not have cisterns could buy their water for one penny a bucket. No one knows where the pumping engine was – it could have been in the Mall Gardens or it could have been where the top station of Clifton Rocks Railway is now. The spring was then purchased by the Bristol Water Company in 1846 along with other springs in the area so that all of Bristol could get clean drinking water (William Budd who lived in Lansdowne Place had associated cholera with dirty water) and so Mr Coates received £13,500 for compensation of loss of revenue and laying the pipes.[16] The Merchants' Society received £18,000 for the riverside springs, machinery and plant.

By 1841, Sion Spring House was just a lodging house and in 1844 it was a lodging house and livery stables. Mr Aitkins' Reading Room and Public Library was removed in 1850.

'We are glad' announced by Mr Bolton (proprietor of the Royal Clifton Spa): 'to find that the Sion Spring Baths are at last finished':

> They will be found very convenient for invalids, and others, who do not wish to descend the hill. Shops are opened in connexion with them for the sale of Stationery, Guide Books, Views, Minerals, and a variety of other articles useful for visitors. A telegraphic communication exists between the two establishments, so that orders for the Hotwells can be left at Clifton, or vice versa.[17]

He seemed to have stopped trading in 1858 when it became a boarding house. Henry Pearce then ran the baths from 1863 until 1870 when St Vincent's Rocks Hotel bought Sion House. In 1871, it became known as St Vincent's Rocks Hotel and Baths when Sion Spring House changed hands. The Hotel ran the baths until 1877. St Vincent's Rocks Hotel was converted to flats in 2002. Being across the road from the Grand Spa Hotel, the proprietor saw them as a competitor when the Grand Spa Hotel was opened.

1.3 Sion Hill tunnel

We know that before the railway was built there were stables on the site. When we started working in the Railway in 2005, we cut a hole in a 14-inch-thick, wartime wall that was filling an alcove in a corridor between the top station and the hotel ballroom. We were amazed to find a 50-ft long, steeply sloped tunnel underneath Sion Hill leading to the entrance of the garden of Spring House before veering sharp left up a slope to a chamber accessible from a manhole cover in the garden. The tunnel had been built in the eighteenth century so was absolutely nothing to do with the Railway. Railway artefacts, however, had been left at the front of the tunnel. It terminates with a left turn and the start of a set of steps with a large concrete slab overhead to block access. There is also a possible junction to the south. The walls are built of rubble

1.10. The Baths in 1833. Spring House is to the right of the Baths. The house on the left with stairs in the middle is Sion House. These buildings became the St Vincent's Rocks Hotel. The entrance of Clifton Rocks Railway is opposite Spring House. (Author)

with brickwork on the arches and corners. The bricks are probably handmade. We also know that white mortar was used in the brickwork rather than grey. White mortar was made with lime at the time Sion Hill and Caledonia Place were being built. We assume it was built when Sion Spring had been bored. The thickness of the walls is unknown, but the tunnel is 52in wide and 69in high. The roof is about 13ft below the pavement at the Railway side and of unknown depth at the other end (due to gradient). The tunnel was blocked off at the other end at an unknown time. There are paving slabs and slate slabs stacked up so the tunnel was presumably used as a storage area. It is possible there was a tunnel under St Vincent's Priory (built 1810) and a deep shaft that could have been a well. There is a junction on the right of the tunnel, which could lead to the Priory, or it may just be an alcove.

This tunnel was not a water culvert as it was not designed to hold water; it is more likely for overflow water. It was built for a specific purpose as it goes from only one house rather than under the pavement. Had it been a culvert, it would have had to be blocked off when the Railway was built otherwise one would have water flooding the passageway. In the early 1800s, culverts were constructed to convey surface water from the streets, and sewage through ash pits. Sewers were laid in Clifton between 1855 and 1857 (Hotwells from 1858 to 1859) to discharge into the Avon. Similarly, it could have been an overflow tunnel for the prolific spring, but again it would have had to be blocked off when the railway was built, and the water from the spring could not have been used to top up the tanks under the railway carriages.

In 1922, we know one of the borings to the spring was 70ft deep with a diameter of 5in and was opposite the entrance of the railway. There was another boring at an unknown location nearby.[18] This was presumably the boring in the Hotel Pump Room for the fountain.

1.4 Turkish baths revival

In 1856, after a break of over fourteen hundred years, the hot dry-air bath was re-introduced into the British Isles. On 3 January 1861, we are informed that the introduction of the Turkish bath into Bristol, by Bartholomew, has been a decided success[19] (at Brunel House College Green). There will be more of Turkish baths in chapter 9, as the spa is revived in Clifton.

1.5 Revival of a Spa in Clifton

A final effort to re-establish a spa began to take place in 1885. On 6 March 'we are informed that a company has been formed and registered, viz. The Clifton Spa

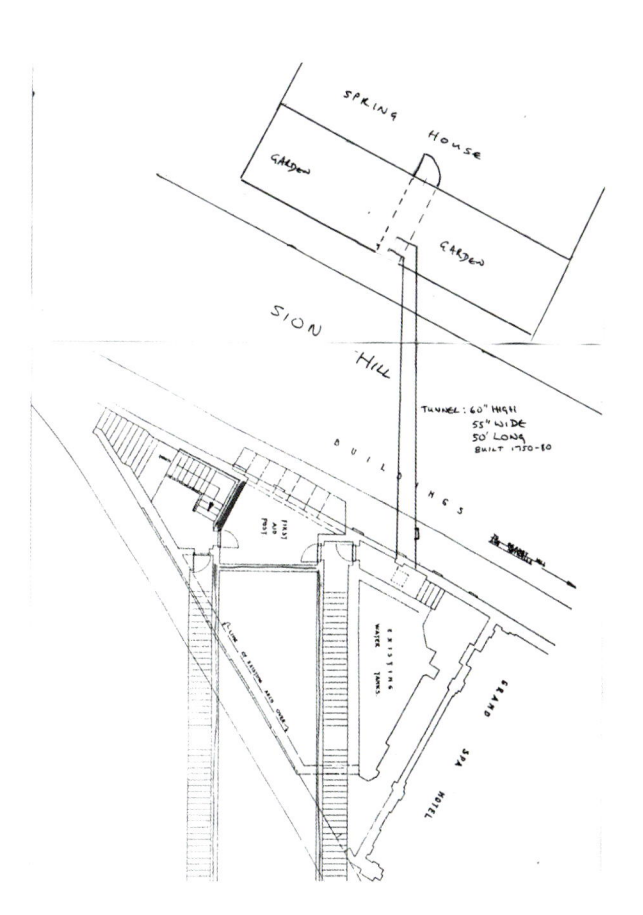

1.11. Map of the Sion Hill tunnel. (Author)

1.12. Entrance to Sion Hill Tunnel from near the top station. Pennant flags had been lifted on the left. Railway artefacts had been left at the front of the tunnel. (Author)

and Pump-room Company (Limited)[20] capital £30,000 in £5 shares; and on 9 April 1885 the *Western Daily Press* (*WDP*) reported:

> visitors to Clifton have frequently complained that they have no opportunity of meeting each other in any assembly room, pump room, or other institution which is open to all and is generally frequented. In this respect Clifton differs from nearly every important watering place, and the result, there can be no doubt is disadvantageous to local interests generally. We mentioned some time ago that a proposal was under consideration for rendering our fashionable suburb more interesting to visitors, and the scheme has since been so far matured as to be set forth in a prospectus which we publish today. A company has been formed for the purpose of adding to the attractions of Clifton by the revival of the Clifton (Hotwells) Spa, the provision of a pump-room, and other accessories, and a glance at the prospectus shows that the undertaking being promoted under very favourable auspices. The Clifton Spa and Pump Company proposes to utilise the Hotwell waters, and for this purpose they have secured the option of purchase of the valuable mansion and the adjoining house [14 and 15] at the end of Prince's Buildings, Clifton, both of which stand on the rocks immediately over the spring, while the grounds of the mansion extend to within twenty yards of its source. The company now propose to provide a pump-room in connection with the mansion in Prince's Buildings, which shall also serve as a reading and writing room for members and visitors, and in which there shall be a fountain of the mineral spring. The adjacent grounds are to be laid out in terraces and walks; a band will play on certain days in front of the pump-room; and the scheme also includes the provision of baths of every description recommended by the Medical Council, which as will be seen, includes the names of several of the leading medical man of Bristol and Clifton. Attention is called in the prospectus to the advantageous position of the property and to the magnificent scenery commanded from the hill, and consideration is also invited to the claims of Clifton as a health resort, it being pointed out that the death rate of Clifton is 25 per cent. below the average of 28 watering places. The directors of the company are Sir Philip Miles, Bart., M.P.; Captain George Bridges, Mr E. Kenyon-Stow, Mr R. W. Butterworth, Mr A. Lyndhurst Deedes, and Mr C.H. Low. and the capital of the company is to be £25,000, divided into 10,000 shares of £2 10s each. The completion of the Spa and pump-room would revive public interest in the hot spring which formerly had so great reputation, and would also provide for visitors a pleasant meeting ground that would be so much appreciated. *The Bristol Mercury* considers that from a financial point of view similar enterprises have flourished, the most recent instance of this being the giant strides towards popularity made by Bath.
>
> The mineral spring is under the control of the Corporation of Bristol. The money for the property, including the mansion, 14 Princes Buildings, the land, buildings, fixtures, fittings and Appurtenances, together with the option to purchase over the adjoining property has been fixed at £7,200.

On 10 April 1885, *Cliftoniensis* writing in the *WDP* was enthusiastic about the idea but wished the quality of the mineral spring to be analysed. The Medical Health Officer reassured him that it had been and remained satisfactory.

On 16 April 1885, a private meeting was held. Mr Munro (later to become involved with the Newnes' scheme, see chapter 2) confirmed the spring was recovered in 1876. The existing spring was under the bed of the river, almost in the centre of the channel. There was a plan to drive a tunnel from the grotto to the Sion Spring (which was nearly under St Vincent's Rocks Hotel). The Sion Spring belonged to the Bristol Water Works Company so the plan was abandoned because there was doubt about obtaining an ample supply of water. There would be no issue in getting the temperature from 80 to 90 degrees. There were also queries as to whether enough money was being raised.

On 14 May 1885, a public notice was made to 'Architects, Engineers and Surveyors': 'PLANS are invited by the Clifton Spa and Pump Room (Limited), for the erecting of a PUMP ROOM AND MINERAL BATHS, including Layering-out of the Grounds. Particulars with Plan.' They offered a premium of twenty guineas for any plans they may select. On 13 June the *WDP* announced that:

> the Clifton Spa Company has met with favourable support with regard the taking of its shares, and that allotment will probably take place at an early date. This would have taken place before now but for the agreement which was entered into by the directors to the effect that no allotment should take place until a requisite number of shares had been applied for to place the company on a sound basis. This protects the first applicants from risk. The architects plans are to be sent in on 1 July, and a number of these gentlemen are competing to carry out the work.

On 17 July 1885, the clerk of the Merchant Venturers reported that by a deed of 1808 the Society was under a perpetual covenant not to build on the side garden of no. 15 Princes Buildings. He was instructed to so inform the Clifton Spa Company. This is where the Pump Room and Clifton Rocks Railway would be built in a few years time.

On 30 October 1885, the clerk of the Merchant Venturers was directed to inform the Clifton Spa Company that their plans could not be considered until the owners of houses in Caledonia Place had consented to waive the covenants as to buildings.

On 15 October 1885, the *WDP* reported that Mr W.V. Gough won on grounds of:

> superiority of design and arrangement of plan: a pump room of similar size to Bath with windows with large balconies along the entire frontage towards the river and ornamental interior, marble pillars, plaster ceiling, alcoves for mineral water fountain supplied direct from the spring, a stage for entertainments; the house to be turned into a boarding house for patients; a storey under the pump room for bathing arrangements approached by a handsome staircase; ladies' and gentlemen's retiring rooms; billiard, smoking and reading rooms all lit by incandescent electric light. Terraces extending down the hill with fountains, tennis lawns and band kiosk for music.

However, on 28 June 1886, in the *WDP* Dr John Macpherson considered that all that was needed was the Observatory being converted into a pump room.

> It only needs the expenditure of half a dredger by the Corporation of Bristol to restore the past repute of the spa, and improve the future prospects of Clifton.

On 25 March 1887 an application of the Clifton Spa Company for the approval of their plans on condition of their getting rid of the restrictive building covenants was declined and it was resolved to maintain those covenants.

On 27 May the Committee of the Merchant Venturers, having viewed the poles and ropes set up by the Clifton Spa Company to show the lines of their proposed building, determined to recommend that the 'restrictive building covenants should be released' and the Company's plans approved, the Company first raising the required capital and guaranteeing to complete the work within a prescribed time, and not to use the building for any other purposes than are permitted at the Bath Pump Room, 'or to obtain a licence for the sale of liquors'. It is not recorded why they changed their mind only two months later.

Not until 9 July 1887 was there a letter confirming that the Merchant Venturers were now prepared to sanction the building (having not given it before). A committee meeting was held on 13 August 1887 confirming this and all that remained was to find the necessary capital. The Bath medical officer then accused the Clifton officer of doctoring the death-rate figures.

On 17 April 1888, according to the *WDP*, the Clifton Spa Company, after overcoming many difficulties, was now in a position to carry out works, and the directors would proceed forthwith. However, on 17 October 1888, the *London Gazette* states the Clifton Spa and Pump Room Company Ltd, was wound up voluntarily. This is a different company from Newnes' company which started buying up property in 1890.

1.6 Plans to build a cliff railway

In 1879, tram routes were sanctioned to operate from St Augustines Parade to the Port Railway Station at Hotwells, but proposals to operate trams from the Victoria Rooms, at the junction of Queens Road and Whiteladies Road, to Clifton Suspension Bridge, were withdrawn since the residents objected to tramlines (this matched experiences elsewhere in the British Isles – tramways were welcomed in lower-class or lower-middle-class districts but not by the inhabitants of more well-to-do districts). To be fair, many of the Clifton roads are narrow, hilly and bendy. In Clifton, according to the *Western Daily Press* of 3 October 1878, it was feared that trams would bring in hordes of 'undesirable' visitors, depress property values, encourage jerry-building, and cause residents to shop in Bristol, to the detriment of local traders. A more vitriolic letter appeared in the *Bristol Mercury* later that month:

> is it not something terrible and most wicked that the disgusting tramway is to bring in the nasty, low inhabitants of Bristol up into our sacred region.

On 23 July 1880 Mr George White proposed an inclined railway from Hotwell Road to near the north end of the Suspension Bridge. According to the Merchants Venturers' Proceedings, the project was not entertained. No reason was stated.

On 16 April 1885, it was reported that Mr George White, secretary of the Bristol Tramways Company, had applied for licences for 65 of the cars belonging to the company. He ran 42 cars a day and 52 were in excellent condition. He did not understand why he had to apply for licences when cabs and omnibuses did not. He was told he was not allowing enough space for passengers – an Act of Parliament stated that all new cars should have 16in sitting space; his cars only

1.13. In 1887 Bristol's first regular horse-bus service started operation, from St Augustines Parade and Clifton Suspension Bridge.[21] (BRO 43207/9/35/40)

supplied 15 or 15½in per passenger. Three cars were refused licences.

On 4 May 1885, in the *Western Daily Press, Pro Bono Publico:*

considers there should be an easy means of access from the lower levels than there is at present provided. Aged and infirm people, who would be sure to take advantage of a reform in this direction, are practically debarred from taking a trip by water or rail simply because they cannot get up the steep hill or afford to pay a high cab fare. This difficulty could be easily met by laying a line of tramway in a slanting direction from the high level to the low, near the Suspension Bridge, or through the grounds of the Spa Company, and by providing two cars; the descending car being weighted a little heavier than a descending one, they would work safely without steam or any costly power. The rails would of course require to be double in the middle of the line to allow the cars to pass.

When the cars are not properly weighted with passengers either ascending or descending to move gently, it would be easy to regulate this by putting in or taking out weights. Only one man for each car would be required, and the expense when not going would be very little. By this means a novel and pleasant little trip could be had from the high to the low level, and vice versa, for, say, 1d or 2d. The opening of such a line would send grist to the mills of the steamboat companies, the railway companies and the Tramway Company, and ought in consequence to receive substantial support from each. To the speculator who takes the hint and puts this affair in motion there is a fortune sure. Tramways running on steep inclines have been in operation for many years at Scarborough, and more recently at Saltburn-by-the Sea.

Scarborough South Cliff Tramway Company Limited was the first funicular railway in the UK, created in 1873. It opened in July 1875 and cost £8,000 to build. On 6 May 1885, in the *WDP Pro Bono Publico:*

has again drawn attention to the feasible goal to construct a lift for the tram-cars and passengers from the Hotwells line to Clifton, so as to connect the present line at Redland and thereby make it almost circular. This would be a great advantage to the centre of Clifton, with its intended spa, while now strangers and others visiting our beautiful Downs and the grandest bridge in the world have to undertake a tiresome walk or pay a costly cab fare. But with the connection of trams it could easily and cheaply accomplished in safety. Anyone standing by the Port and Pier Railway station in Hotwells can see there is a place almost left by nature to construct a strong and safe lift which it only needs a little engineering skill to accomplish. On July 4 1877, you kindly inserted a letter of mine advocating amongst other local improvements, a line to the Hotwells and lift to Clifton. Our Tramway Company tried to carry out the idea, but were strongly opposed in the Town Council. Still I hope to see this improvement carried out. A. D. Collard, Merrywood, Bedminster.

On 27 September 1889, Mr Kincaid applied by letter for leave to construct a passenger lift from near the Hotwells to Clifton Down but the Merchants Venturers' Committee did not approve of the scheme. Messrs.

Broad and Potton made a similar application to the Merchants in September 1890 with a similar result.

1.7 George Newnes, MP enters the scene

Meanwhile George Newnes, MP and publishing magnate together with George Croydon Marks, a hydraulics engineer (see chapter 2), had had great success building a cliff railway at Lynton and wanted to get involved in the new fad of cliff railways.

On 9 September 1890, in the *Exeter and Plymouth Gazette* we hear that Newnes:

> who has just completed a lift at Lynton, has purchased property at Clifton with the idea of making a similar lift close to the Suspension Bridge, to convey passengers from the Hotwells to close to the Observatory Hill. The promoter will have to gain the assent of the Merchant Venturers previous to carrying out his design, and will have to obtain Parliamentary powers. The scheme is not altogether a new one, as it was embodied some years ago in a plan brought before the citizens, but which fell through.

On 26 September 1890, Newnes applied for permission from the Merchant Venturers to:

> make an underground inclined lift from Hotwell Road to the garden of 14 Princes Buildings, the house he proposed to purchase for a Spa Room. Merchants resolved to confer with engineers (which they did the following month).[22]

On 2 October 1890 the *Bristol Mercury* reported:

> Mr G. Newnes, is, we are told, still contemplating the construction of a lift between Clifton Down and the Hotwells. He has, we believe, purchased the property on Sion Hill which was acquired a few years ago for the purpose of establishing a spa, and the railway will start from that point, and is to debouch on the road on the Bristol side of the Suspension Bridge. When the Bristol Tramway Company projected a scheme further down, the Merchant Venturers who own some of the property, opposed it, and they will possibly oppose the present scheme.

On 31 October 1890, the Merchant Venturer engineers' conference having taken place,

> the scheme was discussed and it appearing that the property affected was held on two leases of 1000 years at rents of £17.12s on number 14 and £10 on number 15, it was resolved that consent to the project be given on the following terms:
> That rents be raised to £40 on number 14 and £60 on number 15.
> That works be finished by January 1893.
> That no alteration be made to the exterior elevations of the buildings and that they be used for no other purposes than permitted at the Bath Pumproom.
> that no licence for liquor be obtained.
> That the road in Prince's Buildings be not obstructed.
> That no blasting takes place between 7 p.m. and 7 a.m.

On 4 November 1890:

> The Society of Merchant Venturers have recently had before them more than one application for permission to erect a lift railway to connect the Hotwell road with Clifton [Broad and Potton according to their minutes], but, with their usual jealousy for the natural beauties of our unsurpassable cliffs, they have withheld their consent and sanction until it was clearly demonstrated that the proposals would not injure or mar the scenery, The application that Messrs Osborne, Ward, Vassall and Co. have made on behalf of Mr George Newnes MP, has we are informed; received the unanimous approval of the society. The engineers, Mr G. Croydon. Marks, A.M.I.C.RE, M.I.M.E. of Birmingham, and Mr P. Munro M.S.A., F.S.L, of St Stephen's chambers, Baldwin street, have laid their elaborate plans before the society, and fully demonstrated the feasibility and practicability of their scheme, by attending with the committee upon the ground which Mr Newnes has purchased for the site. It is proposed to cut a tunnel under the garden grounds belonging to No.14, Princes Buildings, Sion Hill, and through this to lay lines of rails for the purpose of carrying cars for the conveyance of passengers. The cars are to be worked upon the principle for, which Royal Letters Patent have been granted to Mr Newnes, Mr Marks, and Mr Jones, and upon a plan similar to the system still working at Lynton, and to be worked also at Torquay and elsewhere. The surface of the ground will not be in any way broken open or interfered with, and no inconvenience of any kind will be caused by the construction of the lift, while a great public service will, it is confidently expected, be rendered: The question of feasibility of the plan of working is not a matter of experiment or speculation, and the experience which has been gained in carrying some 60,000 persons during the season at Lynton will help the engineers further provide for the needs of the public. The Clifton passengers will be taken, from a point near to the present pump in the Hotwell road, and landed on a level with the existing private road passing between the lower and upper grounds of Prince's buildings. By this arrangement no unsightly cutting or embankment is made, and no nuisance or annoyance is caused, seeing that a simple well-lighted

tunnel affords the line of communication, which is within the solid rock beneath the surface of the promoter's grounds, thus avoiding, although at very great expense, everything which might be regarded as objectionable or capable of affording grounds for complaint from anyone. Another most attractive feature is provided for Clifton, as the grounds of the existing house are to be laid out in an artistic and attractive manner, for the recreation of those who may be staying within the new Spa building. These grounds will not be thrown open to those travelling by the lift, or to the public indiscriminately, so that a Spa of a very select character will be formed to meet a long and sorely felt want.

A grand Pump Room and Spa promenade is to be provided as an adjunct to the existing property which will be remodelled and embellished in an attractive manner. The engineers propose to provide everything of the highest possible character to render the Spa one of the handsomest and most attractive in the kingdom. They are convinced, upon the testimony of the highest experts that the waters of Clifton, are capable of effecting cures and all maintaining health in a degree not to be surpassed by Bath, Scarborough, Leamington, or Harrogate. To further tender their work attractive to the residents and visitors of Clifton, they propose to furnish a handsome lounge and reading room for the benefit of those visiting the Spa, The engineers, in their report accompanying the application, assured the Society of Merchant Venturers that nothing should be wanting in any way on their part or on the part of their client Mr Newnes, MP, to further the interests of the visitors and residents, and they desired not only to maintain unblemished, but also to enhance, the beauties and attractions of Clifton. The work of cutting the tunnel will commence in a few weeks, and it is expected that the operation will be completed in about six months, The Spa will be ready at the end of 1891.

Here then are the conditions set by the Merchant Venturers: the Railway would be built in a tunnel to preserve the beauty of the gorge, and that a Pump Room and Spa would be provided at Newnes' expense. These were rather optimistic demands – especially when the other group could not raise even £25,000 to create the Spa. Nor were railway travellers very welcome at theSpa. Construction actually started four months later, and the Spa was not ready at the end of 1894.

two

The People Involved
Investors, Engineers and Master Blasters

2.1. Opening Day medallion 11 March 1893
All medallions from Clifton Rocks Railway 1893 opening day have the initials PM (Philip Munro), GN (George Newnes, whose initials are surrounded by laurel leaves) and GCM – George Croydon Marks. (Author)

2.1 Sir George Newnes (1851-1910)[1] financier

George Newnes was the son of a Congregational minister, born in Matlock Bath. He started work as a haberdasher in Manchester. In 1881, he had an instant publishing success launching a popular penny magazine *Titbits*. Within two hours of its launch, he had sold 5,000 copies, and within ten years it provided him with an annual income of £30,000. The *Strand* magazine came in 1891, and enabled books to be serialised (including Sir Arthur Conan Doyle's *Sherlock Holmes*). With magazines such as *Country Life* and technical books, he became very wealthy. Newnes and his wife had two sons (the eldest died aged eight).

In 1885 he was elected as Liberal Member of Parliament for the newly created constituency of Eastern Cambridgeshire (also called Newmarket). He held the seat for ten years. The *Daily Gazette for Middlesbrough* noted that the new Parliament would contain no less than 34 newspaper proprietors and journalists, the latest being Newnes:

> He must have tender regard for the day he was seized with the inspiration to abandon the calling of a commercial traveller and reproduce in a cheap form the work of other men's brains.

The *Cambridge Independent Press* however, found him in the space of a few weeks, to have:

achieved a most extraordinary popularity and has earned golden opinions from all sorts of people. The rapturous delight and the uncontrollable laughter with which his speech was received is a proof that he is still advancing in popular favour and that he possesses the rare combination of a superb gift of humour with a genuine sympathy with the toiling and suffering poor.

When defeated in 1895, he was offered the safe Liberal seat of Swansea, which he held from 1900 to his retirement at the 1910 general election. In 1895, he became a baronet of Wildcroft, in the parish of Putney, for 'political services' (i.e. for subsidising Liberal newspapers).

Newnes used to visit North Devon and loved Lynton and Lynmouth. He became a key player in development of these towns. In September 1887, he was invited to stay with a friend and was horrified to see the horses and donkeys struggling up Lynmouth Hill. Within 24 hours of arriving, Newnes agreed to put up most of the money for a cliff railway, having met with a recommended engineer, Bob Jones. George Croydon Marks was also in Lynton visiting Bob Jones (a relative of his mother), and so got involved too. Newnes took the matters in hand. A new company, The Lynton and Lynmouth Lift Company, was formed, chaired by Newnes, who attended to financial and legal matters, while Marks and Jones

2.2. Sir George Newnes. (*Life of Sir Charles Newnes*)

concentrated on the design and construction. Newnes boasted that in one month building the railway was begun, and in two years it was opened with thirty pressmen from London to report it. Newnes being a publicist, he was very eager to publicise his work widely. At the time it was the steepest inclined railway in existence. After twenty years of operation it never lost a parcel or a passenger. Newnes noted about his meeting with Jones that:

> The only mechanical part for which I was responsible for was the provision that the man at the wheel should do something before the car could move. The arrangement is that he has to lift the brake off and keep it off in order that the car may proceed. If the brakesman does not do his duty for any reason, the brake acts automatically. [The forerunner of the dead man's handle.] Whatever might go wrong the worst that could happen, up or down, would be that the car would stop.

In June 1888, a patent was filed in the joint names of Newnes, Jones and Marks for the four separate braking systems. Car brakes were controlled by a foot plate gripping the rail, an additional rope wheel brake, and a governor spring. A central cabin was employed to control the brakes on each wheel. The unique safety features – designed so brakes could be applied to either car independently of the other – have been responsible for the railway's remarkable accident-free record.

Schemes which met a public need and likely to make him a handsome profit appealed to him. He built a summer residence in Lynmouth and had his own stop on the railway to get to his house halfway up Hollerday Hill – having purchased the entire hill. He was also responsible for getting the 19-mile Lynmouth and Barnstaple Railway opened in 1898. He funded the Town Hall and he gave this railway to the town. He also gave away large amounts of money to charities such as the Salvation Army, both to ease a conscience troubled by sudden and enormous wealth, and as an expression of his upbringing as the son of deeply religious parents.

He attributed his extraordinary capacity for work of all kinds, to his love of tennis and golfing, and having Saturdays off – according to an article in the Cambridge press[2] in 1892 'Mr Newnes at home. The Journalist's and Politician's domestic life'. He was always on the lookout for new and original ideas. He was also a chain smoker of cigars.

By 1908, his businesses were failing and this adversely affected his health. There were public displays of drunkenness and he lost a fortune gambling on oil and rubber. By 1910 his fortune had gone, he had retired to live in Hollerday House and died of diabetes. His servants and the whole countryside were in mourning for the kind 'Sir George'. The only thing he did not manage to do was to build a pier at Lynmouth, permitting the town to be approached by sea. The Bill to construct it had been passed, and he had paid all expenses related to the Bill, but other work pressures and ill-health meant his rights to build the pier had lapsed. His house burnt down in 1913 in mysterious circumstances: on 5 August, it was reported in the *Devon and Exeter Gazette* that the fire at the unoccupied house may have been caused by the Suffragettes (known for arson of selected empty houses belonging to prominent members of society). Five women, strangers to the town, had obtained the keys from the agent with a view to seeing the sunset from the house. Being 500ft above Lynton, the fire could be seen for miles around. It was difficult to get an efficient water supply for the fire engine so the firemen directed their efforts in protecting the adjoining plantation. It was a very sad ending.

2.2 Sir George Croydon Marks (1858-1938)[3] engineer

At 13, Marks became an apprentice in mechanical engineering at the Royal Arsenal, Woolwich. In 1878, he became a junior draughtsman working on engineering components. The firm worked with the Metropolitan Railway on improved methods of lubrication. In 1882, he joined Tangye Ltd., who made cranes, hoists, steam and gas engines, hydraulic jacks, etc., where he became manager of the Lifting Machinery Department. In 1883-4, he made several visits to Saltburn-by-the-Sea during the construction of a cliff railway operated by water-balance – Tangye supplying a gas engine, water pump and other machinery.

In 1887, he was invited to Lynton, Devon, by local builder Bob Jones, a second-cousin, for discussions with promoters of a new cliff railway which included George Newnes. In Lynton, there was no need to pump back to the top water discharged from cars at the bottom, because a plentiful supply could be piped from the East Lyn River. However, improved brakes were required for the unprecedented 862ft of track, and 1:1.75 gradient. Three days after the meeting, his brother Edward assisted with drawings for four separate braking systems for gripping the rails. This chance meeting with Newnes led to a whole new venture designing cliff railways.

In 1887, Marks obtained sales agencies for Worthington Pumps, the American Elevator Company, and also for engineering insurance. He travelled extensively, advising private undertakings and government. With his energy, enthusiasm and personal charm and integrity, his reputation grew, but financial rewards came slowly and he augmented his income with engineering drawing work, many commissioned to support patent applications. He hoped that his brother, Edward, would join his business. He encouraged him to leave Tangye Ltd in 1887 and take a job as draughtsman at Youngs, Ryland Street Works, manufacturers of lifting tackle and hydraulic machinery.

In 1887, Marks founded a law firm in Birmingham specialising in patent law. Dugald Clerk joined as a partner the following year, so the firm was named Marks & Clerk. Clerk had served an engineering apprenticeship. At the Royal Technical College in Glasgow, he became engaged in the development of the two-stroke internal combustion gas engine, filing his first patent in 1877. Clerk's definitive work on gas and oil engines was first published in 1886. He wrote a paper[4] comparing one of the Crossley 9 engines from Clifton Rocks Railway which had been built in 1892, to a Crossley engine built in 1882. He drew wonderful scaled drawings (see chapter 4) which we have hung up in the bottom station of the Railway. Clearly, since Marks was involved with Clifton Rocks Railway, and it had two brand-new engines, Clerk would have used one as a benchmark. Clerk was also involved in the development of the motor car. Both being engineers, in 1887 Marks and Clerk also

2.3. Sir George Croydon Marks. (Godfrey Marks)

2.4. Youngs jack, found in the rubble at Clifton Rocks Railway. (Author).

launched a new weekly periodical *The Practical Engineer*. Having recognised the potential of patent agency work, Marks and Clerk quickly mastered the complexities of patent law (originated in 1851 and reformed in 1883), and Marks published the *Inventors Pocket Book* in 1888. Marks had filed his first patent in 1888, for lifting-tackle improvements by Youngs, his brother's former employers. Other patents followed, dealing with such diverse subjects as hat stretching, window-blind brackets, mechanical toys, fishing tackle, and Avery scales and balances. Many of these were at Birmingham locations as it was a big hub of engineering industry. The most significant patent was filed in 1888 in the joint names of George Newnes, George Croydon Marks, and the Lynton builder Bob Jones, in which they described the unique features of the Lynton & Lynmouth Cliff Railway braking system. Marks also published definitive books: *Hydraulic Machinery* in 1891, *Notes on the Construction of Cranes and Lifting Machines* in 1892, *Engineering Materials* and *Notes on the Construction and working of Pumps* in 1893.

In 1887, Marks and his wife Margaret visited Bristol to see friends and her sister Martha, who was married to Walter Pearce. Probably Marks as the engineer responsible for the design of the proposed Clifton Rocks railway played some part in the appointment of Pearce who would one day manage the railway. I was contacted by someone researching her family tree: during their research she found a 1911 census entry for her great-grandfather, Mr Walter C.H. Pearce, age 52. He lived at no. 4 Colonnade (adjacent to the Railway) from 1896 until 1929 which is when it was bought by the Clifton Grand Spa and Hydro Company Ltd. (see chapter 9). He was the manager of the Clifton Rocks Railway from 1906 to 1916. His son Stanley (age 27) was an engineer for the Railway. In the 1891 census Walter was an advertising artist so it is quite a career change. Marks was helpful at finding jobs for his relatives, and in this case accommodation. He also liked to employ local people, which is probably how Munro was employed. Munro had been involved on the previous failed Clifton Spa scheme in 1885.

Following the success of the Lynton & Lynmouth Railway, Marks and Newnes looked for other suitable sites to build similar railways, where they could employ the technology they had developed and patented. They succeeded in completing such projects in Bridgnorth, Matlock, Clifton, and Aberystwyth. More later in this chapter.

Marks was awarded the 1893-94 Telford Premium by the Institute of Civil Engineers (for a paper, or series of papers). In 1894, Marks received the first of many commissions from the USA, to file a British patent for innovations in water-tube boiler design. He crossed the Atlantic more than 30 times during the next 35 years, enjoying luxury and making business contacts aboard the great transatlantic liners. In December 1894, the *Bristol Magpie*, described him as:

> of medium height, with dark hair and deep-set eyes, who takes the greatest interest in the management of the (Clifton) Spa concerts, and personally superintends the intricate details of the elaborate baths nearing completion. He is head and shoulders above his fellows in the profession, and extremely genial to those who enjoy his acquaintance.

In 1905, Marks was invited to stand as Liberal candidate for Launceston in the General Election, no doubt encouraged by his friend and mentor Newnes. Although relatively unknown compared with the popular local Conservative candidate, Marks applied all his charm and eloquence to campaigning on Free Trade, amendments to the Education Act, wasteful Government expenditure, unemployment and the Licensing Laws, and was unexpectedly elected to Parliament. He held his Parliamentary seat until it was abolished at the 1918 General Election, when he returned for the new Northern Division of Cornwall. He held that seat until his defeat at the 1924 General Election. In 1929, he left the Liberal Party and joined Ramsay Macdonald's Labour Party.

In 1911, Marks received a knighthood for public and political services at the coronation of George V. In 1929, he became a peer, Baron Marks of Woolwich.

2.3 Cliff railways built by Newnes and Marks[5]

It is clear that from their first chance meeting in April 1887, Newnes and Marks established a great mutual respect and friendship. Both were entrepreneurial, hard working, had similar religious and political backgrounds, and shared common interests in mechanical innovation and technical publishing. They gave each other valuable assistance in a variety of ways. Marks probably became an MP because of his friendship with Newnes. Newnes' contacts also helped Marks.

At Lynmouth, on 30 August 1888, it was announced in the *Western Times* that:

> excavations are being made for the proposed water-balance cliff railway for goods and passengers. Two tramcars will be drawn up and down by the means of two steel wire ropes. The railway will obviate the necessity of traversing the present steep and circuitous lane which

connects Lynton and Lynmouth. The station will be located 500 feet higher than the Esplanade. The total cost of the railway will be about £6,000, and the act of Parliament allows of capital of £10,000. The chief promoter of the scheme was Mr George Newnes, M.P., proprietor of *TitBits*.

It was opened in April 1890 at a cost of £8,000. The railway belonged to a limited company (Lynton and Lynmouth Railway), of which Newnes was the Chairman. It was constructed on a system originally propounded by Bob Jones of Lynton. Marks was the engineer and it is still very successful.

On 24 April 1890, Newnes, having helped the development of Lynton and Lynmouth by connecting them by a cliff railway, undertook to help Babbacombe in a similar way. Plans were prepared for construction of a hydraulic lift from Norclyffe to Redcliffe. This particular plan never came to fruition. Another scheme was implemented later.

Marks read in the *High Peak News* of a proposed inclined scheme in Matlock. He immediately alerted Newnes, who had been born there. Newnes wrote to Matlock in July 1890, pointing out that he was best qualified for such a scheme because of 'certain valuable patents' he held. By the same post, Newnes sent Marks a £1,000 cheque for a survey and detailed proposals for a project which he was prepared to finance. He stated the scheme would cost £20,000 and he could readily find the money himself without appealing to the public. It was reported on 11 March 1892, that the first section of the line was cut and that the erection of the buildings would cost £2,500 and would take 2 months to complete. However, the tramway was opened a year later on 28 March 1893 (17 days after Clifton). It was not a financial success and in 1896 Newnes bought out his fellow shareholders and donated the entire tramway to Matlock Urban District Council, who received it with mixed feelings as they were worried about maintenance costs. This came to a head in 1898 when they had had it for nine months. The public heard the tramway was operating at a loss so they would have to pay for it through increased rates. Passenger receipts were £851 5s 10d. Expenditure including wages of £401 1s 8d, left a balance of £32 5s 10d. Having raised the fares, the tramway made about £70 per year. The Matlock tramway closed in 1927.

Marks read in the *Birmingham Post* in 1890 of a meeting in Bridgnorth about an improved connection between the lower part of the town by the River Severn and the higher part by the ruins of the castle, and he immediately contacted the mayor. After several visits to survey possible sites, he presented to the Town Council, with the support of Newnes' lawyer, several proposals and detailed drawings for a water-balance cliff railway, which was opened on 8 August 1892. It only took eight months to build despite being steepest. On the 26 September 1892, seven weeks later, 60,000 passengers had been carried. This was clearly a non-troublesome one. Funnily enough, the cliff railway has come full circle and is now owned by cousins of Marks.

On 9 September 1890, Newnes purchased property at Clifton with the idea of making a similar lift close to the Suspension Bridge.' The Clifton site was half the height of Lynmouth and Lynton. The lift and the Spa were expected to cost £30,000 and take one year to build. In the end the whole scheme cost £100,000 and took more than two years to complete: two years to build the Railway (£30,000), one year to build the Pump Room and modify no. 15 Princes Buildings (£23,000); two years to build the Turkish Baths and make nos. 12-15 Princes Buildings a hotel (£35,000). The railway went bankrupt in 1908 and was bought by Sir George White.

It was reported in the *Hartlepool Mail* on 10 March 1891 (the day after the construction at Clifton started), that a cliff railway for Durham was proposed to provide a means to surmount the precipitous bank. The cliff railway was to connect the centre of Durham to the railway station, and to provide a track for pedestrians as well as one for cars. Marks prepared the plans for Newnes, but it appears that, despite the publicity, this did not progress.

In 1895, Marks was approached about the development of Aberystwyth as a tourist attraction. He responded with enthusiasm, and the Aberystwyth Improvement Company was formed with Marks as managing director. Newnes was not involved. This, like Clifton, required pumps (Worthington Compound steam pumps) at the bottom to supply the top station with water. Marks patented a method to assist in the deceleration of the cars by reducing the track gradient. The cars were designed by Marks with a 'toast rack' design. This railway cost £60,000 to build owing to the huge amount of rock that had to be removed. Marks was humorously described as 'an authority on how to get up a hill'. No railway was built to the same scheme using the same equipment.

Marks was then appointed consulting engineer to Alfred, the new third Duke of Saxe-Coburg Gotha (son of Prince Albert and Queen Victoria) to construct a funicular railway on the outskirts of Budapest from the west bank of the Danube up to Blocksberg fortress.

Cliff Railway	Date built	Length (overall, rail)	Vertical Height	Grade	Power	Gauge	Water, (car, reservoir)
Lynton and Lynmouth	1887-1890	(890 feet, 862 feet)	500 feet	1 in 1.75	River water (Protected by 1888 Act of Parliament)	3' 9"	(3 ton, 100 gallon)
Bridgnorth	1891-1892	201 feet	111 feet	1 in 1.81	Water and Barker gas engine, now electric	3' 6"	(30,000 gallon reservoir, 2000 gallon tank)
Clifton	1891-1893	(500 feet, 450 feet)	200 feet	1 in 2.223	Water and Crossley gas engine	3' 2 1/2"	probably 3 ton tank approx. 5,000 gallon reservoir top and bottom
Matlock tramway	1892-1893	1 mile			Hauled up by continuously moving cable under road.		none
Aberystwth	1895-1896	(798 feet, 778 feet)	374 feet	1 in 2	Water and steam pump, now electric	4' 10"	4 ton (1000 gallon)

Table 2.1. Comparing cliff railways Note: 670 gallons = 3 ton, 894 gallons = 4 ton

Proposed

Durham 1891

Babbacombe, proposed 1890, built 1922-1926

Lansdown, proposed April 1895, never built. (A proposal in April 2017 to build a cable car from Combe Down to Bath Spa station is getting similar horrified comments.)

The scheme was not completed by Marks but he was invited to the opening in 1896 and awarded the Cross of a Knight of the Ducal Order of Ernestine.

In March 1895, Newnes proposed a cliff railway at Lansdown, Bath. Plans for this were submitted and show that it was proposed to go under no. 17 Camden Crescent. Accompanying the plans was a report by Marks showing it would cost £7,115 for a line on the surface, £12,325 in a tunnel, excluding the cost of purchasing the house. Besides concerns it would not pay, there were others: no one was living at the top of Lansdown Hill; it would add to the congestion of London Street; cause subsidence to the hill; damage house foundations; spoil the beauty of the area; no projections regarding cost of operation or revenue. Only three councillors voted for it. This was to be the last scheme Newnes proposed, and after the problems of Clifton it is surprising he did, especially since it would be built in a tunnel. I guess he considered Bath to be the ultimate spa town to conquer.

2.4 Philip Munro (1833-1911) architect

Munro was the architect and co-engineer of the Clifton Spa Pump Room and Clifton Rocks Railway. He was a Member of the Society of Architects (1887) and a Fellow of the Surveyors' Institute. He was born at Dornoch in Scotland, and educated at Dornoch Grammar School.

At 16, Munro joined the Royal Sappers & Miners (engineers). When he married he was a private, and later, in 1871, a sergeant. He then served articles with, and subsequently acted as assistant to, Mr Alexander Leslie, civil engineer and architect of Aberdeen. Munro managed the War Department's Trigonometrical Survey of Jersey, Guernsey, and Alderney. Later, he was assistant engineer for the Guernsey Harbour Works and was in charge of the fourth division of the Ordnance Survey for eight years, serving under H.M. Board of Works.

In 1859, married to Agnes Kirkpatrick and with children, he was living in Penrith, and later in Rugeley, Staffordshire. They had William Kirkpatrick (the oldest and named after his father-in-law), Janet, and Grace. In 1871 they returned to Scotland living in

Oldfield House, Caithness, with two more boys, James born in Staffordshire and John born in Brampton, Cumberland – so they moved around considerably.

In 1878, Munro commenced practice in Bristol, where he carried out extensive work in connection with the Wiltshire and Berkshire canal, 1870-94, including the bridge over the river Frome at Eastgate. In 1881, aged 47, he and his family moved to Bristol where they resided at 4 Cotham Gardens; in 1901 they moved to 8 Exeter Buildings, Redland.[6] He was now a civil engineer. In 1893, he was appointed by the Board of Agriculture to superintend, carry out, and certify works sanctioned by the Board for Agricultural improvements. He was co-engineer of the Clifton Rocks Railway from 1891 to 1892 (On the railway plans, the bottom station of the railway is called the Munro waiting room after him), and the Clifton Pump Room and Spa from 1893 to 1894 (but not the Turkish Bath or Hotel [1894-98]). Other buildings of his included, Keynsham Paper Mills 1880-82, Burnham-

2.5. Philip Munro. (*Bristol in 1889-99, Contemporary Biographies vol.1*, Pike)

2.6. Munro opening day medallions, and a standard medallion, 1893. (CRR Trust)

on-Sea Town Hall and Market, Burnham Cemetery, and Workmen's Houses for the Smyth family at Ashton Court, 1905. His offices, 'Philip Munro and Sons, Architects and Surveyors', were at 2, St Stephen's Chambers, Baldwin Street, Bristol.

He died aged 78 in 1911 in Bristol. Agnes, his wife, had died a year before aged 84.

We acquired two Munro medallions, one to Mrs William K. Munro (the wife of Philip's son who lived in Nailsea) and one to Mrs Philip Munro. Sale, the manufacturer, is not inscribed on the special silver ones given to the members of the Munro family.

We also acquired a very large heavy clock about 12in square. It has a marble base with a surround made out of limestone rocks that were excavated at the building of the railway, in the shape of a tunnel. The marble base was of the type of marble that would be used to build the Pump Room the following year.

The clock was presented to Mrs W Munro. These mementoes were presented to members of the Munro family by the engineers Croydon Marks and Philip Munro on the opening day of the railway, 11 March 1893. We bought the clock at auction, having met the owner (a retired clock repairer in Frampton Cotterell, South Gloucestershire) several years earlier. He had bought it from a rag and bone man in the 1940s when he saw it on the horse and cart.

Kathleen Nelson, grand-daughter of Philip Munro,

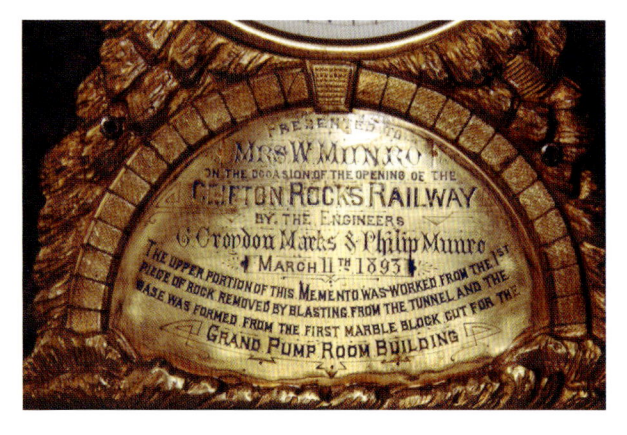

2.7. Left, Munro opening day clock and above 2.8. the inscription, 1893. (CRR Trust)

2.9. Kathleen Nelson, grand-daughter of Philip Munro, with the clock. (Author)

visited us in September 2014, aged 90. She came from Nottingham especially to see us. She had found a standard Clifton Rocks Railway opening day medallion when she was clearing out a drawer. It had been given to her grandfather. We were told he was also given one of two clocks that were made at the time – which was news to us. Unfortunately the clock was no longer in the possession of the family (after her grandfather remarried) so she visited us to look at our clock and medallions.

The one we have was presented to Mrs W. Munro. Kathleen is adamant that two were made, but did not know who received the other clock – presumably Mrs P. Munro her mother. The clock we had was not her family's one as the other one did not have the indent in the top. Janet, her mother, had died young. Kathleen's father (Philip Munro) remarried and there were lots of problems between the two families. When her father died he left things to Kathleen's step-mother who then passed on items to her son. It was very upsetting at the time as many items of very personal value were lost to her family. It was lovely to meet

Kathleen and her family and find out more about missing strands of the story. I then met the great-granddaughter of Philip (Sue Dawson) who visited us in 2016 on a Doors Open day. Kathleen was her aunt so we were able to show her the clock and medallions too, and the façade of the Pump Room. She also saw the houses that Philip had lived in. She thanked me for bringing it all to life.

2.5 George George (1863-1950) gang master and master blaster

A very different character to the others in this chapter, George George was born in 1863 at Hotwells Point Villa, Lower Bridge Valley Road, Bristol and died in 1950 in Bristol. For many years he was one of the leading tunnelling experts in the West as he had amazing blasting skills. His first big job was when he was in charge of one of the sections of the Severn Tunnel (built between 1873 and 1886).

He started work at the age of seven with his grandfather who was a quarryman, in the quarries along the banks of the Avon where he learnt his skill with gunpowder.

He started off as a 'little hammer' (splitting cobbles for the city) which was a hard job, so he became pally with the blaster. There was a storage place for gunpowder opposite Point Villa for the quarries. He never went to school, and could neither read nor write. This did not prevent him from undertaking many major blasting and boring operations. His father, or a charge hand at the quarry, must have promoted his skills. In 1897/1898, family stories say that George George boxed bare knuckles against Bob Fitzsimmons (1896 heavyweight world champion and considered one of the hardest punchers in boxing history) in non-title fights, and won, as the story goes. He was described by his grandson Harold as a 'proper character'. Ray, another grandson said George was exceptional and used his brain.

He was married in 1884. Harriet his wife was different to George. He was a hard man (he had to be); she was a gentle, little lady, 'an angel'. There were nine children in the house. She kept pigs and poultry and the chickens would hatch in her apron. She was overlorded by her husband. She had to do his paper work for him since he was illiterate.

George was survived by four sons and three daughters. His wife died in 1942, soon after they had celebrated their golden wedding. After her death, George went to Bitton to live with his eldest daughter, so this is where the memorial service was. His ashes were spread at Bridge Valley Road. His headstone is

2.10. George George with his wife Harriet. (Harold George)

at Canford cemetery.

A big job was to bore the tunnel through solid rock (carboniferous limestone) for Clifton Rocks Railway (1891-1893) and he was the gang boss.

He also was gang foreman involved in blasting work to cut a track up the cliff for the Aberystwyth Cliff Railway construction in October 1895. Plate 2.11 shows George George in characteristic bowler hat and watch chain on the left. Marks iss there too, below George George. George was presented with a book[7] about Aberystwyth by chief engineer Marks (who signed it in August 1896) and with whom he also worked at Clifton. Harold, his grandson gave the book to his youngest brother Ray, who gave it to us.

From Aberystwyth he went to London where he was in charge of driving a tunnel under the heart of London (1896 to 1900), in which the first London Tube from the Bank of England to Shepherd's Bush was laid. His next job was to build the tunnel to lay the underground cables from the new Chiswick power station to Shepherd's Bush.

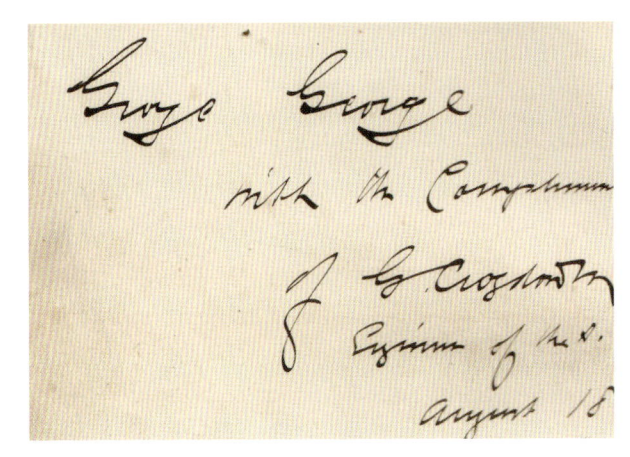

2.11. Left, constructing Aberystwyth Railway.⁶ George is identified by the watch chain on the left-hand side, Marks is below him. (Aberystwyth Guide)

2.12. Above, the inscription in the Aberystwyth book to George from Marks. (Author)

2.13. Point Villa where George lived being demolished. George is the large man on the right yielding the sledgehammer. (Harold George)

He then came back west and for a time worked for the Bristol Docks Authority, blowing rock from the bed and bank of the Avon in the Gorge, to widen and deepen it. George George was exceptional and used his brain to develop his skill so lack of schooling did not hamper him.

He was conductor on the railway for 15 years. As he did not want to wear the Tramways uniform when George White took over, he left.

In 1922, he blasted the sides of the Gorge to lay foundations for the Portway and Bridge Valley Road. This operation involved slicing over 20ft of rock from the sides of the gorge, without causing a landslide.

While he was engaged on this, he was working in sight of his house, which stood where the Portway now runs. It was not long afterwards that he had the job of taking down his own house. It lay just after the exit of Tunnel 2 near the Port and Pier Railway Hotwell Point station (in operation 1917-1922) by the side of the river just past Bridge Valley Road. Point Villa served as an opportunistic refreshment stop for railway passengers. He took his gateposts to Longwell Green where he built his new home.

Hotwell Station (1865-1921) was at the entrance of Tunnel 1 just north of the Suspension Bridge and where George White's trams terminated after 1879.

I met several members of the George family and was privileged to have reunited them having put a photo of Harold (George's grandson) on our website. Their memories are in chapter 8.

2.6 Christopher Albert Hayes (1850-1916)[8]

Hayes was contracted to build the Clifton Rocks Railway tunnel and waiting rooms (contract about £20,000). Hayes was a well-known Bristol firm of builders and contractors known for their craftsmanship. Their offices and works were in Thomas Street,

Bristol, Sanitary Yard Commercial Road, and Timber and Stone Yard Cumberland Basin. He embraced buildings of every description of the architectural and civil engineering trade.

Hayes was born in Clifton, the son of S. Hayes of the Mall, Clifton. He lived at Belmont House, Cumberland Road, and was educated privately. He was apprenticed to his uncle Mr J. Hayes of Bedminster and worked with some of the leading building and contracting firms in London to become a builder and contractor, commencing business in Bristol in 1873. He completed a number of projects for the Bristol Docks Committee (including the frozen meat store and foreign animal wharf), Bristol Corporation (Frome relief culvert and dock and harbour wall and buildings, central electric lighting station), the People's Palace in Baldwin Street, three Lloyds banks and many other buildings. Other contracts he was involved in included: St Katherine's Church in Knowle, Knowle Vicarage, St George's Higher Grade & Technical School and the Bristol Dispensary. He was vice-president of the Bristol Master Builders Association 1894-1895 and president in 1896, a member of the St Kena's Lodge of Freemasons and of the Royal Arch Chapter at Clevedon, formerly member of the old Bristol Incorporation of the Poor, chairman of the Building Committee 1898-99. He was Lord Mayor of Bristol in 1909 and 1910, and presided over the local arrangements for the coronation of Edward VIII, which 'involved him in the reception and entertainment of the Colonial Premiers upon their visit to the City in Coronation Year'. He died on 16 February 1916, aged 65 years at his residence, Salisbury House, Westbury-on-Trym, Bristol.

2.7 Sir George White (1854-1916)[9,10,11]

George White was born in Bristol, the son of a painter and decorator and a domestic servant. In 1869, at the age of 15, he joined a Bristol firm of solicitors Stanley & Wasbrough as a junior clerk. At the age of 18, he was lobbying for Parliamentary bills to seek powers to run a tramway in the city after the Corporation had opposed a horse-drawn tramway and built their own line to Redland. In 1874, the firm was involved in the promotion of the Bristol Tramways Company, following the passage of the Tramways Act 1870, and White played a major part. In 1875, he left the law and established a stockbroking firm, George White & Co and worked with some of the richest, most influential men in Bristol. He became the part-time company secretary of the Bristol Tramways Company at the age of 20 and was paid an annual salary of £150. His father then

2.14. Christopher Hayes. (*Bristol Contemp.Biographies vol 2*)

2.15. Sir George White.

died and he was left to support his mother and younger brother. The 1880s saw the tramway network in Bristol grow. He maintained a good public profile and worked with the local press so that the working class districts of Bristol would see the benefits the trams would give them. In Clifton, a more affluent area, it was feared that the trams would bring 'undesirable' visitors and depress property values, resulting

2.16. BTCC Commercial Vehicle outside the Railway bottom station. (Peter Davey Collection)

in the Clifton route being served only by horse omnibuses. He tried several times from 1877 to 1885 to get a 'lift' built from Hotwell Road by the Port and Pier station where his tram terminated – the Portway was not built until 1922 – up to the Suspension Bridge. But each time the Merchant Venturers refused since it would spoil the beauty of the gorge. In 1892 he became involved in other tramway companies in Dublin, Reading, Middlesbrough and London.

In 1887, the Bristol Tramways Company was merged into the new Bristol Tramways and Carriage Company (BTCC) with White as managing director after the company gained a monopoly on horse-drawn cabs travelling from Bristol Temple Meads station. By 1889, the company owned 876 horses and by 1891 the combined tramways, cab and omnibus company provided 38 journeys per head of Bristol's population per year. As the motor engine developed the use of horses declined. In 1900 only 392 horses were needed, but they lingered on until 1922 when the funeral department (which used black horses) closed down after operating continuously for 36 years. He truly provided transport from the cradle to the grave.

In the 1890s, he was an enthusiastic promoter of electric tramways together with the engineer James Clifton Robinson, to cut the high costs and get rid of the dirty aspects of horse-drawn trams. He started with a line in Old Market and quickly rolled out the electric line to the existing tram system, as well as extending lines further out in the suburbs, which was completed by 1900.

White also got himself involved in greater projects such as expansion of the Bristol & London & South Western Junction Railway, which involved using the tracks that the Great Western Railway ran on for trains from Waterloo. The venture did not come to fruition owing to overwhelming opposition from Great Western Railway, but it still increased his profile and standing. By 1887, he was the largest shareholder in the Bristol Port Railway & Pier Company and launched an ambitious but unsuccessful project to link the City Docks with the Avonmouth Docks (ultimately opened in 1911) to make a direct connection with the Midland Railway at Bristol.

Between 1905 and 1908, he tested and developed a fleet of twelve motorised Thornycroft double-decker buses, with routes starting from Bristol and travelling to depots in Bath, Cheltenham, Gloucester and Weston-super-Mare. He introduced motor taxis to Bristol's streets in 1908 and began to make Bristol buses. In 1912, he finally managed to buy Clifton Rocks Railway after it had gone bankrupt, which then ran until 1934. Bristol Tramways directors purchased it for £1,500 even though it had been constructed at a cost of £30,000. In 1913, White built a motor construction factory in South Bristol capable of building 300 vehicles per year, and by 1914 BTCC was one of the biggest employers in Bristol with 17 tramways services and 15 omnibus services and a fleet of 44 buses, 169 tramcars, 124 taxis and 29 charabancs, plus vans, lorries and commercial vehicles.

White also became interested in heavier-than-air flight, and in 1910 founded the Bristol Aeroplane Company (originally the British & Colonial Aeroplane Company). This set out to produce aircraft on a commercial scale, with premises in Filton, again at the end of his tramway terminus. The company produced 80 Bristol boxkites to great commercial success, many being sold overseas.

There is a replica boxkite in the foyer of the Bristol Museum and Art Gallery (plate 2.17, a photo taken during Bristol Aeroplane Company's 100th celebrations; there is a replica 1919 Bristol Babe on the floor). The Bristol Fighter went into production in 1916, shortly before his death, for use during WWI.

In 1904, White saved the Bristol Royal Infirmary from debts of over £15,000 by increasing the number of subscribed donors and organizing a fundraising carnival. He established a £50,000 fund to build a new hospital building. The new Bristol Royal Infirmary was completed in 1912 (there is a plaque on the building honouring him). A self-made man, White turned to philanthropy and was also a benefactor of the Red Cross and other charities. He was created a baronet on 26 August 1904 for his public service. He received his knighthood through the recommendation of the Prime Minister Arthur Balfour.

His grandson (also Sir George) came to re-open

2.17. Replica Bristol Boxkite and, on the floor, a Bristol Babe. (Author)

Clifton Rocks Railway in May 2005 on our first open day after we had started restoration in March 2005. He said his grandfather had reopened the railway after it had gone bankrupt, his father had closed it when it had gone bankrupt again and he was very privileged to reopen it again. A memorable day.

three

Construction story

Trials and Tribulations

Clifton Rocks Railway is an amazing, large tunnel on a 45% slope, with many historic civil engineering features for everyone to look at close at hand. When constructed, it was the largest tunnel of its kind in the world. The aerial photograph at the beginning of chapter one shows its impressive location. This chapter aims to give an introduction to historical landmarks of tunnelling, surveying, equipment used, and brick lining before describing the unexpected problems that beset Marks and the contractors of the Railway. (Note that the term 'lift' is used in contemporary documentation to avoid confusion between a railway and a tramway since cliff railways were a new concept. Imperial measurements have been used throughout.)

The problems explain why the tunnel took two years to build (March 1891 to March 1893) and cost £30,000, rather than an expected six months and £10,000. It gives an insight into the difficulties of building a tunnel on a steep slope through solid hard rock.

The first cliff railway to be built in a tunnel was at Hastings (about the same length as Clifton but in a shorter tunnel and only two tracks, just like all the other funiculars). Construction of the line was started in January 1889 by a private operator, the Hastings Lift Company, with the intention of an August opening (it finally opened in August 1891 after taking two years and eight months to build). It also originally used a Crossley gas engine to pump the water back up. Objections to plans caused major re-design and expensive delays, pushing up the cost from £10,000 to £16,000. A 363ft tunnel was constructed, using an existing cave, at an inclination of 1 in 3. This should have been easier than Clifton Rocks Railway since it was not as steep, not as wide, and not as long. The constructors were competitors to Newnes and Marks and the company later became Otis (founded in 1853 and still making lifts, escalators and moving walkways). The first operator went bankrupt in 1894, probably as a result of the construction delays and cost overrun.

Building cliff railways was a new Victorian craze, but most were not underground and were two-track rather than four. I am sure that Newnes and Marks would have watched the progress of the tunnel at Hastings and learnt lessons.

A big disadvantage of steep tunnels is that one does not have a view. The Avon Gorge is a wonderful sight to behold – but the Merchant Venturers would only allow a cliff railway to be built in a tunnel, to protect the splendour of the Gorge.

3.1 Surveying and air shafts

Prior to railway tunnels, canal tunnels were the forerunner of underground excavation.[1] The tunnel line was first established using a rudimentary surveying technique of a strong rope pulled over the hill from the entrance to the exit. Borings were then drilled vertically to investigate the ground, and these were then made into larger shafts sunk down to tunnel level to assist in construction. Heavy plumb lines and hanging lit candles were then used to give the line of sight underground. The correct line and level could then be checked. One could use a compass or theodolite (the first was developed in 1576 complete with compass and tripod) to check the angles. In fact plate 2.11 shows George George and Croydon Marks at Aberystwyth with a theodolite in the foreground. One shaft can still clearly be seen half way down the tunnel because there is a change in the pattern of brickwork.

At the ceremony at Clifton Rocks Railway to celebrate work starting in March 1891, the first blast took place one-third of the way down the cliff, dislodging half-a-ton of rock. The shafts then gave working access to the main tunnel works so one could either lift the rubble out or excavate at the bottom of the shaft. The rubble would have been either spread on the surrounding land or a chute inside the tunnel could have been used to get the rubble down to the bottom (a chute was referred to in the Colonnade court case later on in this chapter). The rubble could then have been transported away on a vehicle to help

**Rocks Railway
Sections of Tunnel**

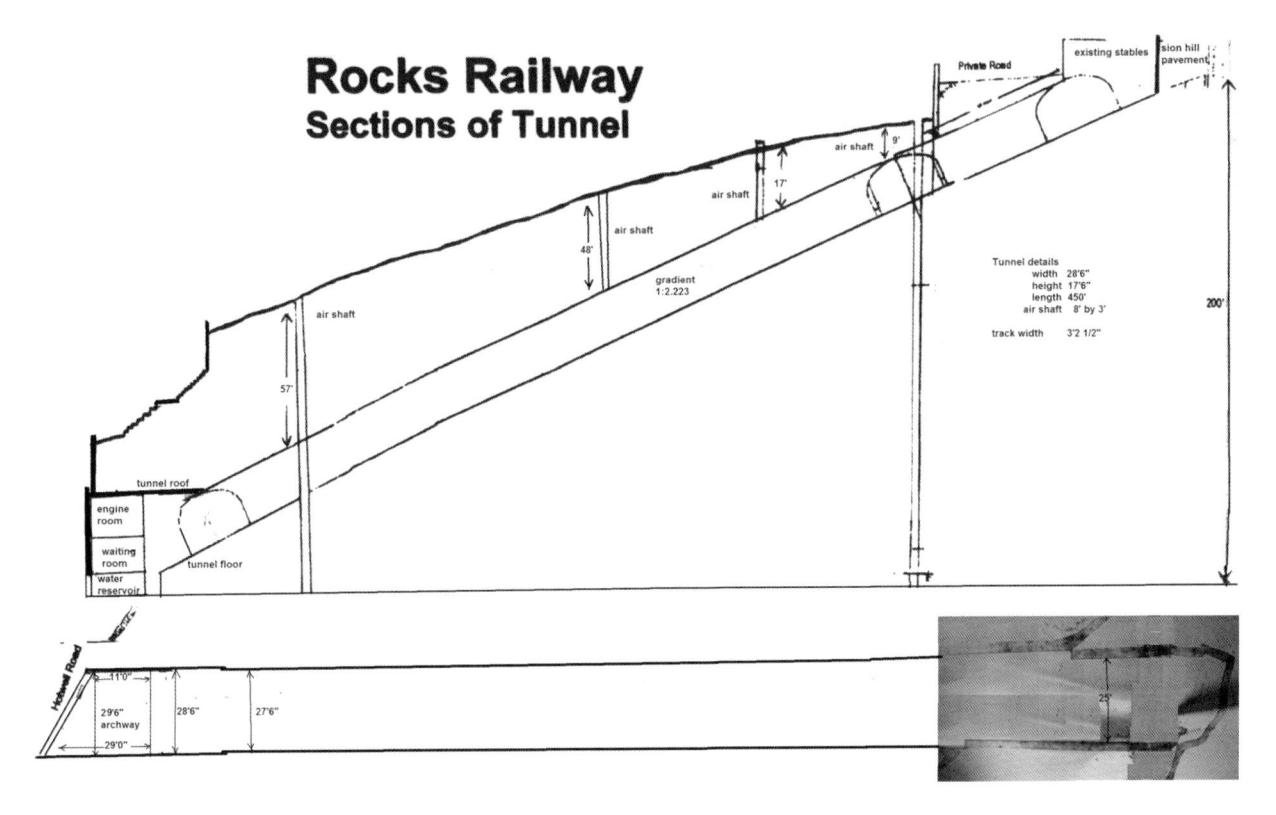

3.1. Tunnel section March 1891.
The 8ft x 3ft shafts can be clearly seen on the Croydon Marks' March 1891 section plan. The bottom shaft is 57ft deep, the next is 48' deep and the next is 17ft. The first blast was at the 17ft shaft. The 60° skew to Hotwell Road and Princes Lane can clearly be seen at the bottom of the section map. The map also shows that there were existing stables at the top station before 1891. (BRO42054.G.Drawer 4/18831)

fill the Belgrave Hill quarries on Clifton Down using Bridge Valley Road or the landing stage could be used to get the rubble away by boat.

3.2 Rock drills and landmark achievements of tunnelling and concrete

A significant proportion of great inventions in tunnelling belong to the nineteenth century. Consider the great skills and abilities of these engineers and their workmen in very poor working conditions and with limited technology, which have resulted in tunnels being in constant use and which have lasted a long time. Modern attempts in looking deeper through the linings have failed to reveal their secrets.[2]

Richard Trevithick (1771-1833) designed a high-pressure steam rock-boring machine, which also lifted and loaded stone for transport. Mining rock drills were not adopted until the last quarter of the nineteenth century well after being invented in 1851 in Boston, USA. They increased the rate of sinking shot holes dramatically, and, being operated by compressed air, also greatly improved ventilation and reduced working temperatures. However they did have a downside of creating deadly sharp dust, noise and white-finger (hand-arm vibration syndrome). Cornish manufacturers pioneered dust suppression by delivering a water spray to the drill bit.

Air drills are referred to in Croydon Marks' paper. He states that:

3.2. Holman air-powered rock drill with spray arrangement, made in Cambourne, Surrey.[3]

37

machine and hand-drilling was employed, but from the many settings-up required, the Author is convinced that the hand-labour was on the whole, as economical as the machine-drilling.

It is said that drillers fresh from constructing the railway in the Rockies came to help out in Clifton Rocks Railway. Locally there were trials of the 1875 Beaumont Diamond Rock drill – the first to be used over substantial distance. In 1875 it was used in Clifton Down tunnel, and in 1875-1886 in the Severn tunnel (which was also the first to use the telegraph during construction).

3.3 Concrete

In the 1870s concrete joist floors were being developed for Lancashire Cotton Spinning Mills which had hundreds of machines per floor.[4] The concrete filler-joist floor was a form of fireproof flooring developed in the second half of the nineteenth century that came to be used quite extensively in industrial and commercial buildings. Iron, and later steel joists embedded in concrete provided a crude form of reinforcing.

In the 1890s the first successful and widely used system of reinforced concrete framing was being developed in Belgium by François Hennebique. It has been surmised that Clifton Rocks Railway could have been an early adopter of Hennebique since both the bottom station floor (over the water tank and supporting three turnstiles) and the bottom station ground floor ceiling (under the two heavy Crossley gas engines on the first floor) needed to be strong. However, since the first building using this technique was constructed in 1892, it is unlikely.

3.4 Tunnelling in Clifton area

There was experience of tunnelling in the area before Clifton Rocks Railway was constructed. It was very challenging because of the very hard undulating oolitic limestone. Details of methods of construction follow, to aid comparison with Clifton Rocks Railway.

3.5. Hotwells Bristol Port Railway and Pier[5] 1865-1921 (north of Suspension Bridge)

The Hotwells to Severn Beach two-track line was built to cater for local traffic and allow large vessels to partly unload and then proceed to Bristol, rather than wait for favourable tides. It is 5¾ miles long and was constructed between February 1863 and March 1865. The location of Hotwell Station can be seen when one looks northwards over the Suspension Bridge parapet, as well as the entrance to tunnel one. It is close to Clifton Rocks Railway with the same kind of rock.

There are two tunnels lined with three layers of bricks between Hotwells and Shirehampton, 14ft 6in wide and 18ft high – narrower than Clifton Rocks Railway. The tunnel roof has a flat profile for protection against rock falls and the first tunnel, at 73 yards long, was used to store archives during WWII; the second tunnel, 175 yards long, was used as an air-raid shelter. Clifton Rocks Railway shelter was described as a palace compared with the Portway tunnel which was wet and overcrowded. The second tunnel was filled with concrete in 2010 when the supporting wall on Bridge Valley Road started to lean out and the tunnel roof split dramatically. Hotwells Halt 1917-1922 (just past Bridge Valley Road) was opened to handle war workers, and was close to where George George the master blaster lived (see chapter two) and learnt his trade in the quarries. Hotwells to Shirehampton was only a spur line and taken out of service in 1921 to build the Portway.

3.5.1 Clifton Down to Avonmouth

This tunnel under Clifton Down was built 1872-1875 using a diamond rock drill boring machine with compressed air (inventor Major Beaumont MP 1875). It was the first successful, soft-rock tunnelling machine. Beaumont did his trials with rock drills here. There is a ventilation shaft at the top of Upper Belgrave Road, and in the gulley in the Avon Gorge. Ponies were used in tunnel construction. The tunnel is still in use today – starting at Clifton Down Station and ending in Shirehampton. The tunnel is 1,768 yards long with a gradient of 1:64. The maximum depth is 160ft below ground level.

3.5.2 Clifton Rocks Railway 1893-1934 (south of the Suspension Bridge)

Croydon Marks in his contemporary paper Cliff Railways[6] which compares the construction and working of Lynton, Bridgnorth and Clifton cliff railways, states

> The whole of the track at Clifton is in a tunnel, the width of which is 28 feet, the length 450 feet, and the height from floor to crown 17 feet; the vertical rise of the cars being 200 feet. There are four cars at Clifton, each running upon its own pair of rails, and there are two sets of upper-track pulleys which are placed parallel with the rails. The tunnel was cut out of solid limestone rock, and presented some difficulty owing to the inclination, which prevented one set of men working with safety above another set when removing the rock. The tunnel is brick-lined from end to end, the timbering being built in where necessary

to save disturbance from its removal.

He made no mention in the paper of a rock fault or having to line the tunnel with bricks as a result of a fatal accident. If only one person died, it would have been a good record for Victorian times.

This tunnel has the Port and Pier Profile (shown on plate 3.13) just like the Hotwells Railway. It has a gradient of 1:2.223. It was built between March 1891 and March 1893 with a maximum depth of 57ft below ground to tunnel roof. It was the widest tunnel in the world at the time – and was widened by 1ft 6in to incorporate a brick lining with four layers of bricks after the accident. The crown had to be increased by one foot. It has a skewed tunnel construction (60° to Hotwell Road) and cost £30,000 rather than £10,000 – a result of widening by brick lining due to the clay fault.

It was very difficult to construct because of the steep incline which would not allow to arrange gangs of men working above each othe. Oolithic limestone is very hard rock, and broke many drills. Steam power was used to supply compressed air to the drills.

3.6 Explosives

George George (see chapter two) learnt his skills as a master blaster in this area using dynamite (which has only been in use since 1867). Tonite explosive was used in Clifton Rocks Railway. The shot holes are still visible from when they had to make the tunnel wider to fit in the brick lining. Tonite was also used in the Severn Tunnel as it gave off less noxious fumes than any other explosive, other than the highly-washed gun cotton. Tonite was also unaffected by cold or moisture and was conveniently packaged. The method used was to drill into the rock to the required depth and a suitable number of tonite packages were placed into the hole together with a fuse. The number of packages was critical; too few and the process would have to be repeated, too many produced a danger of loosening rock outside of the required area. Tonite was patented in 1877 and stopped being used in 1906. It is an explosive that needs to be fired by a detonator. Its name was taken from the Latin *tonat* – 'it thunders'.

In 1880 was the first prototype use of an electrically powered detonator, which made blasting a lot safer. The 30-second delay was not in use until the 1900s.

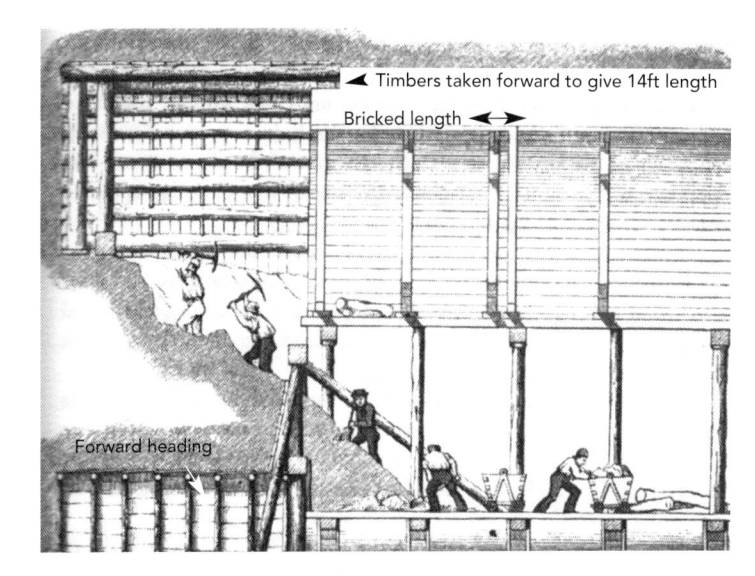

3.3. Constructing a tunnel on a slope. (Simms)
This shows how one would normally tunnel, bricking as you go. Note the timbering. Not the nicest of conditions to work in. Timber supports were needed along the length of the tunnel once it had been excavated. The gradient of the Avon Gorge is so steep that Clifton Rocks Railway had to be built at a skew of 60% to the Hotwell Road – even then it was still 1 in 2.223 which is just under 45%. It was also a very wide tunnel which makes it an amazing feat of engineering.

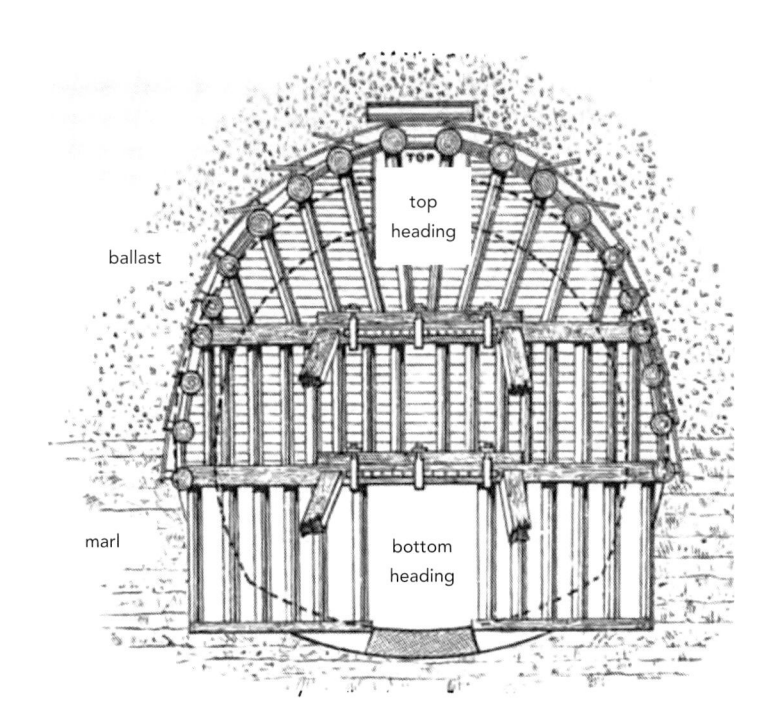

3.4. Timbering support (Walker).[7] There was a full-face, full-length excavation with timber, crown, arch and face support. The men would have worked their way up the sidewalls to complete the very difficult work in the crown. Images of the scaffolding used in Clifton Rocks Railway are shown later.

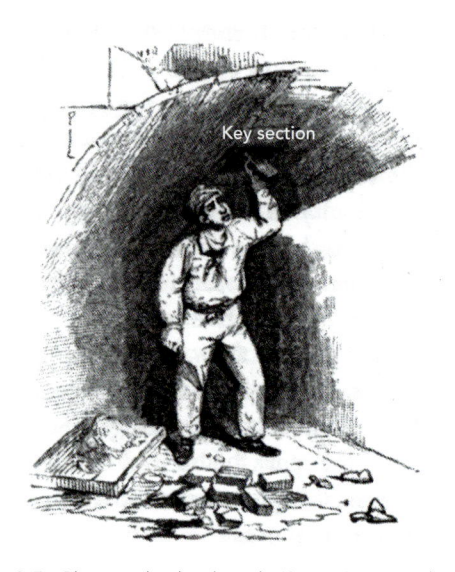

3.5. Closing the brick arch (Simms). Note the key section of brickwork above the bricklayer's head – he is working on an almost-finished length in which the bricks will be held in place by compression.

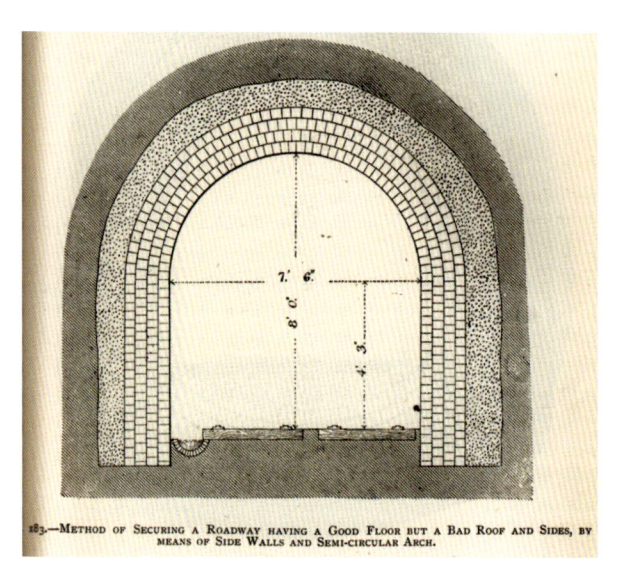

3.6. Four-ring brick arch. (*The Colliery Manager Handbook*, Caleb Pamely. Crosby and Son 1904)

3.7 Skewed brick tunnel construction

A skewed arch is a method of construction that enables an arch to span an obstacle at some angle other than a right angle. A masonry skew requires precise stone-cutting as none of the cuts form right angles. The problem of building skewed arch bridges was addressed by many civil engineers from 1787 onwards.[8]

The development in 1838 of using multi-ring surfaces built to a *forma* (to establish the shape or profile) with bricks, was considerably cheaper than stone. The first canal tunnels were about 9ft 6in wide and with a lining only two rings thick – not enough for the CRR tunnel due to the lateral earth pressures and its huge size.

When canal tunnels were increased to 22ft wide a heading with roof beam supports was needed. The brick linings were built in an arch under the crown supports and were between four and six rings thick. Blechingley[9] cost £71 18s 7d per yard – and used 11,099 bricks per yard, Saltwood £118 per yard and used 10,677 bricks per linear yard, so tunnels were very expensive to build compared with cuttings. The largest span canal tunnel was Netherton Tunnel built in 1858 with a span of 9 yards, which cost £302,000 as opposed to the £238,000 estimate prior to construction. The main reason for the Netherton project being over budget was the extra works necessitated by the condition of the ground through which the tunnel passes.

3.8 Dimensions of the tunnel and brickwork

Interestingly enough, brick masonry work was not included in the contract (see plate 3.7). The whole of the tunnel ended up being widened and having to be brick-lined after an accident in August 1891 (see later). The tunnel had to be increased to a height of 18ft 6in, and a width of 30ft to fit a four-ring brick lining (bricks of 1ft 6in either side) – just when it was almost finished.

Because Clifton Rocks Railway was on a slope, it was difficult to remove rubble. At the bottom station it can be seen in plate 3.8 that the constructors dug in horizontally by about 20ft at the shortest side to 28ft at the longest side due to the 60° skew to the Hotwell Road before working upwards to give themselves room to work and for the waiting room.

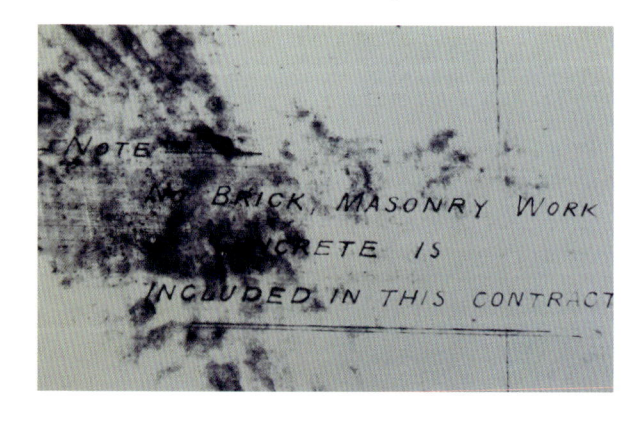

3.7. Marks' 1891 plan expressly affirms that brick masonry work is not included in the contract. (BRO)

3.8. Marks' 1891 plan shows the tunnel to be 27ft 6in wide and 17ft 6in to the crown. (BRO)

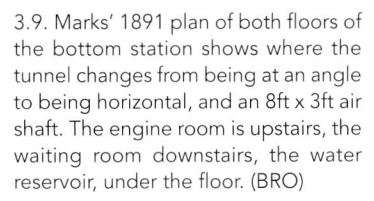

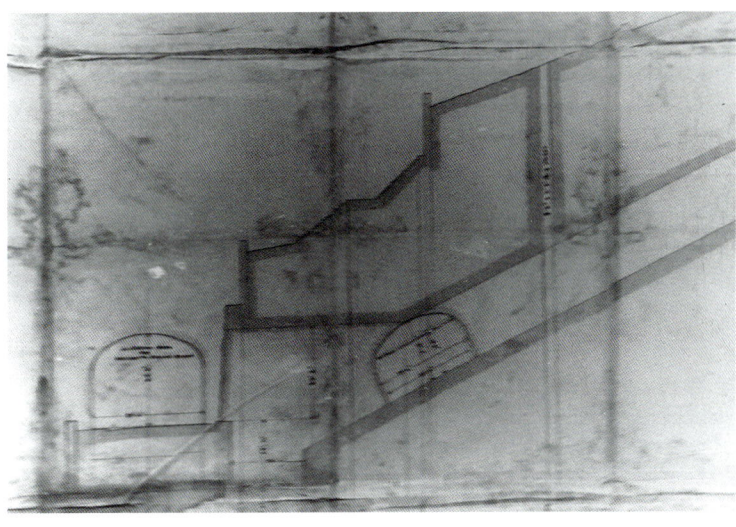

3.9. Marks' 1891 plan of both floors of the bottom station shows where the tunnel changes from being at an angle to being horizontal, and an 8ft x 3ft air shaft. The engine room is upstairs, the waiting room downstairs, the water reservoir, under the floor. (BRO)

There is no slippage of the brickwork at the bottom where the tunnel gradient changes from horizontal to 45% (shown in plate 3.8). The forces would have been at their greatest here so it is very well built – as a visitor said 'Brunel would have been proud of that joint'. Yet near this point was a rock fault and the cause of all the unexpected problems.

The tunnel is skewed at 60° (rather than 90°) to the Hotwell Road and Princes Lane, otherwise it would have been too steep. This tunnel uses Brunel's skewed tunnel method which is a clever solution to building bridges rather than the regular right-angled method. Skewed bricks can clearly be seen when looking at the tunnel entrance at the top station. As a strengthening device (the tunnel is not far below the surface of Princes Lane), the next layer of bricks would be laid at right angles to the first layer, and so on. This creates stronger edge arches where the span is greater. Having gone past Tuffleigh House, the bricks change to being laid parallel to the wall of the tunnel. This would have been the hardest part to lay because each brick had to be cut to fit (see plate 3.22). The skewed section is 15ft 10in long on the north side.

3.10. Skew shown at the top station tunnel entrance. The bricks are laid at right angles to the tunnel wall above, rather than parallel to the walls of the tunnel. (Author)

3.11. Tunnel roof. Conventionally laid bricks to left, parallel to wall, skewed bricks to right. Beyond Tuffleigh House, west of Princes Lane. See plate 3.22 to see what happens when the skewed bricks join the tunnel wall. See plate 3.23 to see what happens when conventionally laid bricks join the tunnel wall.

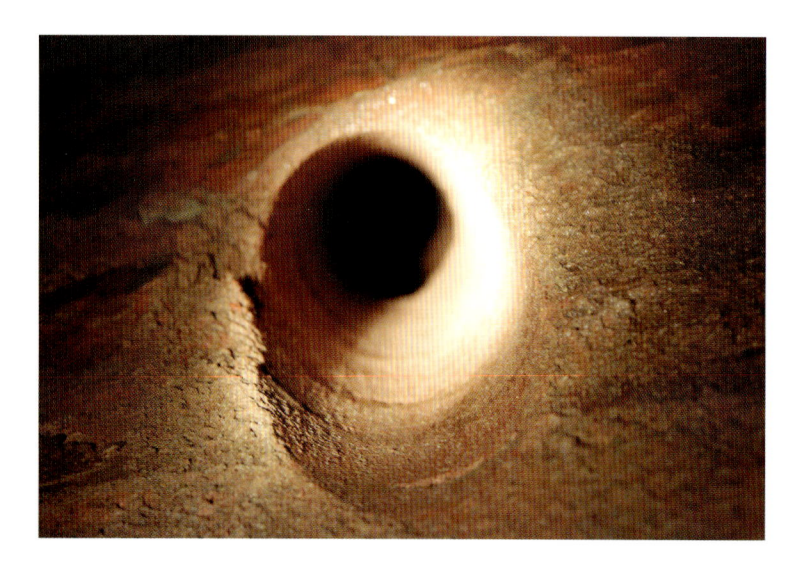

3.12. Roof section bore hole shows four rings of bricks. There were four rings of bricks because of the extreme width of the tunnel and laid as a result of the accident you will read about later. This bore hole is a result of testing stability of the rock above the Railway in 1959 by Skempton[10] (see chapter 14). It is near the shaft that is 17ft deep. (Author)

3.13. Marks' plan shows the the Port and Pier Railway profile along the Gorge (see chapter 5). It has a flat profile because it was not load bearing but merely to stop rock falls landing on the cars. It is also easier to build a flat profile. (BRO)

3.14. Tunnel roof. An air shaft (the 48ft deep shaft) can be seen here. It would have been used to lift the rubble out during construction. Note the flat profile. (Author)

3.15. 14ft tunnel roof brickwork, end of length. The roof brickwork was laid in 12ft to 14ft lengths starting from the bottom, using a proforma, as there are lines to be seen crossing the tunnel roof at regular intervals. (Author)

43

3.9 Bonding in brickwork

Bonding is the arrangement or pattern of bricks in a wall. Each unit should overlap the unit below by at least one quarter of a unit's length, and sufficient bonding bricks should be provided to prevent the wall splitting apart. Common bond patterns are Flemish, Stretcher, English and English Garden Wall. Brickwork in the tunnel would have been very heavy, especially over such a width, so good, strong bonding was required.

English bond is used for the wall of tunnel, and is made of alternating courses of stretchers (the narrow, long side of brick exposed) and headers (the short side). This produces a solid wall that is a full brick in depth. English bond is easy to lay and is the strongest bond. It is the preferred bonding pattern for bridges, viaducts, embankment walls, wartime shelters, and other civil engineering architectures. The Clifton Rocks Railway tunnel lining is four-bricks deep, so it is very strong.

Stretcher bond is used for curved part of tunnel roof and is one of the most common bonds. It is easy to lay with little waste and composed entirely of stretchers set in rows, offset by half a brick. Bricks were used end-on in the curved roof to aid compression.

3.16. English bond brickwork. (www.i-brick.com)

3.17. Stretcher bond brickwork. (www.i-brick.com)

3.9.1 Bricks and brickwork in Clifton Rocks

Several types of bricks are in the tunnel. Engineering bricks would have been used in the tunnel for their strength, durability and water resistance. Probably Cattybrook bricks were used, as in the Severn Tunnel. These are 9in x 4in x 3in. They do not appear to have frogs, and are very dense bricks. They weigh 8lb. The bricks are hand-made, heavy, fine-grained and one can see marks where they have been dried on the ground.

It is noticeable that the bricks are laid horizontally to create the tunnel support walls, and had to be cut at angles since the roof walls match the angle of the tunnel (plates 3.22 and 3.23).

Larger Malago colliery bricks have been used in the top and bottom station and are 9in x $4^{3}/_{8}$in x 3in. They weigh 6lb 12oz.

The bricks used in the war-time structures are London Phorpres; they are 9in x $4^{1}/_{4}$in x 3in and weigh 5lb 12oz (pressed four times and efficient to make).

Where there is a large amount of rock sticking out, there was no point in removing the strong rock to replace by brick, so it is very noticeable that buttresses have been built underneath and brickwork has been carefully built around the protruding rock. A rock plan is shown later in plate 3.28. In plate 3.18 the layers of

3.18. Protruding hard rock. (Author)

rock can be seen to be fractured. The railway is in a tunnel and is part of a transport system, and the car moves quickly, the tunnel being lit by gas light, and there are skylights at the top so there was no point in making the tunnel look pristine.

Malago colliery bricks from Bedminster were used in the bottom station above the sliding doors and for the chimney flue in the top station. They are not as good quality bricks as in the tunnel, and they are only used for small structures.

In the ballroom corridor larger bricks were used – 9in x 4½in x 3in where visible. This bond has three courses of stretchers between every course of headers

3.19. Malago colliery brick. (Author)

3.20. Brickwork along the back corridor. The Pump Room is behind this wall. (Author)

(as can be seen in plate 3.20). It could be standard English Garden Wall bond, or it could be American common bond. A bigger brick makes for a thicker (and thus more insulating) wall.

3.21. Internal buttress brickwork. (Author)

Note the angled bricks in the top floor of bottom station to help as extra bracing and strength – an internal buttress. Further along this wall, all the bricks are laid horizontally. Bricks are 9in x 4in x 3in.

3.22. Skewed roof bricks meeting tunnel wall. Each brick has been cut to accommodate the curve of the roof. This photograph was taken on the north side of the tunnel near the top. Look back at plate 3.11 to see how the skewed roof bricks meet the parallel roof bricks. (Author)

There is a skewed roof at the top of tunnel. The bricks are cut at steeper angles. Same size bricks are used in the wall and tunnel roof, but both wall and roof bricks have to be cut at the joint. The supporting bricks are laid horizontally, the roof bricks at a steep angle. Each joining brick has to be cut at an angle. Bricks are 9in x 4in x 3in but because it was hard to use this length brick at this angle and this curve, each had to be chamfered to fit so some have been reduced by as much as one inch in length. This has been done for the first 15ft 10in on the north side. There is a slight overhang of the adjacent upper brick.

3.23. Brickwork where there is a not a skewed roof, so the roof bricks are laid parallel to the wall. Note bricks cut at different angles. This photograph was taken on the south side. (Author)

Where there is not a skewed roof, the supporting bricks are laid horizontally and the roof bricks follow the line of the tunnel (compare plate 3.23 with 3.22). Only horizontal joining bricks need to be cut at an angle so are easier to lay than the skewed brickwork.

3.10 Construction starts

The story is told by contemporary newspapers and Merchant Venturers' reports. Note that the Merchants owned the site upon which the Railway would be built; note also that early reports were optimistic.

> 17 November 1890: We hear that the works in connection with the proposed lift railway from the Hotwells to Clifton Downs will be commenced in about a fortnight or three weeks. The cost of making the line and of constructing the spa at Sion Hill is estimated at about £30,000 and the whole of the capital will be provided by Mr Newnes MP, the spirited promoter of the undertaking.

> [In the *WDP*] 6 January 1891: Mr Newnes, MP, is to carry out the Clifton Lift and Spa with the business-like promptness which characterises him; and it will be understood that no time has been lost when it is stated that the tenders for the works are to be sent in by the end of the week. The lift will be the subject of one contract; and the Spa, with all its necessary pumping arrangements will be comprised in another; and the whole scheme is expected to cost about £30,000. Not a day's delay will be allowed unnecessarily to interfere with the progress of the undertaking; and the public will therefore look forward to seeing Clifton in the enjoyment of these evidences of a visitor's public spirit.

> 20 January 1891: Mr P. Munro the engineer of the proposed lift railway applied for the temporary use of some ground at the foot of the zig-zag for the storage of boring material etc. The request was granted for a period of six months.

> [The *Bristol Mercury*] 22 January 1891: the tender of Mr C. A. Hayes [see chapter two] Bedminster, for the construction of the Clifton Rocks Railway has been accepted by Mr George Newnes, M.P. The work of tunnelling from Sion Hill down to the Hotwell road is to be commenced immediately, and will be carried on night and day until its completion which is expected to be in June next. There were a large number of tenders and it is gratifying that a Bristol man has been successful in obtaining the contract; that the work will be executed by Bristol men and that it will occasion the expenditure of a large sum of money in Bristol. Mr Philip Munro is the resident engineer.

The Railway would not be completed in June 1892 and was not opened officially until March 1893. The writer wrongly describes Newnes as a Bristol man and, though one of the conditions set by the Merchant Venturers in 1890 was that 'no blasting takes place between 7 p.m. and 7 a.m.', the article states tunnelling will be carried out night and day.

> 17 February 1891: There have been rumours flying about as to a hitch in connection with the construction of the lift railway from the Hotwell to Clifton Down but we hear that the promoter, Mr Newnes MP and the Merchant Venturers are in complete accord. It is true that a little matter has arisen with regards to one portion of the undertaking [see 27 February], but it is hoped that this will be satisfactorily arranged in the course of a few days. The work, it is believed, will be started in a very short time, and the Mayoress (Lady Wathen) is likely to take a prominent part in the ceremony with which the tunnel will be commenced.

> [The *Bristol Mercury*] 27 February 1891: New plans were submitted to the Merchant Venturers on behalf of Mr Newnes of the Pump Room and of the Lift station to obviate the necessity of obtaining the assent of the owners of Caledonia Place to an alteration of the covenants [see 28 February]. The same were approved.

> 28 February 1891: The works in connection with the construction of a lift railway from Hotwells to Clifton Down and the building of a spa on a site on Sion Hill will shortly be commenced. The 'first shot' of the blasting operations necessary for the making of the tunnel will be fired on Saturday next and it is hoped that the Mayoress

(Lady Wathen) will be present at the initial ceremony. The projector of the Clifton Rocks Railway and Spa is Mr George Newnes MP, and he has enlisted the services of two talented engineers, Mr G. Croydon Marks of Birmingham and Mr Phillip Munro, St Stephen's Chambers, Baldwin Street. It appears that one of the inhabitants of Caledonia Place objects to the construction of the edifice on the ground that it would contravene the covenants on which he had purchased the property and would interfere with his vested rights of light and air. Mr Newnes made endeavours to come to an arrangement with the objecting proprietor, but without avail, and now alternative plans have been submitted to the Merchant Venturers Society, who are the owners of the site, and these have been accepted. The elevation which the lower scheme offers to the front is not of a striking character, seeing that a height of five feet only precluded any attempts at ornamentation.

This is why the entrance to the Railway and Pump Room are in Princes Lane, and there is only a 5ft wall along Sion Hill.

On the same day it was announced that a contract to lay a tramway in Matlock had been tendered by the newly-formed Matlock Tramway Company. It would cost between £8,000 and £9,000 and be completed within five months of the contract being signed. Directors were Newnes and Marks amongst others – gluttons for punishment.

2 March 1891: Mr C. Hayes, the contractor for the Clifton Rocks Railway, will commence operations at once, and it is expected that the railway will be ready for opening early in August. The contract was to have expired in June, but there have been two or three weeks delay arising from a cause over which he had no control.

3.11 Firing of the first shot, March 1891

[*WDP* 3 March 1891] We have been informed by Messrs G. Croydon Marks and Philip Munro, that the preliminary ceremony in connection with the above railway will be performed on Saturday next, March 7th, at 3.30pm, the Mayoress of Bristol (Lady Wathen) will fire the first blasting by electricity, and military bands have been engaged to enliven the proceedings and to interest the spectators. An engineer's enclosure will be provided for members of the Corporation, while ample accommodation remains for the public in the private road and upon the Downs next adjoining. This initial ceremony is one of the most important events that have occurred in recent times to this city, and we have no doubt that the citizens will not be wanting in showing their appreciation of the enterprise of Mr Newnes MP and will also accord their approbation to the indomitable energy, courage and skill of Messrs Marks and Newnes in thus far overcoming all difficulties. Mr and Mrs Newnes will, we understand, be present at the ceremony.

7 March 1891: The older part of Clifton – the district of it which suggests thoughts, not of adjacent Bristol, but of Bath or Cheltenham – was the most pleasurable twitter of excitement on Saturday afternoon. There was a gay display of bunting, servants were at windows and area gates as if a smart wedding was being celebrated just a few doors off, and on all hands there were signs of rejoicing. The occasion was, of course, the first blasting of the railway which is to carry passengers from the valley to the top of the hill and supersede the breathless climb up the Zigzag. It is the population of Bristol which must be looked to make the undertaking pay, and their interest was shown by the thousands who were massed upon the Downs, along the Suspension Bridge, and wherever a view of the proceedings could be obtained. No doubt was entertained that the railway will be very profitable, for as Mr Newnes justly says, the natural beauties of Clifton are so great that a cheap means of approach is sure to be popular. The scheme also includes a spa, to revive the fame of the Hotwell water. That will certainly be a slower development. It is currently believed that the bulk of the supply has been choked up, and at any rate the Hotwells have been erased for many years from the list of watering places, though at one time it dreamed of rivalling Bath.

The promoter of the railway and spa is Mr George Newnes MP whose enterprise in establishing a line between Lynmouth and Lynton is a marked success. It is proposed to cut a tunnel from Hotwell Road to the upper ground facing St Vincent's Rocks Hotel and within the tunnel to lay four lines of rails in order that two cars may ascend while two others descend. The motive power will be water, used on a principle protected by law letters patent. It is intended to utilise the water over and over again by means of two storage tanks or reservoirs which will be built on the upper and lower end of the line respectively. Powerful pumps, driven by a gas engine will force the water from the lower to the upper tank, and this will be the only expenditure of power necessary. The weight of the water itself being the agent for actually raising the cars. As is well known in this system, every provision has been made for safety, and special improvements have been introduced with a view of, if possible, having everything duplicated throughout. There will be no delay in carrying out any the work in hand, and several portions will be pushed forward simultaneously in order that time may be economised. The entrance to the railway at the Hotwell Road will be in the form of a handsome castellated building [must have changed plan] cut from the rock, while the upper end of the line nothing but a graceful archway leading to a waiting room, will be, seen, so that in no sense it is asserted, can it be said that the railway will be an eyesore or a spoliation of the scenery. The engineers engaged in the works are Mr G. Croydon Marks of Birmingham and the contractor is Mr C. A. Hayes.

Saturday's ceremony is fixed for 3:30 and for some time before that hour people began to assemble various points commanding a view of the proceedings being quickly crowded. Visitors responding to the invitation of the engineers found accommodation in the grounds attached to Princes Buildings where a platform had been erected and decorated with bunting, and reception room had been elegantly furnished by Messrs. Trapnell and Gane [cabinet makers] of College Green. During the interval preceding the function of the day the united bands of Bristol Artillery and Rifles stationed on the lawn rendered an excellent program of music: There was some uncertainty as to the weather but the slight showers that fell early in the afternoon had apparently very little effect on the attendance, and by the hour the shot firing was to timed to take place there was a dense crowd on the green adjacent to the engineer's enclosure while considerable gatherings were to be seen lower down the hill and on the Suspension Bridge in the distance A slight shower fell a few minutes before half past three o'clock but afterwards the sky cleared and the weather occasioned no further trouble. With commendable punctuality the Mayoress Lady Wathen who had promised to fire the shot took her place on the platform escorted by Mr G. Newnes MP etc. No time was lost in disposing of the business of the afternoon. The whole affair lasted no more than 15 minutes. Lady Wathen was conducted to the front of the platform where the necessary firing apparatus was fixed upon a table in full view of a large number of spectators and a few moments were spent imparting requisite instructions to the Mayoress. After this an exchange of signals with handkerchiefs indicated that everything was ready and without further delay a shot was fired. This was done by Lady Wathen pressing a key which made an electrical connection completing the circuit upon which a fuse in the distance exploded a large charger of tonite the most powerful agent known, and much safer in working than ordinary material.

The explosion occurred about a third of the way down the cliffs [see plate 3.1, the second air shaft from the right would have been too close to Tuffleigh House. Even though the railway is being built in a tunnel, this was one of the four intermediate 8ft x 3ft shafts needed for construction, both to enable excavation at the bottom of each shaft and lift the rubble out] and dislodged about half a ton of rock.

Simultaneously the band struck up the National Anthem and the crowd applauded enthusiastically. Cheers were called for Lady Wathen on the success of her effort and Mr Newnes observed that it 'was simply splendid'. The inaugural explosion was followed by the discharge of a number of signet rockets some of which travelled 1000 feet into the air and speedily there were replies from vessels in the harbour and river which continued at intervals for some time. The singular shriek of a fog siren was particularly noticeable especially when it occurred during the brief spell in speech making which comprised part of the program. The penetrating sound completely smothered one of two remarks, but the speakers were in no way perturbed and the interruptions had only the result of provoking some merriment.

After the shot firing, Mr Newton presented to the Mayoress a large inkstand, which had been cut out of the rocks by a special operation. The rock was cut so as to form a miniature representation of the Avon Gorge at St Vincent's Rocks, and on the lower portion, in which was the inkwell there was a silver plate bearing the following inscription: 'Presented to Lady Wathen, Mayoress of Bristol, on the occasion of her firing the first blasting charge for the Clifton Rocks Railway, March 7th 1891. George Newnes, MP. Chairman: G. Croydon Marks and Phillip Munro engineers.' The polishing was done by Messrs. Leo Bros. and the silver mountings by the Goldsmith's Alliance. The tray upon which the inkstand was rested was of marble, and accompanying the gift were several silver penholders. In making the presentation Mr Newnes expressed the hope that the gift, which was offered on behalf of the engineers, the contractor, and himself would be an interesting memento of that day's proceedings which the Mayoress had so gracefully adorned with her presence. He explained that the stone of which the inkstand was made had been cut from the rocks, and he added that they were all extremely obliged to her for the willingness she had shown in accepting the invitation to take part in today's ceremony. He thanked her in the most graceful manner in which she had performed the duties allotted to her (applause). [We have never seen any images of the inkstand.]

The Mayor who was asked to reply, said it afforded him very much pleasure to return the sincere and hearty thanks of the Mayoress for the very handsome present given her in commemoration of the event of the afternoon. He was delighted to welcome Mr Newnes amongst them (applause). Mr Newnes was a gentleman who had had the courage to come to the midst of a large an influential city like Bristol for the purpose of promoting an undertaking for the benefit of Clifton and the immediate neighbourhood. Clifton in years gone by occupied a very prominent position as regarded the faith in the Hotwells and other springs. He found on reflecting to the report issued in 1780 of a very imminent doctor that the Hotwells was one of the most fashionable and crowded places in the kingdom and that visitors from all parts of the United Kingdom and elsewhere were sent there and received considerable benefit from the use of the waters, notably in the care of consumption and other diseases of that class. In fact it was reported that the influx of visitors was so numerous that at times, especially in summer, that lodgings frequently could not be got or could be secured only for exceedingly high prices. He hoped that the result of the undertaking they were all inaugurating that afternoon would be to bring from all parts visitors desirous of benefitting from the use of the noted waters of the Hotwells. He remembered when there was in use a pump room at the bottom of the hill immediately beneath the point on which they were standing. It had only been done away with, he believed through the improvements necessary. for the conduct of the navigation of the river in connection with the Docks estate. Judging from the principle about to be adopted pumping the water of this valuable spring to the rooms to be provided by and by, it was apparent that the original character and value of the water would in no way be impaired, and it would be a boon to many residents (applause).

Subsequently Mr Marks presented Mrs. Newnes with a neat paperweight cut from the rocks as an interesting memento [see plate 3.24] and similar gifts were handed by Mr Newnes to the Mayor, Mrs. Marks and Mrs. Munro. Each paperweight bore a silver plate, on which was engraved the inscription: 'Made from the first rock cut in the construction of the Clifton Rocks Railway March 7 1891'. A luncheon was given in The Clifton Rocks Hotel.

This was undoubtedly a massive occasion for the citizens of Bristol and for Newnes. He clearly saw this as flagship project, a major competitor to Bath and other spa towns. He was always a man with an eye to the main chance. Many a paddle steamer carried tourists and trade between South Wales, Bristol and Lynmouth. Building a superb spa at Clifton with a cliff railway station and a paddle steamer stop could bring more trade to Lynmouth, if promoted. Lynton might then attract more trade and not miss out on the economic growth and social vibrancy of Lynmouth. This in turn could attract even more tourists to the Lynmouth and Lynton and the North Devon area.

The *Bristol Mercury* confirmed that Newnes hoped the railway might be open in August (as did Hayes the constructor) – only six months later. Optimistic, but

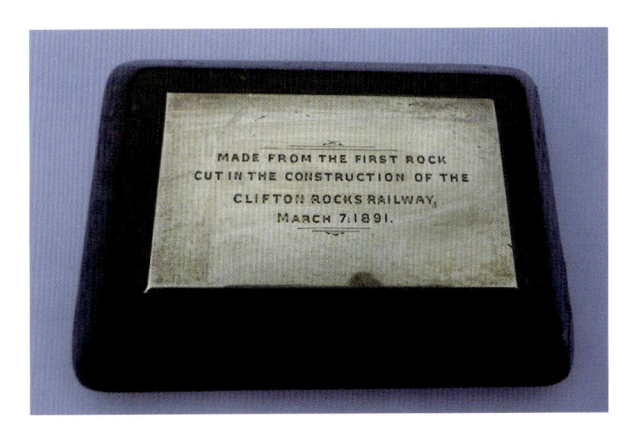

3.24. We were lucky enough to obtain a paperweight at auction, but we do not know what happened to the inkstand. (CRR Trust)

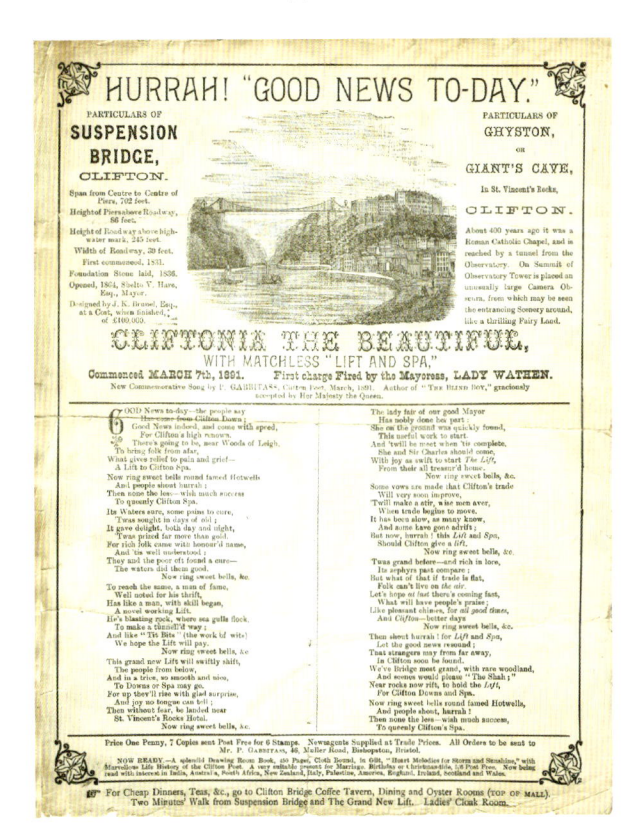

3.25. 'Cliftonia the Beautiful', a poem by P. Gabbitass to celebrate the occasion. (CRR Trust)

then all the other railways had been built overland so were easier to build. It is staggering that he was planning several other cliff railways at the same time.

When built, the Clifton tunnel would be the largest of its kind in the world; it would also be the only one to have four tracks – either for heavy traffic or different classes of ticket.

Blasting occured three times a day – once after breakfast, once after dinner and a third in the evening.

On 24 April 1891 it was reported that the Master of the Merchant Venturers had permitted Mr Newnes to begin operations for making the 'Lift', Newnes having deposited £1,000 as a guarantee for the execution of the deeds, which were in course of preparation. This was over one month after the first blasting.

3.12 Disaster strikes

Disaster struck on 29 August 1891 when James Willis (a timberman's labourer) was crushed to death by a rock fall. William Watts, aged 22, of 238 Hotwell Road was seriously hurt and several other men were hit by falling debris. On 1 September the *Bristol Mercury* reported that Croydon Marks and Hayes were at the inquest. The inquest and subsequent claim for compensation give a good insight into construction and into the checks the foreman and Croydon Marks carried out, their experience, and the types of men employed (engineer, contractor, foreman, timberman, timberman's labourer, miner). In total around 200 hands were employed at any one time.

At the time of the accident Croydon Marks was on site doing his periodic inspection as was Hayes the contractor:

> Yesterday the city coroner held an inquest on the body of the unfortunate man James Willis, who lost his life on Saturday morning last whilst working on the Clifton Rocks Railway.
>
> The coroner, in opening the proceedings said that the subject of their inquiry was a poor man who had met with a fatal accident whilst engaged on excavating at the Clifton Rocks Railway, Hotwells. The questions they had to consider were whether the accident was due to the negligence of some persons in not taking proper precautions, or whether it was one of those sad occurrences which no one could have foreseen.
>
> The wife of the deceased though present was too ill to give evidence and identification was proved by Mary Jane Richards of 5 Charlton Street, Barton Hill, who deposed

3.26. Note the winch with two operators and children watching. The man with the watch chain standing on the scaffolding is probably Hayes. George George with his watch chain and hat is on the right, the gang of men with simple tools, the massive amount of scaffolding. They were engaged in a process called timbering (vertical faces of the excavations need supporting by means of timber, to prevent rock and soil from falling in and injuring the workmen or the work upon which they are engaged). (Official Description of the Clifton Rocks Railway)

the unfortunate man was 39 years of age and had resided at 60 Richmond Street, Barton Hill.

George Wigmore, foreman of the workings in connection with the Clifton Rocks Railway, stated that on Saturday morning last the deceased with several other men, was engaged to dig a tunnel about 45 feet from the bottom entrance when suddenly a huge mass of rock fell [30ft back in the workings], burying the deceased and injuring several other men.

The location of the fall was west of where the tunnel changes from being at an angle to being horizontal (see plate 3.9).

The coroner 'When did you blast the rocks last?
Witness 'The previous night
The coroner 'Is the roof supported after blasting operations?
Witness 'Yes, sir. We fix the supports as we go along.
The coroner 'Do you think the recent wet weather has had anything to do with the fall?
Witness 'Yes sir. I have noticed water dripping through joints in the rocks.
A juryman 'How was the roof supported?
Witness 'We use large trees, about 18 feet in length. They supported the roof from a sill on the side of the rock.'
The Coroner 'Have you had any experience in this work?
Witness 'Yes sir; I have worked under Mr Walker at the Severn Tunnel and several other large undertakings.
Mr George Croydon Marks, consulting engineer to the Clifton Rocks Railway Company, produced plans and photographs of the workings, and explained in detail the formation of the rocks. He said the sides of the tunnel were composed of limestone, granite and sandstone, whilst a few feet above the solid rock was a layer of clay, which had occasioned the fall. The rock which had fallen was in three pieces, one measuring 4 feet in length, 40 inches wide, and 16 inches deep, and the piece circular in configuration was 7 feet in circumference, whilst a thin piece was 3 feet 6 inches wide, 4 feet 8 inches deep and 15 feet in length. He had broken off several pieces and weighed and estimated the weight of the three pieces to be 27 tons. With regard the timbering, the instructions given were to fill up all the way through the tunnel in shoring [placing timber supports under a huge piece of rock], and that had been carefully done.

A juryman 'Whom were the instructions given to?
Witness 'To Mr Hayes, the contractor, and to the foreman. They have been faithfully carried out. I have been there myself every week, and my colleague every day.
A juryman 'Is this the first clay that has been found?
Witness 'No. We have had several slips of rock, but no one has been injured. It would have been impossible for anyone to tell there was clay above eight feet of solid rock. When the men heard the rock come down two of them jumped to the left and escaped but the deceased jumped to the right and came in contact with the falling rock.
A juryman 'Might not a further fall of rock occur?
Witness 'There may be another slip, but they could not take any further precautions. They used the largest trees that could be obtained, and there was so much timber used that persons had taken photographs of the workings out of curiosity[11] [see plates 3.26 and 3.27]. About the end of the week they would meet the cutting above and then all danger would be over.

No further evidence was taken and the Coroner said there was no doubt but that the sad affair was purely accidental for the evidence showed that every means had been

3.27. Scaffolding of top station arch.

adopted to prevent the -rocks-falling. The verdict was accidental death.

3.13 Compensation for the victim's wife
On 13 January 1892 there was a report in the *Western Daily Press* regarding compensation for the widow of Mr James Willis (a timberman's labourer). She had been married for 10 years. She had claimed £210.12s. The claim was under the Employers' Liability Act 1880. Before that it was impossible for a worker to hold his employer responsible for injuries caused by his foreman or another worker's negligence. The insurance company had not responded. She was not allowed to claim more than three years' wages. His weekly wage was 27 shillings (daily 5 shillings; 5½ pence per hour, thus working 11 hours per day), though James often earned 30 shillings. It was contended that the men complained to George Wigmore, the foreman, that the roof was improperly supported, and in consequence some props were put up but not enough. Sometimes the props fell during blasting. A back-ground miner considered that as one drove the heading that the roof should be timbered. A timberman was of the opinion there should have been nine bars to support the roof at that point but there was only four, and one was bound to another by a chain. The sill was the piece of timber that the props should rest on, and on the day it was being taken in. The props were removed to allow it to pass and the roof fell in. He had not heard the foreman suggest that more rock should be cut out at the side to allow the sill to pass.

A miner confirmed that the men had dragged the sill which caused the props to shake and the roof then fell in. The accident was caused by the removal of the middle props. Another witness thought that the 'roof sounded like a drum'. Mr Wigmore said the ground was more treacherous than expected.

The rock had broken away from a spar-joint. Three days before Mr Wigmore had ordered three more bars and props. He had wanted the side to be cut away instead of moving the props, but one of the men thought it easier to remove the props. The judge was surprised he had given in to them and said 'The men are accustomed to using their spades, and not their heads'. The roof had fallen in about one hour after removing the props. Croydon Marks said that the spar joint in the rock would not be noticeable. The primary cause of the fall was a fault in the rock about 8ft up. The timber would have been sufficient but for the fault.

The judge was concerned whether proper precautions were taken and thought the foreman should not have waited for men to ask for timber to be put up. 'Why should the men suggest danger? It is for the foreman to see there is no danger'.

Addressing the judge on the question of damages, the counsel for the defendant Mr Hayes, considered she had asked for too much. There was no negligence, rather an error of judgment. His average weekly earnings were 28 shillings 3 pence. He admitted nothing could compensate for the loss of a husband. The judge said 'I think we generally flatter ourselves that that is so.' It was reported there was then laughter in court. The woman only received £1 per week for support so was not entitled to ask for a proportion of earnings. She was only given £156 rather than £210-12s. It was Mr Hayes, duty:

> to see that some care was taken of the lives of those working under him. The men were working with their spades and could not be expected to be constantly looking up to see if the roof was going to tumble in.

The Employers' Liability Act was replaced by the Workmen's Compensation Act 1897, which removed the requirement that the injured party prove who was responsible for the injury – instead they needed only to show that the injury had occurred on the job.

3.14 Job advertisements
It is interesting to watch the progress of construction by looking at job adverts so:

> 5 September 1891-12 October 1891, masons and bricklayers are wanted (as a result of the bed of clay being found where solid rock was expected).
> 21 November-23 November 1891, more bricklayers are wanted.
> 26 January 1892-28 January 1892, pennant stone cutters wanted.
> 5 March 1892, stone cutters wanted.
> 20 April 1892-24 May 1892, plasterers wanted.
> 12 September 1892, machinery, drills and plant used at the railway for sale by Hayes the contractor. No reasonable offer refused.
> 1 March 1893, Baxters stone crusher nearly new and a boring machine for sale by Hayes the contractor.

3.15 Decision to line the tunnel with brick
On 23 September 1891 the *WDP* reported that:

> The Clifton Rocks railway is within measurable distance, and Munro, the engineer, expects that in about seven weeks all will be ready for the Venture to be opened' [It

would be a further 18 months.] 'It was originally hoped to finish work by the end of September, but unexpected difficulties occurred in the construction of the tunnel. Beds of clay and marl were found where solid rock had been calculated on, and consequently the roofs had to be shored up. The defective portions will be made secure by brickwork The workings are at an inclination of 1 in 2.223, and this fact will give some idea of the difficult nature of the undertaking. The heading has been finished, and the contractors are at present enlarging the tunnel to a height of 18 feet 6 inches, and a width of 30 feet. Diamond boring drills are at work, and with the aid of these two holes 2^1/$_2$in in diameter and 12in deep, can be bored every minute.

This was skilled, dusty, noisy work and dangerous in a tunnel. We are not sure how many holes had to be drilled and the distance apart. Drilling unnecessary holes is wasted effort. The finished width of the tunnel varies as can be seen on plate 3.28 from 27ft to 25ft 11in. It had been widened to allow for four courses of bricks in cement and the sides lined with 18-inch brickwork. The height was increased by 1ft 6in and the tunnel was arched throughout.

There is still evidence of shot holes (see plate 3.29) in rocks jutting out in various places on both sides of the tunnel, but more on the south side than the west about half way down. Where the rock was deemed to be solid (see plate 3.18), it was pointless replacing it by brick. There are many buttresses to be seen, and bricks filling in gaps in the projecting stone. The size varies from as much as 44ft x 5ft 7in to 2ft 9in x 1ft 8in – with the bigger outcrops on the south side. There are shot holes inside the tunnel and long ones outside (see plate 3.30).

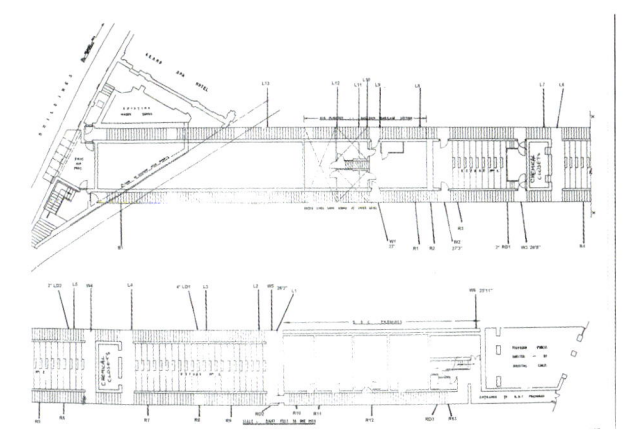

The majority of the visible drill holes are 1^1/$_2$in but two at location LD2 on plate 3.28 are 2in and are triangular rather than round so for some reason they may have been opened up further by a chisel. There are four drill holes in this location so perhaps it was a particularly hard bit of rock.

3.29. Shot hole half way down the tunnel where the rock is protruding. Some are vertical, some at an angle, some round, some oval (indicating trouble in drilling), and some are triangular (made by chisel). (Author)

3.28. Measured rock plan showing location of rock sticking out in the walls, and drill holes. (BRO, modified by Jon Picken)

3.30. Shot hole visible by Hotwell Road entrance, some vertical and some at an angle. (Author)

Interestingly there appears to be 51% of rock still sticking out on the south side (13 outcrops), and 36% on the north side (13 outcrops) when one considers the rock above the steps (with the wall five inches above the steps out of a total height of 7ft. The steps tend to be cut round the rock so there is rock showing under the steps too. The curved tunnel roof is brick lined with no rock visible.

> [*WDP* 23 September 1891] The work has been going on at Zion Hill at the top and Hotwell Road at the bottom, and also at three intermediate shafts. The latter were at a depth of 57', 48', and 17' respectively. All the plant – engines, carriages, rails and other appliances is ready, the waiting room at the top is nearly completed about ten or twelve days will suffice to get the railway in working order.

So near and yet so far.

On 30 October 1891, the owner of Ghyston House (Tuffleigh House) went to the Merchant Venturers to order the surveyor to remove an obstruction which prevented the occupiers of the Colonnade from their right of access to Princes Lane. It was also reported that Newnes' tunnel was above the level of that lane. The clerk of the Merchant Venturers was directed to require the surveyor to remove the obstruction.

On 19 November 1891, the contractors were no longer being specific about a date:

> It was generally expected that the lift railway between Hotwells and the higher levels of Clifton would be finished this autumn but progress has been a good deal impeded by beds of marl and clay which have been encountered in driving the tunnel. However the whole work is considerably advanced, and the line will be opened in the new year. There are still 70 to 80 men engaged in various parts of the line [I was told by a bricklayer's granddaughter that 12 bricklayers were employed]. The tunnel has been driven the necessary distance and the contractor's employees are busily engaged in enlarging it to the requisite height and width and lining it with brick. Then there is some work to be done at either end of the railway, but that should not occupy much time. Everything necessary in connection with the actual working of the line has been prepared so that there shall be no time lost at the last moment getting things shipshape and making a start. Four carriages have been built in Manchester and are ready for use: each carriage has five seats and will hold 25 passengers. No arrangements have yet been made for engaging the staff to work the line – but that is a small matter: nor has anything definitely been decided upon in the way of dispatching the first car on its journey through St Vincent's Rocks.

On 17 March 1892:

> the work is now in measurable distance of completion. Patience and energy have overcome all obstacles. Next week in all probability a start will be made with the erection of the engines. Confidently expected to be ready by Whitsuntide. [Actually one year to go to opening.]

On 9 June 1892 'It is now expected that the Clifton Rocks Railway will be finished by the end of August'. They advertised the plant for sale in September so they must have finished the drilling.

On 19 October 1892 it was reported in the *Western Daily Press* that the cars appeared to be of a different design to that of November 1891 – they were originally going to hold 25 people on 5 seats – they would now hold 18 on two seats with passengers facing each other.

3.16 Claim for damages

On 12 November 1892 there arose evidence of a different problem, when contractor Mr Hayes was taken to court by Mrs Hudson, a resident of the Colonnade. (This was after the blasting has stopped in September.) Mrs Hudson claimed for £44 1s 5d, for damage done to her four houses on the Colonnade, nos. 3, 4. 5 and 6, in the course of operations. The claim also included items for trespass, damage to fruit trees, and money spent on repairing the damage done. The plaintiff said that

> a good deal of blasting was necessary and the concussion broke many of the windows on her property. As the rock and debris were taken out of the tunnel it was thrown into a sort of shaft until in due course of time became choked with the result that stones and earth came through into the gardens of these houses, and pieces of rock weighing as much as 50lb or 60lb bounded over the garden wall onto the roof, causing considerable damage. The chief damage complained of was to a large brick wall, which in consequence to the weight of earth and rock against it had bulged considerably. In order to repair this properly it would be necessary to pull it down and rebuild it, which would cost between £25 and £30. Complaints were made to Mr Hayes, and he promised to send a man to do what was necessary to the roofs, but he let some time elapse before he did this. There had been considerable correspondence. Evidence was given by Mrs Hudson and Mr Wyllie (her brother) who lived at number 5. The case was

adjourned for a site visit. Mr Hayes, giving evidence said that Mr Newnes was the owner of the wall, it being the retaining wall of the upper portion of the land purchased by him. It was in a state very similar to its present condition. With regard to the claim for broken glass, it was contended it was not broken by stones from the tunnel. His Honour said that there were several points more or less complicated. As to the item of £3 5s for labour and materials for the repair of the damage, he was of the opinion that the plaintiff had not proved her case. On the item of £5 17s 6d for repapering ceilings, glass for the greenhouse and other work he considered the action should have been taken by the occupier for trespass instead of by the owner. With regard to repapering, the damage was too remote to return a verdict, as to rebuilding the wall, trespass did not arise. He considered the acts of ownership, found for the plaintiff, and assessed the damage at £15. He gave a verdict of £2 sufficient for the fruit tree. So £17 rather than £44 to the plaintiff and costs borne by the defendant. I am sure she must have been really upset by disruption of the blasting for a year longer than expected.

On 1 December 1892 the *WDP* reported 'The lift is expected to be open for traffic almost immediately'. Reasons for the delay are suggested in 3.18.

3.17 Electric signals

It was reported that, on 10 December 1892, a contract was secured for the erection of the electric signals by Messrs King, Mendbam & Co, Western Electrical Works, Bristol and London. The arrangements to be made include means for signalling from the lower to the top the number of passengers to be carried up the railway each journey, replying and starting signals and telephones. More details are to be found in chapter 4 and some artefacts in chapter 7 (including a telegraphic depth signaller held in M shed (on Bristol

3.31. We found scratched lettering in the bottom reservoir by one of the water chutes H King – a place which would not be normally seen as it was under the floor. (Author)

3.32. Telegraph wire pulley. Along the wall at 2' intervals are the pulleys that would have supported the telegraph wire between the top and bottom station.

docks) that is purported to be from the top station).

3.18 First test journey

The first journey was experimentally tested on 7 February 1893 by Marks 'and his energetic staff of assistants'. It was confirmed that Newnes must have had his patience tried, and the engineers had been harassed and worried by their arduous responsibility. Hayes had had difficulties in getting his men to work for any length of time in the tunnel. The dangers of removing thousands of tons of loose stone, rock and shale threatening to bury them can be understood. Daylight could be seen passing through from the upper to lower waiting room due to the steepness of the tunnel. It had been decided to have four lines rather than two due to the amount of people using the zig-zag in the summer. The railway would shorten journey times since they could get off the Hotwell Road tram and go up to Clifton by the railway.

> To the business and professional man, whose time means money, this saving will no doubt be appreciated and to those whose duty and pleasure it is to stay at home preparing for the return of the sterner members of the family, we can readily believe this new and quick route to the city will be one they will not be slow to patronise. The work, so far as the cutting, building and constructive operations are concerned, is now practically finished, that still remaining to be done being simply details in the manner of fittings, such as the erection of screens, specially designed turnstiles, and a hundred and one other trifles required to please the eye and minister to the comfort and convenience of the passengers.

The other test journeys are to be found in chapter 4.
The steel rails are firmly secured to concrete and

larch timber sleepers, firmly bedded in the rock bed of the tunnel. One can only see six rails now (see the photographs in the next chapter), since the outside two had stairs built on top during WWII. The larch sleepers can be seen, and the maintenance steps up the middle. The gap between the maintenance steps tracks is 2ft 3⅝in. The gap between a pair of tracks is 2ft 3in. At the top, the gap between the larch sleepers is 2ft 8in, then 4ft, then 6ft thereafter.

Finally, on 11 March 1893 the railway was officially opened. The story continues in chapter 4.

3.33. Four-in. fang pin to hold the rail down on larch sleepers. (Author)

3.34. Below, current track (from 1912). Flat surface width is 2¼in, base 4in, height 5in. (Author)

four

Changes
in Architecture and Operation

This chapter covers the period from February 1893 to the Railway's closure in September 1934, and what happened afterwards until war broke out. It shows how the railway looked when it first opened, the staff, how the cars were powered, safety features, opening day, the company and its demise, passenger figures, sale to Bristol Tramways, and demise.

It is worth noting that most funicular railways have two cars running. To have four cars in a tunnel explains why the track of 3ft 3in is so narrow – it is so expensive and difficult to build a big tunnel. The high expectations of 1,000 passengers per hour could have been a justification. Lynton and Lynmouth funicular railway copes with two cars but has doubled the length of the car. It is useful to be able to have a spare system so that one can maintain the cables, engines, etc., without stopping the passenger service.

It is also worth noting that having to reuse the water from the top reservoir – pumping it back from the bottom reservoir using a gas engine – makes life more difficult too. With too much water you flood the floor, with too little you need enough passengers to make the cars move, to make the heaviest car go down. So you have to monitor water levels more, and not pump water unnecessarily. However, one advantage of reusing water is that the reservoirs do not need to be cleaned of river sediment, unlike at Lynmouth.

4.1 Façades

The top station when constructed originally had a stone wall, just like there is now, along the elevated side of Princes Buildings. This is due to the Merchants asking George Newnes to submit plans in February 1891 to reduce height of new buildings. This enabled passengers to 'land on a level with the existing private road passing between the lower and upper grounds of Prince's buildings.'

This meant the only indication that there was a railway was the sign at the junction of Sion Hill and Princes Lane. It was intentionally far more discreet

4.1. Showing Newnes' sign at the junction of Princes Lane private land and Princes Buildings. (Author's collection)

than the bottom station. The railings that are there now (see plate 4.31) were introduced in George White's time in 1912. Visitors used the steps at the junction to enter and exit to go to the Downs, Suspension Bridge, Zoo, shops etc. To go to the Pump Room they used a subway from the turnstile section. This is another possible reason why four cars are provided rather than the normal two – to keep the visitors segregated, as well as to cope with peak traffic. Note that plate 4.1 shows the top station with a glazed roof. There is not a lamp-post in sight in Sion Hill. The large electric posts for the main Clifton thoroughfares that one still sees nearby, were not introduced until 1898.

The bottom façade is larger and more ornate than the top façade to make it eye catching to passing visitors. This is because there were good transport links – see chapter 5 – and because covenants required a modest façade at top station. The pavement was considerably wider than today. There is a typical gas railway-lamp outside the entrance. The lower station itself is constructed inside the rock, and was finished

4.2. Princes Lane façade. Note the ornate Pump Room to the right, the austere decoration of the Railway top station to the left. It is constructed of ashlar oolitic limestone (Bath stone) exterior face in Egyptian style. Some of the face stone has now been displaced. The middle section with the door was used as a retiring room and toilets for those enjoying concerts. A plan is shown in the chapter 9, plate 9.14. (Author)

4.3. 1894 drawing showing contrast of plain Railway exterior and ornate Pump Room down Princes Lane.[1] There is some Railway railing in Sion Hill. (Official Description of the Railway)

4.4 An early image of the Hotwell Road façade taken in 1893. Electric trams have not arrived yet, only horse trams and horse and fly. There is only an ice-cream cart outside. (Pete Davey Collection)

with the façade erected flush with the rock face. The façade was of rubble construction using grey pennant stone, locally mined and used extensively in the less expensive houses being built in Bristol in the 1800s. Quoins and architraves were of Bath stone with decorative gargoyles above the three entrance arches. The two windows, one on either side, remain undecorated.

There is a distinctive tiled canopy for protection against the weather with decoration very reminiscent of Victorian railway buildings. There are six smaller arches above the upper floor windows giving a superb view across the river.

The ground floor contained three turnstiles (as in the top station) and a pay box (now removed) set in a floor of red six-inch tiles (like the top station, some were inscribed Adamantine and have lasted well). The walls of this room were lined from floor to ceiling with vertical pine match boarding which is still there. On the first floor were the two Crossley engines and pumping equipment, one small toilet (left side front) and some staff facilities. The first floor had a rounded ceiling just like the tunnel. The gas supply to the engines is on the right hand side. The reservoir is below the floor.

Postcards with Newnes' façade are more numerous than postcards with the later 1912 George White façade, and can be dated by the lamp posts and horse-drawn or electric trams.

4.2 Interiors

4.5. Another view of bottom station before electric trams were introduced in 1900. There is a gas lamp post on the left from the 1860s. There seem to be more vendors outside, since there is now a weighing machine. Note the signage above the balcony. Observe the bottom station dug into the rock. One can observe bore holes for dynamite even now left of the grotto (plate 3.30). Plates in chapter 5 show trams and all the wires and posts, taken from the same position. (Author's collection)

4.7. Top station, Newnes' interior. Three turnstiles, central control cabin (water could be controlled from here too – note the wheel inside). The valves providing water for each car are hidden by wooden casing (one per car). Two cars at the top. The glass roof can be clearly seen supported by beams (which are still there). The water pipe returning to the top reservoir can be seen in the middle at the back. The pillars to the left are still there and support the infrastructure underneath Sion Hill pavement. At the back can be seen the subway to the Pump Room with a sign above advertising Apartments to let. The ornate sign between the two men says 'Out'. The signs to the right are difficult to read. The bottom one is for a Bavarian Ales and Stout Brewery in Shepton Mallet. (Peter Davey collection)

4.6. Signage on the glass: SWEET STALL INSIDE, WAY IN at the centre, WAY OUT on the other two entrances. There are so many vending machines it is hard to avoid them, but they would have all brought in revenue. The weighing machine is identified by the big dial. Memories in chapter 8 recall some of the vending machines. The lamp above the centre is inscribed Clifton Rocks Railway. The rectangular board in the middle is for 'Results' (perhaps football?) and also advertises Diadem Flour. There would have been a cigarette machine somewhere. Rock falls may have necessitated the fencing. (Ed Scammell collection)

4.8. Curved brick arch corridor leading from the Railway to the Pump Room. (Author)

Plate 4.8 shows a corridor from the turnstile section. 'There will be another entrance to the Pump Room though the subway from the Lift Railway'. One of the conditions that the Merchants set in November 1890 was that the:

> grounds of the existing house were for the recreation of those who may be staying within the new Spa building and not be thrown open to those travelling by the lift, or to the public indiscriminately, so that a Spa of a very select character will be formed.

Possibly the people on the southern-most pair of cars were intended to go to the Pump Room if four cars were running. The subway is as original, and is a curved brick arch corridor. It is adjacent to the top reservoir (on the right on the same level).

4.9. Bottom station interior, known on the plans as the Munro Waiting Room. The sketch is by Loxton (1857-1922), a prolific Bristol illustrator. Note the sliding door on the right hand and the light at the end of the tunnel. (Bristol Central Library)

4.3 Employees of the Railway

Walter Pearce (shown on the 1911 census) was the manager who served the longest, and his grandchildren got in touch with us. In the 1891 census he was an advertising artist so it is quite a career change. Croydon Marks wife's sister Martha was married to him. Marks was known for getting jobs for relatives and played some part in the appointment of Pearce who would one day manage the railway. He conveniently lived at 4 the Colonnade (otherwise known as 418 Hotwell Road) from 1896 until 1929. He was manager of Clifton Rocks Railway from 1906 to 1916, but he received £7.50 commission from Mr Woolley an engineer (see below and chapter 6) in 1901 to replace eight axle shafts. He clearly was associated with the railway from very early on before he became a manager. Usually one got a house at the same time as becoming a station manager in Victorian times to make sure one was on hand when there are any problems. The station manager was responsible for the management of the other employees, and for safety and the efficient running of the station. In 1905 Clifton Rocks Railway had an address in Princes Buildings, at the changeover of managers. Walter Cowling lived at 420 Hotwell Road (no. 3 The Colonnade). After 1924 there did not appear to be a station manager – perhaps if they were losing money they may have made the post redundant. Mr Price, who owned the Hotel from 1918, then appears to be the contact.

Station managers 1902 to 1923
1902-1905 H.I. James
1906-1907 Walter William Pearce
1908-1913 Walter William Pearce (in receivership)
1914-1916 Walter William Pearce (BTCC)
1917-1923 Walter Cowling

When the Railway first opened, smart uniforms were provided for seven men (see the *Mercury* report for 11 March 1893). If there were four cars running, there would have been four conductors required. There was at least one resident engineer; one station manager who could have acted as a conductor; at least one ticket collector and a minimum of two conductors.

We have learnt the names of more employees from people contacting us to see what we might know about their ancestors, see table 4.1. The average wage of a tram conductor in 1906 was £70 (£147 in 1924) per year, ticket collector £64 (£166 in 1924), and railway mechanic £82.[2]

William Watling	Brakesman		Morgan Court/Cumberland Court, Hotwell Road
Richard Tiley	Conductor		
George Webley	Brakesman (on his marriage certificate)	1906	20 West Mall and Caledonia Place, Clifton (born 1876)
Maggs	Conductor (rode on platform). Made redundant		
Mr Poole	Superintendant (plate 4.34)		
Stanley Pearce	Engineer		4 Colonade, Hotwell Road (census 1911)
James Woolley,	Engineer	1911-22	Son of Alfred Woolley (see chapter 6)
Thomas Rosewell	Brakesman (on census 1901 and 1911)	1901-1912	10 Westfield Place, Clifton. Lodged with another brakesman (buried
	Engineer	1913-1924	Arnos Vale with 6 CRR workmates bearing the coffin) 1881-1924. Brother of Portland Lodge.
George Hemmings	Brakesman	Aged 25 in 1929	8 Chapter Street, St Pauls (on marriage certificate)
George George	Brakesman		Point Villa, Hotwell Road (was gangmaster during construction)
Frank Coates	Brakesman	1925	Ambra Vale West, Cliftonwood
Wilf Perry	Gateman and ticket collector	1905-1914	(born 1884)

Table 4.1. Employees. Note, brakesman and conductor are the same job.

4.4 Lighting

Although the tunnel was lit by daylight at both top and bottom, this was supplemented by gas lamps installed down the tunnel length.

A visitor would descend the steps with simple wrought iron balustrading and polished timber bannisters, from either entrance to the top station onto a small platform which was below upper street level. In fact the platform extends under the pavement in Sion Hill. A series of small arched-roof vaults between the rock face and a substantial steel beam, itself supported on cast iron columns, support the pavement of the street above.

4.10. Glass prisms in a skylight of the top station. (Mike Edwards)

The remainder of the 15-ft wide platform not recessed under the pavement was covered with an awning of small glass panels in iron frames (thus pavement lights). Most of these glass panels are still in place. Many are purple and prism-shaped (to deviate the light) rather than the familiar flat glass blocks.

4.5 The cars

Each car consisted of an upper passenger section, with a triangular chassis angled to suit the gradient of the tunnel (see plate 4.12). The passenger section resembled the horse-drawn tramcars to be found operating during the 1890s on the City tramway, and are believed to have been constructed in Birmingham by George F. Milnes and Co (who took over Starbuck in 1886), and who built tramcars for the City.[3] Each car could accommodate 18 seated passengers and had sliding doors at either end, the door at the end facing the river opening onto a small platform upon which the brakesman or attendant rode alongside the brake control. The car was lit by oil. Cars were painted light blue and white with gold lining when new, but were later re-painted in colours similar to the Bristol Tramway Company (royal blue and ivory). The cars were mounted by four leaf springs onto the chassis which were built by Gimsons of Leicester.[4]

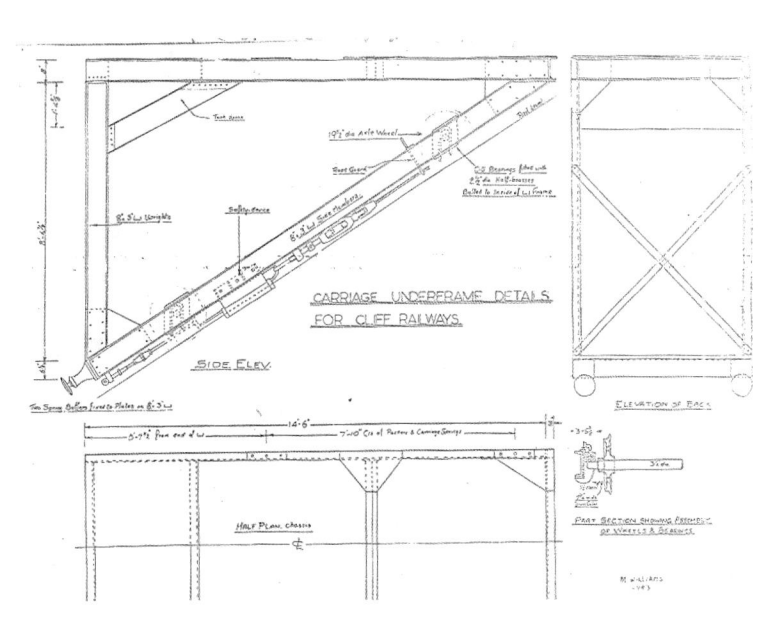

4.11. Diagram of chassis drawn by M Williams, Junior Chamber, 1983.[3]

The four chassis were part of a batch of six ordered on 7 March 1892 and delivered in December, 1892. The other two were for the Bridgnorth/Castle-Hill Railway, those for the Clifton Rocks being designated by a C prefix on the detail drawing. The Bridgnorth equipment had a wider gauge at 3ft 8ins. The chassis were constructed of 8in x 3in steel channel section 14ft 6in long, carried on four wheels. The only chassis item not manufactured at Leicester were the axles which were supplied by J.H. Lloyd and Co. The axles ran in brass bearings fitted onto cast iron housings bolted onto the top side of the bottom chassis member. The cars were handed left and right.[4]

4.6 Operation of the cars

The engineers for this unique little railway Messrs G. Croydon Marks, A.M.I.C.E., M.I.M.E., and Philip Munro M.S.A., F.S.I. have apparently done all that is possible to meet the sentiments of the passengers, for nothing has been left to chance in the arrangements which have been made for safety, and for controlling the working of the cars. It is a matter of surprise to the visitor inspecting this line for the first time to observe the absolute and complete control the drivers have over the cars, and the easy motion and absence from vibration which attends the journey, makes the traffic upon this line a very popular one.

It is on the funicular system, and is to be worked on the counter-balancing principle, the additional weight being obtained from water tanks supplied with water by the Bristol Water Company. There are four sets of rails of a gauge of three feet [actually three feet two and a half inches], and four cars will be working at each operation, two descending and two ascending. Each car will carry 18 passengers.

The principle of working the cars is a simple and economical one, consisting of what is known as the 'water balance' combined with the multiple hydraulic brake controlling appliances, introduced by Mr G. Croydon Marks. The car under-frame had its tank secured within it, the car being secured by spring-bearers to the channel-iron members of the upper table of the frame. For convenience of illustration, the car is shown with the lower and upper station platforms close together. The rope-wheel is carried upon a strong bed-plate, having also an upper tie-plate passing right across it to independent anchor-blocks at the side.

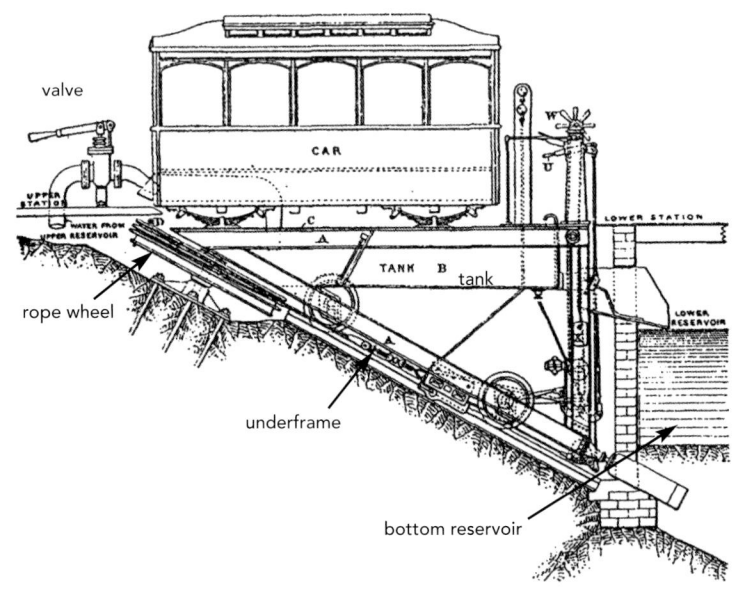

4.12 The general arrangement of the Clifton Rocks Railway car.[5] (Marks)

The valve puts the water into the top car tank (the tank at Lynton and Lynmouth holds three tons of water [670 gallons, plate 4.14] and this would be a similar size. The frame was 8ft 6in high, 14ft 6in long and 3ft wide). Water is let out of the car tank at the bottom into the bottom reservoir when the car has completed the down journey. The capacity of the top water reservoir (see plate 4.13) is estimated to be 5,134 gallons, the bottom to be 9,779 gallons.

4.13. Top reservoir, shows the pipe which puts the water back, having been pumped from the bottom. (Author)

4.14. Bottom reservoir. To the left is a one-way clack valve to get the water from the bottom reservoir to the pumps on the first floor. The pillar can be seen on the ground floor and supports the ground floor ceiling. (Author)

Plate 4.14 shows the bottom reservoir which holds the water ready to be pumped back to the top reservoir by a Crossley 9hp engine and pumps (there is a duplicate system so one can be on stand by).[5] This is crucial because if the water was not pumped back to the top reservoir, no cars could run. More details and images of clack valves can be found in chapter 7.

4.15. Water chute in the bottom station (one per car). Three are still in situ, the fourth was destroyed during WWII conversion to make the new BBC entrance. See plate 4.12. (Author)

4.7 Water balance principle

Gravity (1:2 slope) and weight (people + water), plus a good braking system to prevent a hard landing are the key elements of the water balance principle.

> The two vehicles move together. When one is at the top, water is allowed to flow from an upper reservoir erected near the top station into a tank-like body, arranged immediately underneath the ordinary passenger car. The passengers enter also in the top car, while the bottom car carries no water, but passengers only on the up journey. The difference in weight between the two cars, and their complement of passengers is made up by the amount of water which is allowed to flow into the upper car tank in order that it may overbalance the weight of its bottom companion car; and when the top car, with its water load and passengers, arrives at the bottom station, while the passengers are leaving the car, the water is also automatically flowing out of the under tank-body into a reservoir arranged beneath the waiting room of the bottom station. At the same time that the water flows out from the lower car at the bottom station, the attendant at the upper station allows water to flow into the car that has then arrived at the upper station, so as to be ready to repeat the journey.

4.16. A Lego model has been constructed by the author to show how the heavier car going down brings the lighter car (people and no water) up. The two cars are connected by rope – hence the term funicular. (Author)

4.17. The two pairs of cars work independently since there are two cable wheels at the top – one per pair to enable four cars. The left-hand car is possibly going down and so the extreme left car will be nearly at the top. The conductor is holding the dead man's handle. The right-hand car may be going up and the other of the pair will be coming down. It appears to be empty so there will be people in the top car to justify its coming down. The connecting cables can be seen on the right hand car which run on rollers to avoid dragging on the ground. Each car is mounted on leaf springs on the chassis holding the water tank under each car. The badge on the side of the car says Clifton Rocks Railway and are in the Newnes' livery. Maintenance steps can be seen in the middle. This photograph also shows how much rock is protruding from the walls of the tunnel. The water pipe can be seen on the right-hand side, taking the water back up to the top reservoir. (Image donated to CRRT)

4.8 The cables and wheels

'Two steel wire cables or ropes pass from one car to the other, these two ropes being of a strength such that the combined weight of both cars can be safely sustained by one rope alone, the combined ropes being thirty times stronger than the load that has to be put upon them.'[1] If one cable breaks and all the brakes fail then the cars will not run away.

4.18. The two cables and fittings (including a cable roller) can clearly be seen for each cable wheel. A grease patch is visible to the right where the car would have come to rest. (Author)

4.19. View from the top station. The wheels are securely bolted down. The maintenance steps can be seen between the two wheels. The war-time wall and stairs have been built on the outside tracks which is why only six tracks can be seen today. The larch sleepers are embedded in concrete. (Author)

4.20. Cable rollers made out of cast iron with axles mounted in small bearings support the ropes at intervals of about 30in and stop them dragging on the ground. Note the grooves caused by the cables. (Author)

ton. Four cars need two cable wheels as can be seen in the pictures. The inside diameter of a cable wheel was 6ft, the outside diameter 6ft 4in.

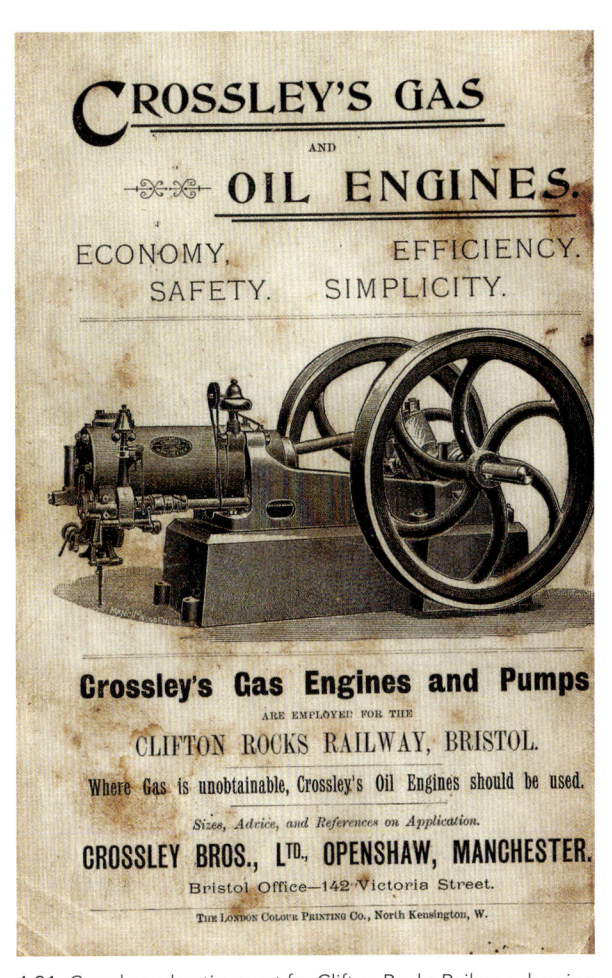

4.21. Crossley advertisement for Clifton Rocks Railway, showing a 12hp engine, not a 9hp. (Official Description)

The cables were made by Cradock (his advertisement can be seen in chapter 6) and are $1\frac{1}{2}$in in diameter. Using an engineering table[6] this would have a breaking strength of 64.4 tons for best patent steel wire rope for winding (72 tons for specially-improved patent steel), and a weight of 392 lb/100ft, 1,960lb or .89

4.9 Engines, pumps and workshop equipment

Dugald Clerk (Croydon Marks' business partner) wrote a paper,[7] in which he compared one of the Crossley 9 engines from Clifton Rocks Railway which had been built in 1892, to a Crossley engine built in 1882 and functioning in Birmingham. (Hot-tube igni-

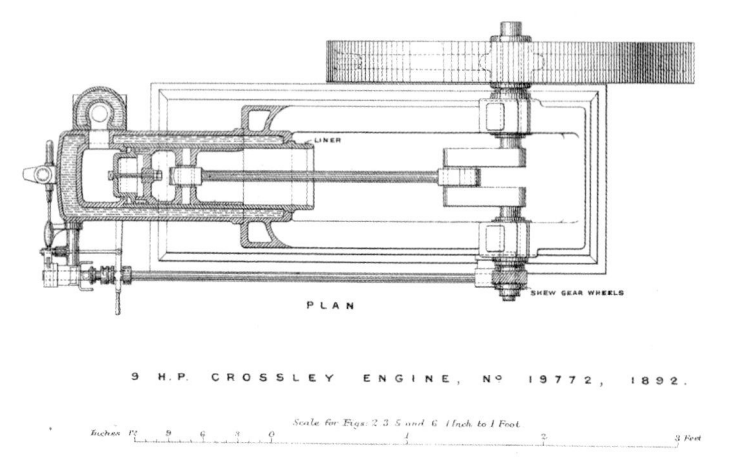

4.22. Clerk's drawing of one of the 9hp Crossley engines. ('Recent Developments in Gas Engines')

tion was found to be better than flame side-valve ignition. The valves and ports had been improved so it gained more pressure.) There are some wonderful scaled drawings (plate 4.22) which we have hung up in the bottom station of the Railway. Clearly, since Marks was involved with Clifton Rocks Railway, and it had a brand new engine, he would have used this engine as a benchmark.

A pair of self-starting Crossley 9hp gas engines (we are not sure who made the pumps) erected in the bottom station first floor (the engine room) were employed for pumping the water back again from the lower reservoir to the top reservoir, so that the water that was needed to work this line was reused. The cost of the motive power for working being that required to drive the gas-engines for pumping the water.

From a condition survey done in 1940, when the tunnel was to be converted for wartime use, we know that in the engine room there was: a large twin-cylinder hydraulic pump; an electric motor with silent chain on the pump; a smaller three-cylinder pump; wells greased and preserved by, and an electric motor. Reported in fair condition: all pipe work and non return valves; a timber work bench with one drawer, fixed to floor; five electric light pendants complete with enamelled iron shades in rusty condition; fittings and conduit; a glazed timber screen across; a timber cupboard; and portable grindstone in poor condition.

4.10 The brakes

There are no fewer than five ways of stopping the cars, more than in Lynmouth. The brakes are thoroughly described in the 'Official Description of the Clifton Rocks Railway' and 'Marks on Cliff Railways'. The rails are flat bottomed in section and have good bulb heads to enable the gripping brakes to work against as deep a surface as possible. There were duplicate brakes using hydraulic pressure on both rails of the line. The brakes were activated by a footbrake, automatically by a speed governor, automatically if either of the cables broke, and by the dead man's handle. The conductor in the top station could also stop the cars.

With the provisions for contingencies, appliances for safety, precautionary mechanisms for making assurance doubly sure which Messrs Marks and Munro have introduced, the most nervous may clearly dismiss every element of fear as to the absolute safety of the Clifton Rocks Railway.

4.10.1 Cable failure

There were duplicate brakes if both ropes break.

An emergency brake is also arranged upon each car for the purpose of seizing the rails in the event of a stretching or breaking of either of the ropes, these brakes acting through independent levers and springs, which are always in readiness to arrest the cars upon a severance or slackening of any of their connections.

This was completely automatic. It consisted of two large cast-iron wedges with serrated faces mounted inside either rail, on the ends of steel arms, both of which were pivoted on a heavy steel cross-member. Both ends of the cross-member enclosed the rails. The free ends of the arms were attached to the two cables, and a large coil spring was incorporated, held under compression by the tension in the ropes. In the event of a cable failure, the load on the spring would be released and the spring pivot the arm, jamming the wedge between rail and cross member. Deflection of the rail was prevented by the outer ends of the cross-member.

First of all, the steel wire ropes from which each car receives its motion, and upon which the thoughts of all visitors instinctively centre. These steel ropes are double, when in fact one would be safe beyond all reason of possible strain for carrying the whole load; each of these double ropes is attached independently to the car safety grippers in such a way that should one rope stretch or become in any way disabled, then all the grippers would seize hold of the rails and securely lock the cars, thus preventing any further movement.

4.10.2 Speed governor

There were duplicate brakes to stop the car from going too fast (the speed governor). 'The governor brake checks the speed should either of the attendants be mindful of their duty, and try to allow the cars to travel too quickly.'

The governor is driven by the main wheels. These brakes had shoes which pressed down on the top surface of the rail and actually lifted the car off the rail by $1/16$in, thereby relying on the weight of the car to give maximum friction between the rail and the brake shoes. The speed governing brake acted independently of the controlling brake to cause the hydraulic pressure to be increased, and the rails to be gripped, should the recommended speed be exceeded.

> The hydraulic brakes, on one side of the car for gripping one rail were independent of those upon the other rail of the same track; they were each fed by a separate and distinct accumulator and weighted ram, and each had a separate gauge-glass, both systems however being connected to one common conductor's hand wheel. The brake cylinders were fitted with piston-rams with drawback springs provided within for causing the rams to leave the rails upon the pressure being withdrawn from the cylinders when starting the car from rest.

4.10.3 Deadman's handle

> The emergency brake is ready for application should any unforeseen circumstance transpire; the controlling brakes continually try to stop the car from movement it starts up and down the track; and ridiculous it may seem, it is yet the only duty of the attendant riding on the cars to prevent them stopping of themselves. Thus it will be seen the cars normally stand still at any position on the rails, and if the attendants relax their hold upon the small controlling wheel which they have to manipulate, or if they carelessly discharge their duties, then the cars are immediately brought to a standstill, and remain standing until those in charge do that which they are expected to do so.

This is the brake that the brakeman manipulates to *keep* the car from stopping, rather than making it stop.

On the railway, the brakes are permanently on, operated by a large water accumulator via the conductor's hand wheel. If the brakesman lets go or leaves the car unattended, the car will stop together with the other car. The action will increase the pressure in the hydraulics and cause the hydraulic gripping brakes to

4.23. One of four governor drives that were found. Removed when cars winched down during war. (Author)

clamp each side of the crown of the rail making it impossible to move. This is why images of cars in the tunnel are shown with no driver, since the car can not run away as there is no one holding onto the wheel.

> It is only when both brakesmen wish the cars to go that they can move the act of turning the hand wheel on each car, raises a weight which is normally pressing on to a cylinder of water for the purpose of conveying pressure to the hydraulic cylinder connected to each brake-gripper. The water under pressure travels through duplicate independent copper pipes and exerts, except when the brakesman raises the weight, an enormous pressure upon the rails; and during the journey of the cars, should either of the brakesmen release their hold upon the hand wheels then both of the cars would gradually stop. The simplicity of the arrangements introduced, together with the entire duplication of the mechanism which cause nothing to be dependent upon one single portion of the mechanism or upon one single connection, leaves nothing unprovided that experience can suggest for ensuring the fullest and most absolute reliability of the unique safeguards of the cars upon the Clifton Rocks Railway.[5]

There was also a specially-designed compensating brake-band which was applied to the surface of each upper wheel, the brake being applied by means of a hand-wheel controlled from the central cabin in the top station.

> The action of the brakes, on account of the neglect or inattention of the conductor being to increase the pressure in the hydraulic mechanism and thus cause the

4.24. Handwheel from the central cabin to stop the cars. (Chris Bull)

4.25. Hydraulic gripper brake visible on rails at top station. (Author)

hydraulic gripping brakes [see plate 4.25] to instantly seize the rails on which the cars travel.[6]

Thus the design incorporated fail-safe systems which were made simple and which were also duplicated throughout the assembly – a very reliable design concept.

> From numerous tests made by the author, after the cars had been running a few weeks, it was found that a pressure of 250 lbs per square inch arrested the cars when an excess load on one car over the other existed to the extent of 15cwt, while exerting a slight pressure by the hand upon the windlass wheel, this pressure was raised immediately to 1,000 lbs per sq inch.[7]

4.11 First journeys
The first journeys are described in very different ways. Safety was clearly of paramount importance to reassure everyone, then the cars are compared with trams and omnibuses to help passengers familiarise themselves. Even opening day seems to have come as a surprise and the contractor is still finishing last minute details. Every newspaper held different information so one would have to have read several to appreciate the full story. The official opening day was published in an impressive number of journals from all over the country: Gloucester, Birmingham, Newcastle, Cardiff, Bath, Exeter, Sunderland, Portsmouth, Somerset, Wells, Nottingham, Lincolnshire, Durham, Gloucestershire, Yeovil, Edinburgh, Lancaster, Cheltenham, Taunton, Wells and more.

The first experimental journey was described in the *WDP* for 8 February 1893:

> The sensation of travelling up is not at all as expected, in fact there is no sensation to be described. The question more than once as to when would the car go even when it had started on its steep climbing journey. The four cars are in effect, neatly designed and nicely hung small tramcars, minus horses and outside passengers, so that the motion is not one that can be described other than it seems as though one were riding in a tram car when the horses had been removed. Messrs Marks and Munro have left nothing to chance or hope in their splendidly equipped little line; the system of brakes and safeguards provided upon each car, being in the eyes of the casual visitor, almost extravagant and unnecessary. But from the experience obtained by Mr Marks elsewhere, we suppose he has found it wise to minister to the fears of the patrons by making these cars veritable homes of safety for their occupants.

4.12 Opening day announcements
Not surprisingly, there was a huge amount of advance press coverage, and coverage about the day itself. The following excerpts have been pruned to remove duplication. On 11 March 1893 the *Bristol Mercury* reported:

> The Clifton Rocks Railway, will be informally opened this morning, the cars starting with any passengers who choose to present themselves at nine o'clock and will continue running throughout the day. We hear that after Saturday the cars will start every morning at 6 o'clock. The time occupied has been more than at first anticipated, owing to unexpected difficulties presenting themselves to the contractor, Mr C. A Hayes, and the engineers, Messrs. Croydon Marks, and P. Munro, in the nature of the work. The tunnel which extends from the Hotwell Road to Sion Hill, a distance of little over 500 feet, having been blasted not entirely out of solid rock, but treacherous conglomerate and marl, which required more care than the solid beds of rock. Although it was March lst 1891 when Lady

4.26. Opening day notice in the *Western Daily Press* and *Bristol Mercury*.

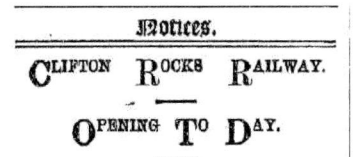

Notices.

CLIFTON ROCKS RAILWAY.

OPENING TO DAY.

IN ORDER TO PREVENT EXCESSIVE CROWDING, AND FOR THE GREATER CONVENIENCE OF THOSE DESIROUS OF TRAVELLING ON THE OPENING DAY, it has been decided to Charge a SPECIAL FARE OF 4D, and to present to each Passenger paying this Special Fare, as a personal Memento of the occasion,

A GILDED MEDAL,

expressly Designed and struck at very great cost. The number of Medals is limited, and after they have all been distributed, other Passengers will be carried on payment of 2D FOR THE UP JOURNEY TO CLIFTON DOWNS, and 1D FOR THE DOWN JOURNEY.

FIRST CAR STARTS AT 9 A.M. TO-DAY.

ON AND AFTER THURSDAY, the 16th instant, THROUGH TICKETS will be issued to and from the DRAWBRIDGE and CLIFTON, via the HOTWELLS TRAMS and the ROCKS RAILWAY.   1909

Wathen, the Mayoress performed the initial ceremony of firing the first charge for blasting, the actual tunnel has occupied only eighteen months, with a gradient of 1 in 2.223. This of course is altogether an exception to those accustomed to use anything in the way of railways, but as a lift railway, the engineers consider it a good comfortable 'slope', while steep enough to admit a good shaft of light descending down from the heights above to the waiting room in the Hotwell Road.

In order that its equipment should be perfect, Mr Newnes through his engineers arranged instead of one car to have four cars on the line, two going up and two going down simultaneously. The cars, which are fitted somewhat after the style of ordinary tram cars, are square and barely one minute is occupied in the ascent. For those accustomed to travel on cliff railways in this remarkably steep ascent, we may add that the greatest care has been taken to provide in every way for the safety of the cars. [The brake arrangements are so contrived that if both steel ropes snapped both cars could be stopped even on a gradient of one in two.]

The equipment of the cars is elaborate and complete and the engineers and men attending upon the railway – seven in number – are provided with smart uniforms (made by Durie and Co, of High St.) of dark blue serge with scarlet facings, their patrol jackets looking very neat and finished. Those of the engineers are enriched with gold lace, and the peak caps are of blue with scarlet band above, on which in scarlet are the letters C.R.R.

Durie seemed to have regular business receiving £3 10s at the end of October and £10 1s in April (see table 6.3, chapter 6) so presumably some clothing items were replaced regularly or maybe new staff needed fitting out. The 1907 Army and Navy priced a double-breasted motor coat at £4, peaked hat at 6s 6d and lined gauntlets at 9s 6d.

The cars, like the conductors' tunics and hats, had all the freshness of recent make, and with their white and blue and gold decorations were as attractive in appearance as one desires.

4.13 Signalling and communication

Electrical signalling is arranged so as to enable one man to signal his companion the number of passengers he has taken. Telephones also enable conversation to pass from end to end. The machinery has been constructed by Messrs Gimson, of Leicester, under the personal direction and to the patent design of Croydon Marks and his associates. Arrangements for the traffic today have been made with a view of avoiding crushing.

In *The Engineer* of 1893, further details are given about signalling by Messrs King, Mendham & Co of Bristol:

The signals are received in succession upon a signal board, and electric bell rings at each fresh indication and the engineer is thus able to provide for sufficient water being admitted into the tank to take a car down, and when it is ready a signal is sent down intimating that the car is starting. The object of the signals is to establish the necessary balance in working the cars. To take a car down, and when it is ready, a signal is sent down intimating that the car is starting.[8]

4.14 Plans for opening day and the commemorative medallion

The issue of the commemoration medal to a limited number of passengers paying increased fare should prove successful in enabling the first-day passengers to obtain the memento without fear of annoyance of discomfort. In order to prevent excessive crowding on the opening day, a charge of fourpence will be made, and each passenger will be presented with an especially designed gilded medal, which will form an interesting memento of the occasion. After a limited number of medals have been issued the other passengers will be charged twopence for the up and one penny for the down journey. After Thursday next, through tickets will be issued to and from the Hotwell tram and Clifton via the Rocks Railway. The medal referred to is of artistic design.

What strikes me is how much in advance the medallion would have to be ordered, and when they decided the date. Clearly the medallion would have to have the correct date or none at all (safer).

On 11 March 1983, the *Western Daily Press* reported:

The Bristol Tramways Company have proved their foresight by entering into an agreement whereby passengers

69

4.27. The commemorative medallion was a 'pretty little medal shaped like a Maltese cross'. On the one side was a wonderful stamped representation of a four track railway with one car on a 1 in 2 slope, and around it the words: 'The Clifton Rocks Railway, commenced 1891, completed 1893'. On the arms of the cross were the initials G.N. of Mr Newnes, G.C.M. of Croydon Marks, the engineer, and P.M. of Mr Munro, the local engineer, who worked with Croydon Marks. G.N was embellished with a laurel leaf crown – a symbol of victory. (Author)

4.28. On the reverse were the words: 'Issued to the passengers on the opening day, March 11, 1893'. If one also looks closely at the wonderfully detailed medallion the letters 'Sale Birm' appears at the bottom. Sale was a well-established Birmingham firm, manufacturing and supplying medals, coins, tokens and badges, since 1862. Sale suffered a fire in 2009 and the building at 393 Summer Lane was demolished. The Clifton Rocks Railway Trust has obtained several medallions. It is impossible to know how many were made, but probably fewer than 6000. (Author)

will shortly be booked through to Clifton from St Augustines Bridge to the Hotwells, and thence via the Rocks Railway, thus effecting a saving of time. The construction of this inclined railway has been a matter of great difficulty, and a vast amount of money has been spent by Newnes, beyond what was contemplated, owing to the faulty nature of the rock and other contingent troubles. The cars have been running daily for a few weeks, in order to make the attendants particularly familiar with their novel duties. A stranger on examining the details of the mechanical safeguards and elaborate contrivances for safety which Marks and Munro have introduced, ventured almost in question the necessity for such a complete multiplication of provisions for doing the same thing, viz to keep the car from running away. The gradient of Park Street is considered steep, rising about one foot in ten feet at some places, but when it is remembered that the Clifton Rocks Railway rises one foot in less than two feet, the difficulties of the task undertaken, and now so ably completed, can be somewhat better appreciated.

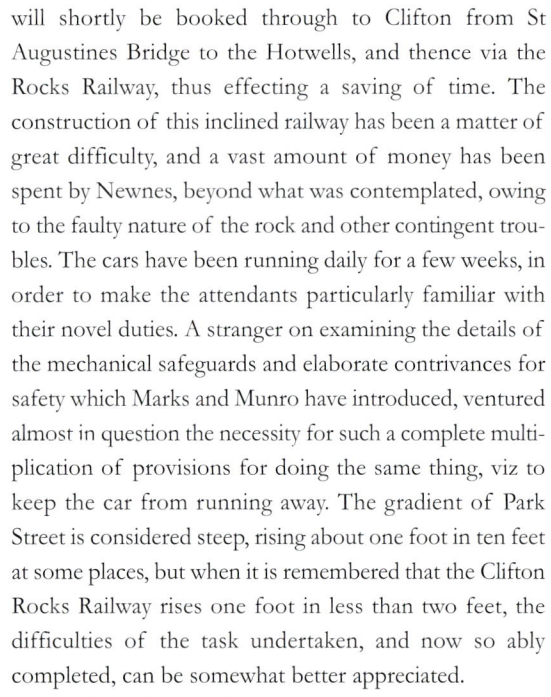

The work has been well thought out, the arrangement of turnstiles for ingress and egress of passengers being designed to prevent holiday crushes.

4.15 Opening day reports
On 11 March 1893 the *Evening News* noted:

The Clifton Rocks Railway was opened this morning. The hour was early; Bristol is not fully awake before 9 o'clock, and comparatively few people had ventured through the frosty mist which gave promise of soon being cleared away by the warmth of unclouded sun. By and by the station doors were open, and the groups of would-be passengers waited patiently by the turnstile forming a barrier across the waiting-room at the bottom of the lift. The room is a light one, with walls and ceiling covered with varnished pine, and a partition of similar material divides the apartment from the tunnel beyond. The opening was so sudden as to come as a surprise to the contractor himself, Mr C.A. Hayes, and the workmen were still employed in many a little fitting, making ready for permanent working. Then the conductors appeared, smart in their fresh uniform, and looking not unlike brand new postmen, and the telephones communicating from top to bottom from top to bottom were heard ringing, and then the first load of passengers were allowed to take their set. Each was supplied with a pretty little medal shaped like a Maltese cross to commemorate the occasion. There was not much time before taking a seat ready for ascent to closely examine the surroundings. The cars, so like those seen daily on the tramway system of Bristol, might almost be mistaken for them at a glance. They have no top or outside seats and hold eighteen passengers apiece. They like the conductors' tunics and hats, had all the freshness of recent make, and with their white and blue decorations were as attractive in appearance as one desires. When an omnibus goes up or down an exceptionally steep hill the level of the vehicle is altered by the inclination of the road and passengers know what it is like to lurch over against those sitting by their side. If, therefore, an attempt was made to run an ordinary tramcar up the rocks railway one might expect to see the occupants tumbling over each other in rather awkward fashion. All that has to be done is to mount the vehicle on a base which shall counterbalance the steepness of the inclination, and thus the floor and seats always maintain their horizontal position.

The first lot of passengers were seated and the start made with about as much motion as is felt in travelling along one of the Bristol tram lines. Half way up the descending car, equally full of those who had presented themselves at the Clifton Down station, and in <u>forty seconds</u> the journey was finished and the up tram load found themselves in a sunny room built on the hillside, and reached by a flight of steps from the road just opposite the St Vincent's Hotel. By the time one had a look round quite a stream of passengers had presented themselves, and soon the two up and two down cars were busily engaged. It is intended the fare shall be 2d up and 1d down, and on Thursday next there is a through booking arrangement in force on the Hotwells tram line, so that passengers may travel by this route and be landed at Clifton Down. Passengers can also alight at the Hotwells entrance to the Rocks Railway and simply get from one car into the other and be hoisted to Sion Hill without trouble; the through fares Sion Hill to Temple Meads 4d and 3d.

Impressive integrated transport. Incidentally a person today is on average three stones heavier than in Victorian times so, if run today, the cars would require less counterbalance water ballast to go up and down and would have a quicker turnaround (though the governor would have stopped the cars from going quicker).

Other comments:

The line will prove a great boon to those not having time or inclination to mount the zig-zag. [Bath]

Its successful completion now unites the upper portions of Clifton with the lower levels of Bristol. The line is unique in its arrangements for dealing with general and with holiday traffic, on account of the crowds who constantly make their way to the Downs. [Cheltenham]

The Railway is in full working order and is attracting large numbers of passengers: everything is well arranged, and the public is much pleased with the great convenience afforded. Clifton is at last waking up in earnest and beginning to appreciate its largely increasing number of visitors; certainly the apathy hitherto shown by the residents to the requirements and attraction of strangers has been unaccountable, but now it appears *Nous avons changé tout cela*. We have changed all that. [Bath]

On 13 March 1893 the *Bristol Mercury* reported:

A considerable crowd of people – not at all deterred by the special charge of 4d made to prevent overcrowding – gathered outside the waiting room of the Hotwell Road before 9 o'clock, anxious to be in the first or at least one of the cars going from the river bank to the heights above, and to obtain one of the artistically designed medals presented to each passenger as a memento of the opening day of the railway. [The *Clifton Chronicle* reckons only a dozen or so were present.] The hoarding round the exterior of the waiting room had been removed, and a good view of the exterior of the building could be obtained. At the other end – the Clifton end – there were also a few people waiting for the opening of the line. The doors were opened shortly after nine o'clock, and at first there was a slight crush at the turnstiles for the honour of riding in the first car, but the effort of a stalwart Clifton police sergeant soon restored order. For the first hour or so the cars were all filled very quickly, and the powers of the attendants were severely taxed, but after the first crush the work was carried on very steadily throughout the day.

It was noticeable that almost everyone who rode made the return journey. Entering into the waiting room at the Hotwells, the passenger finds himself in a light, cheerful compartment, the walls and ceiling being panelled. Immediately facing the entrance is a handsome railing extending across the room, with three turnstiles in it, and behind these are three doors – a small one each side and a larger one in the centre – leading out onto a small platform from which the cars start. From this platform the whole length of the line can be seen, and the gradient appears very steep. The platform and waiting room at the Clifton end can be dimly discerned in the distance by the light of the lamps at the top. In the waiting room at the Hotwells is a large ticket box. The cars are exactly like ordinary tramcars, except for size, and there being no seats on top.

The journey is performed in little <u>under a minute</u>, and the cars travel at a very good pace and easy and without oscillation. The Clifton waiting room, which is smaller than the other one, is fitted up in a similar style to that of Hotwells. The entrance to it is by a broad flight of stone stairs divided in the centre, one side for exit and one for entrance. The cars are built upon large triangular framework so they are perfectly level, and the motion going up is like that of a tram car, but the down motion is somewhat different, the car seeming to sink gently to the bottom. The tunnel being at so steep a gradient permits of daylight passing through from the Hotwells to Sion Hill, and at no place is the line in total darkness. There are gas jets at intervals on the walls of the tunnel, and the cars are provided with lamps. The amateur photographer of course put in an appearance on Saturday and took several views of the railway and cars. The number of passengers carried was 6,220. The cars did not run at all yesterday, in order to give the men, who had been working late all week, absolute rest before commencing regular work. The Rocks Railway opens today at 8:30am and runs until 9:00pm. On Sundays the first car will start at 2:30pm. The Grand Pump Room and Spa will now be pushed forward with energy by Marks, Munro and son, who have completed all arrangements in connection with this interesting work.

Owing to having to attend church and Sunday school in the morning, one would not start until 2:30, even though now, to us, the whole of Sunday is a leisure day and to ride on the car would have been a special Sunday treat. Newnes was also the son of a Congregational minister so would have conformed to Sunday opening times.

On opening day, if there were 6,220 passengers to board four cars, 18 at a time, there would have been 86 trips at 8.72 minutes each over $12^{1}/_{2}$ hours.

On 14 March 1893 in the *Bristol Mercury*, it was admitted there could have been 8,000 passengers (111 trips at 6.76 minutes each) had there not been the customary little delays in connection with opening a new undertaking.

[In the *Bristol Mercury*] One of these delays was caused in the afternoon by some chips getting into the water and choking the suction pump used for pumping the ballast water into the tank at the bottom of each car. The extra weight of water thus pumped into the tank is sufficient to cause the car at the top of the tunnel to draw up the other by its weight as it descends, and the delay caused by the suction pump getting choked lost some passengers, as the officials only worked two sets of rails and two cars instead of four during the delay. The arrangements when working smoothly, as they now will, are made for the steep and romantic little line to carry a thousand passengers per hour; and what this will mean on Bank Holidays must be left to actual experience.

But a more interesting little adventure occurred later on in the evening, about seven o'clock, when a car crowded with passengers, some of who were standing up in the centre after all the seats had been filled, suddenly came to a standstill after travelling up about 80 or 100 feet of the steep line of rails. Much has been made by the o'er timid, and there have been exaggerated statements as to the ladies getting in a panic and insisting on getting out and walking; and it was even stated that most of the passengers got out and walked. But the whole affair was very simple indeed, and only showed how thoroughly well the automatic safety grippers hold to the rails when once the cars are brought to a standstill. When the men ascended to the car and used a crowbar to try to start the wheels, it seemed so tightly they gripped that 'wild horses could not move them'. At the sight of the crowbars we are officially informed only two passengers – a gentleman and his wife – 'got out to walk, or rather do the steep climb of one foot in every two'. The engineer then discovered that the man in charge had forgotten to let the water out of the tank when he had reached the bottom of the tunnel railway on his journey down; and as he had retained it in his tank in addition to an extra load of passengers the top car even with its load of water, had not sufficient weight to move up the overweighted lower car. The plugs of the tank were removed, the water let out, and up the car glided as smoothly as possible – and the passengers who had remained in laughed at the timid, and at the man who had forgotten all about the water. The man had been specially trained for the work, but at the start 'one can't remember everything'. How embarrassing.

Saturday's trips were made with only one or two of those slight hitches that are bound to attend a new enterprise of this sort and revealed the fact that the arrangements are almost as complete as it was possible to make them. We only need to point out two necessary alterations. One is that the exceedingly awkward slope at the corner of the pavement near the top approach should be altered in some way, and the other that instead of the doors of the cars opening outwards as at present they should slide sideways as in the ordinary tram cars or the lift railway at Hastings. The cars themselves are comfortable and airy, each seating eighteen persons, and are lighted with oil lamps, but the tunnel is so steep that the daylight can penetrate from top to bottom. Later on we believe it is the intention to have the electric light installed. Over 6000 persons used the lift on the first day, and a large number of these, by travelling early and paying double fare, were able to secure a pleasing little memento.

4.16 Finances

One year later, on 28 April 1894, the *Gloucester Journal* reported that the Railway has had remarkable success.

The traffic has become so great as to make it a matter of wonders that such an apparently indispensable public convenience should have been left to an enterprising visitor to personally carry out at his own private cost. The traffic during the twelve months ending with March 19th last, reached the total of 427,492 persons; and at holiday times so great has been the demand that the whole of the four cars are called into constant requisition. The construction of the tunnel, and the equipment of this inclined railway has far exceeded in cost the original estimates for the work; while the length of time required to carry out the scheme led many to imagine that it would be one more of the many abandoned undertakings which have started with every prospect of success. Since the opening of the line on March 11th 1893, the traffic and management has been entirely in the charge and under the personal direction of Mr Croydon Marks, whose system of mechanism is adopted for working the line. The safety of the travelling has been placed beyond all doubt, not only in theory, but by results of a years hard working. A limited company is to be formed forthwith. The holding of the chairman (the vendor Mr Newnes) will be in ordinary shares [this shows Newnes' confidence in the undertaking], while the public are invited to subscribe cumulative preference shares and debenture stock. All costs, charges and expenses of forming the company will be borne by Newnes.

A company, 'The Clifton Rocks Railway Limited', was formed in the Spring of 1894 for managing the enterprise. Mr G. Newnes M.P., was chairman; Mr P. Fussell director, and Mr G. Croydon Marks, managing director and engineer. Mr A.A. Yeatman, secretary; and Messrs. Osborne Ward, Vassell. & Co., solicitors.

Subscriptions were invited for £10,000 in 6 percent cumulative preference shares of £5 each. The company has been formed to acquire Clifton Rocks Railway. Mr Newnes takes all the ordinary shares, the consideration money to be paid him being £30,080, viz, £20,000 in cash and £10,000 in ordinary shares. The subscription list opened on 23 April and closed on 3rd May 1894. There were 6 subscribers each taking one preference share (Newnes, A.A. Yeatman – CRR secretary and chartered accountant, A Henry Johnson secretary Essex, F. A Boyer clerk Leytonstone, Essex, Charles Harrison publisher Clapton, Jonas Beves clerk Highgate, London, Sidney Herbert Nugent accountant clerk Islington)

Total receipts £2,460 11s 9d;

Working expenses (all out-goings, repairs) £874 5s 0d;

Net profit £1,586 6s 9d

Increased Revenue of £150 per year would be received for advertisements in the cars and stations, but in future rent, directors' fees would need to be paid (which would not exceed the increased revenue. The net profit was sufficient to pay a full year's interest on 4 1/2% debenture stock of £450, 6% on cumulative preference shares £600, and £1,050 dividend on ordinary shares £586 6s 9d, 1,586 6s 9d.

Mr Newnes has undertaken to demise the railway and stations for the term of 900 years, at the yearly rent of £50 with a covenant to indemnify the Company against a superior rent to be paid to the Society of Merchant Venturers.

Debenture stock would be secured as a first charge on the company's undertaking, the interest paid half-yearly and redeemable at any time after 10 March 1914, at par, on six months' notice from the company.

There is a covenant requiring the company to maintain the tunnel with the lines, stations etc., fit for use as a railway tunnel. This caused problems to those who wanted to adapt the railway for wartime use, as will be seen later.

It appears that no profit was held back for reserves, with all excess income used to pay dividends and interest on preference shares. As Newnes owned all 10,000 shares, he would be in for the entire £536 dividend payment.

Income seems to tie up: if 427,492 passengers paid an average of 1.5 pence per journey, that is 641,238 pennies, £2,671.83 in income. This almost matches the £2,460 11s in the accounts. More people did go up than down, so that could account for the difference. They never really considered other forms of income, like advertising. It shows how reliant the whole set-up was on getting enough passengers. Clearly expenses would increase as the equipment began to wear out.

On 29 August 1894, it was reported that 2,000 preference shares and 2,000 ordinary shares at £5 each have been taken up, and the amount of calls £9,970. The Railway had earned more than sufficient clear profit to pay the interest on debenture stock, and dividend on the preference shares up to 10 March 1895.

Looking at the summary of capital and shares, there were 39 subscribers to preference shares. Newnes had 1900 ordinary shares and 1467 preference shares, his occupation listed as Baronet. Marks had 100 ordinary shares with occupation listed as Chief Engineer.

On 3 June 1899 the summary of capital and shares stated there was nominal capital of £30,000 divided into 6,000 shares of £5 each, 2,000 preference shares taken up, 2,000 ordinary shares taken up, and £5 that had been called up on each of 2,000 preference shares. There was a total debt of debenture shares of £10,000.

Looking at the summary[10] of capital and shares, there were 43 subscribers to preference shares. Newnes had 1900 ordinary shares and 1267 preference shares; Marks had 100 ordinary shares.

On 3 June 1904 there were 39 subscribers to preference shares. Newnes still had 1900 ordinary shares and 1267 preference shares. Marks had 100 ordinary shares. Lily Marks (Croydon Marks' sister-in-law) had 70 preference shares. The directors are listed as Philip Fussell J.P. of Kingswood Hill (director of West Gloucester Water Company); George Peckett of Hyde Lodge, Clifton (director of the Avonside Locomotive Works); and Alexander Alfred Yeatman (secretary).

On 25 December 1907 there were 38 subscribers to preference shares. Newnes still had 1,900 ordinary shares and 1,267 preference shares. Marks still had 100 ordinary shares. Lily Marks still had 70 preference shares.

By 1908, George Newnes' businesses were failing and this affected his health. By 1910 his fortune had gone, he had retired to live in Hollerday House in Lynmouth where he died on 10 June 1910, of diabetes.

4.17 Receivership

On 17 July 1908 A.A. Yeatman and William Welsford Ward solicitor from Falmouth appointed themselves receivers and managers of a trust deed of debenture stock (dated November 1908).[10] The next day £151 5s 4d in account was transferred to the receiver.

On 1 December 1908, a receiver was appointed. On 11 December 1911 Bruntworsh and Black paid £21 0s 9d for an option to purchase for two months, but this did not seem to come to anything. Then, on 24 October 1912 it was announced that Clifton Rocks Railway had been purchased by Bristol Tramways and Carriage Company (BTCC):

> who intend to work a new service of motor omnibuses through old Clifton in connection therewith. Motor omnibuses to and from other districts in the city will make the railway at both upper and lower levels their point of arrival and departure, and in the summer a large excursion traffic will certainly be secured. New rolling stock will be introduced, the tunnel brilliantly lighted during the day, and other improvements made, so that it will form quite an attractive portion of the new communication between Bristol and Clifton. We understand the Tramways Company will enter into possession next month.

On 30 November 1912 the last week's wages paid £5 5s 4d and on 20 December, BTCC paid the purchase balance of £1352 9s 7d. At the end of the month the Clifton Rocks Railway closed.

As part of the the Bristol Tramways directors' report of February 1913 it was announced that the undertaking of Clifton Rocks Railway Company had been

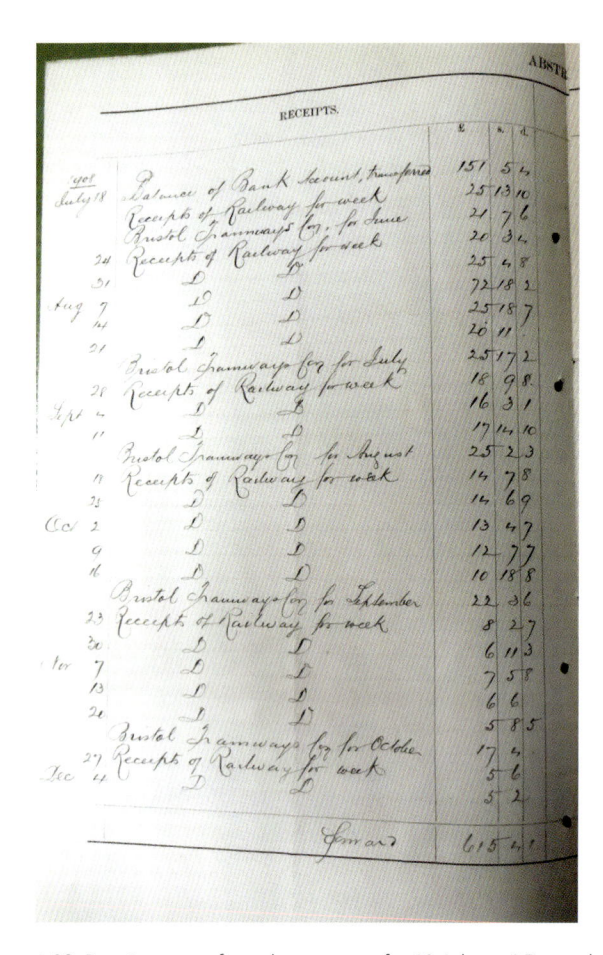

4.29. Receipts page from the accounts for 18 July to 4 December 1908, the year the company went into receivership. (National Archives Kew, 28/162 BT 31/5845/41016)

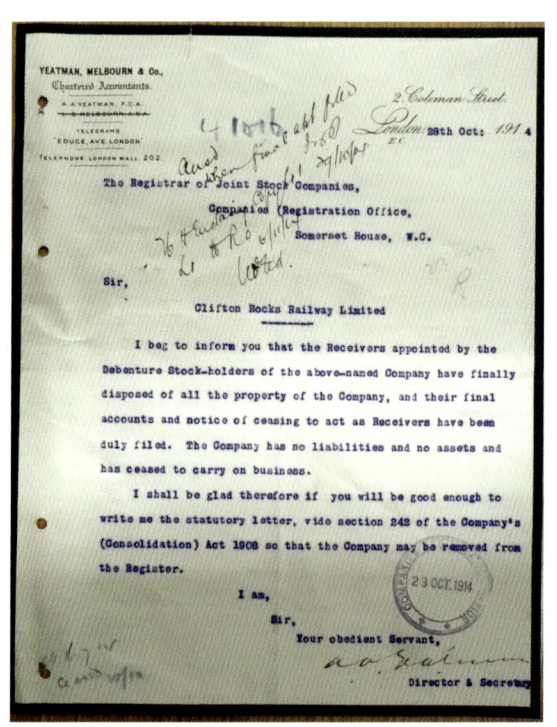

4.30. Letter to Company Registration Office confirming disposal of assets of Clifton Rocks Railway Limited. (National Archives Kew, 28/162 BT 31/5845/41016)

Table 4.2.
Yearly receivership income in decimal money.

item	July-Dec 1908	1909	1910	1911	1912 ceased running	1913	total
A Pole advert (printer)					£18.63	£15.63	£34.26
balance of account				£2.37			£2.37
Bruntworsh and Black option to purchase 2mth				£21.04			£21.04
BTCC		£122.97	£171.40	£160.88	£109.30	£74.08	£638.62
BTCC purchase balance					£1,352.48		£1,352.48
gas refund					£31.05	£7.66	£38.71
Hewitt list of debt holders					£0.10		£0.10
interest			£0.98	£1.95	£2.13	£3.75	£8.80
Newnes advert		£80.00	£11.58	£115.49		£6.09	£213.16
receipt		£381.98	£692.24	£699.51	£652.98	£441.32	£2,868.03
transfer fee			£0.26		£0.25		£0.51
Total		£584.95	£876.45	£977.83	£787.81	£1,917.90	£33.14 £5,178.08

Table 4.3.
Yearly receivership expenditure in decimal money.

item_type	July-Dec 1908	1909	1910	1911	1912 ceased running	1913	total
unknown			£11.17	£4.80	£1.34		£17.30
expenses	£1.50		£4.20	£7.92	£9.49	£8.00	£31.10
finance	£15.38	£302.63	£199.69	£27.70	£1,027.00	£332.35	£1,904.74
insurance		£21.25	£21.45	£22.30	£21.31		£86.31
maintenance	£7.58	£42.30	£79.71	£61.47	£146.50		£337.56
remuneration	£23.25	£52.50	£52.50	£52.50	£52.50	£21.00	£254.25
rent					£32.27		£32.27
utility	£208.18	£151.63	£228.92	£232.06	£174.25		£995.03
wages	£190.23	£387.00	£413.00	£400.40	£373.05		£1,763.68
Total spent	£446.10	£957.31	£1,010.64	£809.14	£1,837.71- £1325 (debenture repay)	£361.35	£5,422.25
Total income	£584.95	£876.45	£977.83	£787.81	£1,917.90- £1,500 (sale)		
Net	138.85	-80.86	-32.81	-21.33	-94.81		

purchased for £1,500, having cost £30,000 to construct.

The railway can be profitably worked by the company and made available as a convenient link for those passengers who travel by the vehicles now or at any future time in and around Clifton and Hotwells.

By 28 October 1914 Clifton Rocks Railway Limited was dissolved, having paid off the debenture holders £1,325 at the end of 1913 and all property disposed of. It is still called Clifton Rocks Railway but an offshoot of BTCC.

Yearly receivership income
Tables 4.2 and 4.3 show receivership income and expenditure for 1908-1913. Compare this with 1894 when total receipts were £2,460 11s 9d and working expenses including all out-goings, repairs £874 5s 0d making net profit £1,586 6s 9d (excluding interest on shares and debenture stock) so half the income, and few repairs and more expenditure.

The wages per week varied between £7 14s and £8 3s, so the seven men earned about £1 per week. Wages were clearly a big part of expenditure. (National figures in 1906 were £1 5s for ticket collectors, £1 7s 7d for conductors, and £1 11s 6d for mechanics) so the men must have earned less than the national average and certainly no more than seven could have been employed. Gas was also an expensive cost since the engines were running on it. Every three months there was a bill for between £15 (in January since less passengers) and £18 (in October) – an average of just over £60 per year. Rates were £85 every 6 months which was even more expensive but they were reduced by about £10 each year so were £54 in 1912. The debenture stock interest was high too, and had to be paid off in 1913.

It is curious that on 6 May 1912 that Clifton Grand Spa Hydro charged rent of £16 0s 2d – for the first time – after the Railway went into receivership.

4.31. The top station now has railings rather than a wall along Princes Buildings, a new entrance sign – the Newnes one at the junction has been removed. The main entrance is now in Princes Buildings. This makes the Railway far more prominent to visitors. Compare with plate 4.1. The railings confirm that BTCC now own the railway. In the 1960s the right hand section of the railings was removed so the Trust only had space to put back 'Clifton Rocks Railway'. The George White ornamental sign was replicated in 2016, see plate 15.55. (Peter Davey collection)

4.18 Change of façade by BTCC in 1913
Subsequent expenditure by BTCC in 1913 was: £3,328. This includes repairs, painting and reupholstering cars. They also changed both the top and bottom façades, the rails and appear to have changed the turnstiles.

Net profit
1894 £1,586; 1914 £1,824; 1919 £73; 1913-1920 may be £5,426 total; 1920 £34; 1925-27: *Losses* at an average of £500 p.a.

4.32. Photographed in the 1950s[11] (Cedric Barker), this section of railing has since been removed. Note the Scissor Gate from George Newnes' time. The signboard is missing.

4.33. Top station. Even the turnstiles changed. Note the uniform. The man to the right is clearly a conductor/ brakesman since he has a ticket machine. The partitions have changed, the valves are in view behind them. The doorway behind leading to the subway to the Pump Room appears to be shut. Visitors to the Pump Room were expected to use the main entrance in Princes Lane. Compare with plate 4.7. (Peter Davey collection)

4.34. The windows, lighting and entrance have changed. There is new signage. No vending machines. Superintendent Poole stands in front of the station with BTCC Inspector Wall. BTCC employed inspectors who liaised with the Railway staff. The panoramic view from the first floor still exists but the canopy has gone. Compare with plate 4.5. and 4.6.
(Peter Davey collection)

4.35. The bottom station in 2005. It is in original condition except that the back wall should not be there, nor the wall on the left (converted to a blast absorption area during WWII). Visitors got onto the cars at the tall entrance. The ticket office was on the right with sliding door. Three turnstiles and quarry tile floor as at top station. (Author)

4.36. Arriving at the bottom station. This is the only contemporary view of the bottom station we have. The conductors' ticket clicker can be seen. The tunnel is wet with signs of limestone seepage.
(Peter Davey collection)

	1925	1926	1927
Wages	£1,093	£1,071	£1,075
Repairs and maintenance	£293	£303	£322
Electric current	£44	£55	£55
Cleaning and Lubrication	£9	£11	£9
Gas and Water	£53	£33	£34
Telephone	£9	£11	£11
Punch Hire and tickets	£11	£10	£11
Rates	£174	£183	£195!
Ground rent	£50	£50	£50
Fire insurance	£17	£18	£18
Total	**£1,753**	**£1,745**	**£1,780**
Passenger receipt	**£1,336**	**£1,294**	**£1,252**

Table 4.4.

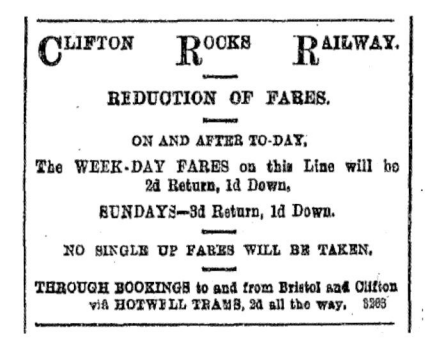

4.37. 28 April 1893: After only six weeks of running the weekday ticket prices are reduced to 2d return and 1d down. Interesting that no single up fares will be taken, especially when in 1928/9 it was recorded that twice as many people wanted to go up than down.

4.19 Operating expenditure 1925-27[12]

Note from the above table that wages have more than doubled since 1912, electricity and telephone have been added and repairs have quadrupled.

On 13 January 1928 a Rating Appeal reveals the following figures: administration £100p.a., depreciation not less than £200, interest on capital £350, so losing over £1,000p.a. The working account over the past six years showed an average loss of £549p.a, with capital expenditure £7,000 and a purchase price of £1,552. The Rates Officer thought there could be a reduction, but not considerable. It was suggested that if the Assessment Committee were not prepared to accept the owners' views, they should carry the appeal to Court. The Officer said he would point out the Railway was a useful feature in the amenities of the City and that as it was unprofitable there was a risk it might close. He asked if he could state that if it was not relieved from rates it would definitely be closed, but he was told that he could go no further than to state the continuance of the present rating would be a contributory cause to being closed. We are not sure whether the decision was to reduce to £200 gross, £132 net.

4.20 Passenger figures and prices
11 March 1893:

> Opening day: the number of passengers carried was 6,220. It is intended the fare shall be 2d up and 1d down. Passengers can also alight at the Hotwells entrance to the Rocks Railway and simply get from one car into the other and be hoisted to Sion Hill without trouble; the through fares Sion Hill to Temple Meads 4d and 3d.
>
> Easter Monday [6 April 1893]: our tramcars carried over 60,000 passengers; the aggregate journeys covered 3.580 miles. Fifteen thousand went on the Hotwell line only, and no less than eight thousand used the Clifton Rocks Lift Railway. The receipts of the lift were over £80.
>
> The Railway opens at 8:30am and runs until 9:00pm. On Sundays the first car starts at 2:30pm.

By May 1893, the Railway has been open for traffic between six and seven weeks during which time 111,400 persons had been carried. During the first six months over 330,400 persons passed through the turnstiles.

By March 1894, during the first twelve months, 427,492 passengers were carried upon the line (demonstrating a dramatic decline in the second six months), without accident and the net profit was £1,586 6s 9d. The Railway had become a regular means of travel between Bristol and Clifton. On Bank Holidays 1,000 passengers per hour could be carried. By October 1894 it was said that the Railway had earned more than sufficient clear profit in the last half-year to pay the interest on the debenture stock and dividend on the preference shares up to 10 March next, the end of the company's financial year.

On Easter Monday, 15 April 1895 the band of the Royal Engineers gave two concerts and Mr Walter George ('an entertainer who is gifted with considerable humour and drollery') appeared on both occasions. The Railway carried 11,020 passengers.

On 2 July 1895 the first advertisement appears for night journeys to connect with trams for Grand Spa concerts. Up to now the Spa entertainments had avoided mentioning the Railway as presumably the wrong kind of people might have attended the concerts. (Remember one of the conditions that the Merchants had placed upon the Railway being built was that people travelling on the Railway should not mingle with people using the Spa.)

On 18 January 1896, it was suggested that in the summer season the Railway gives the 'readiest possible access to the steamboat excursionist'.

On 14 April 1898 the first of the series of daily evening concerts was given at the Clifton Grand Spa Hydro (this was after the Pump Room had reopened and the Hotel and Turkish Baths had been built) and there was a fair attendance. Concerts were also held on Saturday afternoons. The Spa company had arranged for a series of monthly tickets, and visitors arriving at Hotwells and buying their sixpenny ticket at the offices at the railway would be conveyed up to the Spa for free, and receive a programme and be provided with a seat at the concert. The Bristol Artillery Band played. They were now encouraging railway visitors to concerts.

Using the income figures from the accounts above, one can estimate the number of passengers. It appears clear that twice as many people used the Railway on its own rather than as a part of a trip to the Tramways Centre. On the assumption that BTCC paid 1d back to the Railway as part of the 2d fare, then £1 represents 240 (240d per £) passengers. On the assumption twice as many went up the Railway paying 2d than 1d down (as shown by the 1920s figures), then each £1 represents 96 up and 48 down.

On 26 March 1913, Easter Monday, only two of the four cars were running, the other two at Brislington works undergoing repairs, painting and reupholstering. Since it would be impossible to get the car out, it seems more likely that just the bodies were removed by taking them to pieces. No fewer than 4,525 passengers were carried between Hotwells and Clifton. Since the Tramways owned the Railway the up fare had been reduced from 2d to 1d and this has resulted in a daily increase of receipts as compared with those of last year. By contrast, a total of 214,856 BTCC passengers were carried, using 233 tramcars and 14 buses.

A joint tram and Railway ticket was pink, early BTCC tickets were buff coloured. We found some buff BTCC tickets in the turnstile when it was retrieved.

On 3 June 1914 at the International Exhibition held on Ashton Meadows in the Bower Ashton area on Whitsun Bank Holiday, 7,761 were carried on Clifton Rocks Railway, twice as many as in 1913. Total BTCC passengers 315, 919 using 233 tramcars and 27 buses.

4.38 Ticket from BTCC times after fares reduced. Depending on whether the round trip started at the Centre, one would either go up or down.

4.39. Specific ticket for direction since different fare.

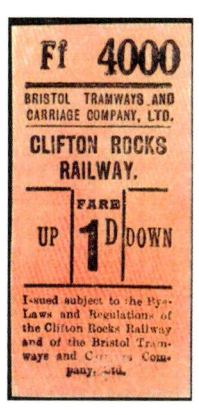

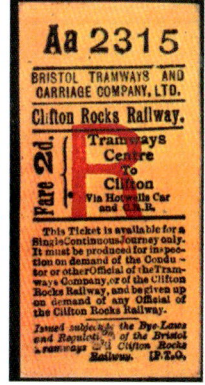

4.40. 3d round trip going up. Note the round trip tickets require the bottom portion to be given up. This would help with distribution of income to BTCC and CRR.

4.41. 3d round ticket going down. (Peter Davey collection)

yr	BTCC	Estimated CRR passengers	receipt	Estimated CRR passengers	Estimated total CRR passengers
July-Dec 1908	£122.97	29,513	£381.98	55,005	84,518
1909	£171.40	41,136	£692.24	99,648	140,784
1910	£160.88	38,611	£699.51	100,733	139,344
1911	£109.30	26,232	£652.98	94,029	120,261
1912	£74.08	17,779	£441.32	63,550	81,329
Easter week 1912	£1?	240	£15.59	2,245	2,500

Table 4.5. Estimated passenger traffic, 1908-Easter 1912. A big fall from 1893-1895.

The financial situation deteriorated even further after 1922, when the Portway was constructed. This required the closure and demolition of the Port and Pier Railway spur, and the Hotwell Road being widened outside the Railway and becoming a major road. By 1925 passenger receipts stood at £1,336, expenditure was £1,753. In 1926 there were 55,890 down passengers and 112,564 up passengers, with 382 mailcarts and bicycles going up and 283 down; a total of 169,119 passengers raising £1,294 9s 7d. But with expenditure at £1,745 this constituted a loss of £451. Despite the higher fare for up traffic, there were twice as many up passengers. It is staggering that the fare to travel up stayed the same from 1893, with only the down fare increasing.

Between 1926 and 1934 the Railway opened at 8:45am and ran until 9:15pm; Sundays from 2:30pm to 9:15pm. In 1927, 53,458 passengers paid 1^1/$_2$d to travel down, 109,426 passengers paid 2d to travel up. Along with mail carts and bicycles there was a total of 163,449 passengers raising £1,252 5s 2d, but with expenditure of £1,780 and a loss of £528.

4.21 Other news after the opening

[2 May 1893] Yesterday, at the Bristol Police Court, Henry Fellows, a respectably dressed youth, was brought up in the custody of Police-constable 29B, who stated that early that morning the accused gave himself up to him saying he had only 2d and was unable to get a night's lodging. Replying to the Magistrates clerk, the boy said the firm for which he worked [H.B. Sale – see plate 4.28] struck the medals for the Clifton Rocks Railway, which he wanted to see. Accordingly on Saturday he appropriated his weeks wages (4s 6d) and came by train to Bristol, sleeping the night in Hotwell Road. The Bench remanded Fellows to the workhouse until tomorrow for further enquiry to be made.

Sadly no more of the story was reported – I hope he saw the railway before going back to Birmingham.

On 4 July 1893, Philip Munro applied on behalf of George Newnes to lease a plot of land between the Colonnade and the Railway (see plate 4.42). It was decided to ask the applicants to submit tracings showing the exact nature of the building proposed to be erected, and the purpose to which it was to be used.

On 21 October 1893, a Bristol Corporation report noted they had had an application from Mr George Newnes the owner, to purchase a plot of ground

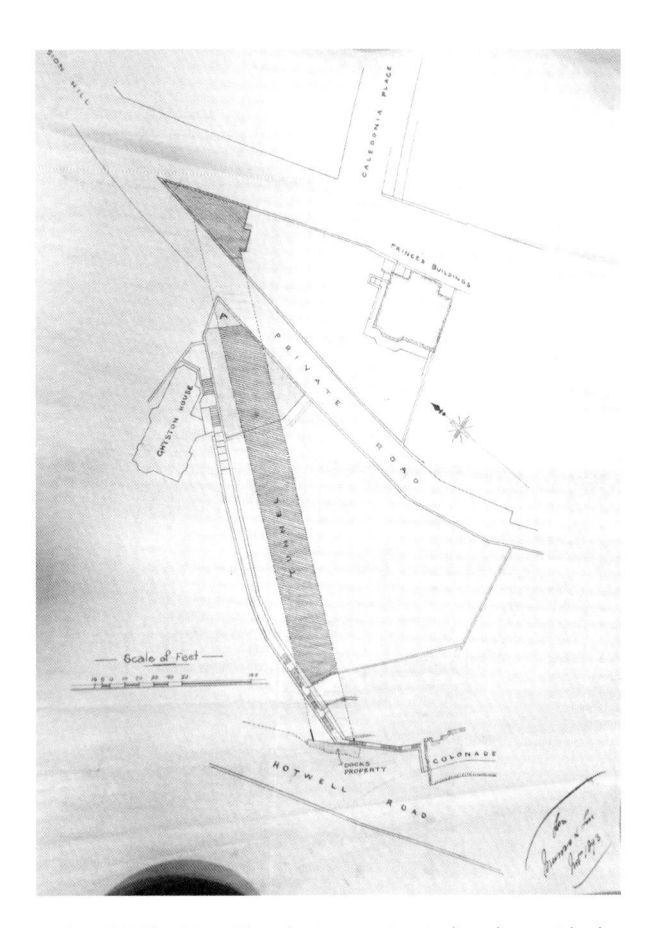

4.42. 1893 Plot Map. The plot in question is directly outside the entrance of the railway, and judging by the photographs on the bottom station was to be used for stalls and slot machines (plates 4.4-6). We are not sure when permission was granted judging by the discussion in 1893 to 1895, but the stalls do not encroach either side of the bottom. The map also shows steps going over the tunnel. (BRO 39735/8)

about 11 square yards in area exclusive of what is covered by the overhang of the roof at the entrance to Clifton Rocks Railway, the use of which was granted to him by arrangement with the committee in 1891, at a cost of £3 per annum, and they recommended the Corporation to sell the plot of land for the sum of £120, Mr Newnes paying all costs and expenses in connection with such sale, including the consent of the Local Government Board.

On 31 October 1893, some members of the Corporation after a two-hour discussion felt a disinclination to part with any part of the rocks at the Hotwells, though Mr George Newnes offered £120 for twelve yards in connection with the entrance to his Railway. The report of the Council was rather astonishing. A plan was provided on request showing where the land was situated. It was part of the salvage purchased under the Bristol Dock Act 1855, for widening the road. It was a small piece of rocky ground close to the entrance of the Clifton Rocks Railway, and which had been occupied by the proprietor, Mr Newnes, whilst constructing it. It was part of the piece included in the entrance to the tunnel. It had nothing to do with the larger piece of land to the side. The committee could not turn it to any other use, and the price was thought to be a satisfactory one. An Alderman thought they could trust the committee with a small transaction like this, otherwise it was no use electing committees if the business was to be discussed all over again in the Council. There had been too much of parting with city land in the past unless every member of the Council knew what they were selling. It was queried whether negotiations were going on for making a refreshment place, but that was for a site situated further away in the direction of the Colonnade. There was discussion about its value to Newnes, whether the Docks Committee should sell any city property which it would be in the interests of the citizens to keep, someone objected to the doctrine preached by Sir George Edwards, that the committee were to arrogate to themselves the rights of the Council. It was quite clear there were other negotiations with Mr Newnes, and they ought not to place it in that gentleman's power to do what he liked with this piece of land, which it was proposed to let him have at a very moderate price.

> Alderman Pethick said some years ago there was a class of man who were described as 'straining at gnats and swallowing a camel', and he thought their lineal descendants must have found their way into the Council. The Sanitary Committee appeared to be able to do what they liked, but the Docks Committee must do nothing. Mr Bastow thought Alderman Pethick was one of those men who had often swallowed a camel, and that morning he seemed to have taken the opportunity of swallowing a gnat. He for one could never accept the doctrine that committees were independent of the Council.

The Docks Committee had refused the offer, no negotiations were now pending with regard to the adjoining land and there would be greater check on expenditure by the sanitary committee. Twenty-two voted against and 16 for.

On 10 August 1895 the Committee recommended to grant Newnes a lease on 83 square yards near the Colonnade and adjoining the Railway for a term of 75 years from 24 June 1895, at a rent of £15 per annum in consideration of erecting a building according to an approved plan. The lease contained a covenant that the building could not be used to sell refreshments. At the same meeting the Docks Committee reported on their failure to get the steamers to agree to get passengers to pay poll tax (having only collected £2 8s each vessel), so clearly they were now keen to get revenue from wherever they could.

On 9 April 1896, there was a death on the opening day of the Snowdon Railway. The newspapers reported that they had humbly followed the pioneering Swiss engineers.

> Every district does not, like Clifton and like Lynton, possess a Sir George Newnes with the intelligence to comprehend and the liberality to grasp occasion by the hand and doing so to all but create and found new towns or villages.

Marks defended the water balance system as being far safer than the rack system:

> One car only runs upon each pair of rails, and is connected by steel wire cables with the car upon the adjacent rails, and thus one car can not move without the other, the ascent or descent of the one permitting the descent or the ascent of the other. No rack is required with such a system, for the cars, in addition to resting upon the rails laid upon the inclined track are also suspended by the connecting cables passing round the wheels at the top of the track.

On 13 January 1899 a quantity of rock suddenly fell from St Vincent's Rocks, close to the Clifton Rocks Railway. The stone fell into the middle of the road, and completely blocked traffic for about half an hour. The docks foreman was informed of the fact, and immediately had the obstruction removed. He also caused to be moved a quantity of rock which although it had not fallen he considered dangerous.

In 1905 advertisements appeared stating that one of the features of a house to let in Clifton is its proximity to the Grand Spa Hydro (with low charges for all kinds of bath and treatment), and Clifton Rocks Railway Lift connecting with the electric railway (ie. a tram) to Bristol.

On 10 February 1906 a 13-year-old boy was charged with stealing chocolate from an automatic machine at Clifton Rocks Railway by using metal discs. Forty discs representing 3s 6d worth of chocolate, were inserted by him and two others who escaped between Saturday night and Sunday afternoon. He was ordered to be birched.

On 22 February 1908 there was a slight accident at the Railway with no serious injuries, despite initial rumours to the contrary.

> While one of the two lines were being worked a car appeared to get beyond proper control, both cars – the one ascending and the other descending – reached their destinations with considerable velocity. The supposition is that one of the brakes did not act in the customary manner. Fortunately there was only one passenger, and he sustained only a cut on the hand from a falling piece of glass. When the ascending car reached the top station the impact was sufficient to break the glass in the office, and the man who was in charge there naturally beat a hasty retreat. The driver, or conductor who stands in front of the car when it is making a journey, sustained some shock: but the conductor of the descending car – a portion of the footboard of which was splintered – escaped without injury. The passenger who received a cut was in the ascending car and was attended to by a local doctor. Our representative was officially informed that the matter was a trivial one, and that the railway will be open for traffic today as usual.

On 15 April 1914, it was reported that since the Tramways Company acquired Clifton Rocks Railway, that an improvement has been effected in the appearance of the frontage, and the cars.

Between 1915-1922, it was noted that donations to the Bristol Royal Infirmary (George White's major charity) were made by Clifton Rocks Railway.

4.21.1 Hotel and Railway up for sale

In May 1922, it was announced the Hotel would be sold by auction on 6 July 1922 by Mr and Mrs Price who intended to retire if sold. As well as the Hotel, included in the sale were the Terraced Gardens, Tuffleigh House, the Hotel Garage and the Grand Pump Room (now converted into the Spa Cinema), and the leasehold reversion of the Clifton Rocks Railway, held under a sub lease for 872 years unexpired, at the yearly rent of £50 payable to the vendors. The site was not sold.

Six years later, on 25 February 1928, the site was sold by Mr Price to the Grand Hotel in Broad Street.

On 1 June 1928, 32 children from Chewton Mendip School were taken on an education visit to Bristol. The went to Fry's in Union Street, Bristol Museum, Cabot Tower, Suspension Bridge and were particularly interested in the mechanism of Clifton Rocks Railway.

4.21.2 Railway closure and neglect

On 1 October 1934 the Railway quietly closed as a after years of deficit.

In April 1935 nine boys broke in and did a considerable amount of damage costing £7 to repair, with broken window panes in the cars and windows overlooking the road way, and oval glasses in lead lights in the drivers cabin, broken roof panes, cushions thrown out, and a broken 14-rung ladder.

In March 1937 Bristol Tramways ascertained that it was their responsibility to maintain the tunnel and property. Since it was on a lease it was impractical to abandon it. Next month is is reported that:

> The re-opening of the Clifton Rocks Railway would be very acceptable to a number of people, not only visitors, but those who reside in the Cotham, Cheltenham Road and Westbury districts, who by means of the bus to the Suspension Bridge, and via the Rocks Railway would have easy access to the pleasure steamers, thus avoiding a tedious journey through the city. It would also be used by thousands during the Coronation celebrations. The question is often asked why was the Rocks Railway closed, and the reply is that it was more or less a white elephant. The public did not appreciate the enterprise and it was discontinued as a going concern. Whether it could be made to pay now is problematical.

On 1 July 1938 it was reported that Mr Packer wanted to turn the premises into a petrol filling station.

In May 1939 we read:

> Since the Clifton Rocks Railway was closed down a few years ago the entrances to this very convenient lift have become shabby and unsightly through utter neglect. This is particularly true of the upper entrance, adjoining the Clifton Spa Hotel, where the structure is more extensive. Here, owing to lack of paint, the metal work shows many signs of rusting, and so decayed is the vertical end of the canopy at the foot of the steps that the glass has completely fallen away; while one of the glass signs has likewise disappeared from its frame, probably from the same cause. On the other hand, the cracking of the canopy roof is no doubt accounted for by the stone-throwing activities of irresponsible louts and guttersnipes, whose mischievous pranks are likely to continue so long as the station remains closed and deserted.
> But whatever the future of the railway may be, it is surely somebody's duty to do all they can to prevent exposed and visible properties at both ends of the lift from becoming permanent additions to the city's many eye sores of which the derelict areas left by roughly finished demolition

schemes present such flagrant example. What is even more desirable is that restoration of these structures should be followed by the re-opening of the whole system, if only during the summer months. for there can be little doubt that, although not perhaps a profitable undertaking in itself, it provided a useful means of transport for those who wished to avoid the arduous physical exertions of clambering up one or other of the steep ascents from Hotwells to Clifton, between which two districts there was and still is, no direct bus service.

There is still no direct bus service in 2017.

In March 1940 the Ministry of Works (for Imperial Airways) arranged an annual tenancy at £100 per annum so it could be converted to an air-raid shelter. That story continues in the chapter 10 with a whole new life for the Railway. The surrender of the whole to Bristol Corporation was sealed by Company 23 April 1941 but the date left open and eventually agreed 1 July 1942.

five

Other Transport Links

The many transport links in Hotwell Road made the Railway viable. This chapter describes each link, and what happened to them.

5.1 Port and Pier Railway

The building of the Suspension Bridge (completed 1864) saw a resurgence of tourism to the area and was accompanied by new transport opportunities at the foot of the gorge. In 1865 the Bristol Port Railway & Pier Company opened its line – not connected to the main line network – from Avonmouth to Hotwells (the latter station confusingly originally being named Clifton so renamed to Hotwells in 1891) located just a short way beyond the Suspension Bridge.

The company realised it needed connection to the network, if only for goods traffic to and from the docks, and during the 1870s the Clifton Extension Railway was built, which has a long tunnel going from Clifton Down Station under Upper Belgrave Road and ending up at Sea Walls. The Hotwells station was so cramped that ingenuity was called for in train operation, the lines originally converged on a turntable at the south end of the station; engines would change tracks on this, and draw their carriages further into the platform by a rope from the adjacent track. The Board of Trade did not like this arrangement and the turntable was removed in 1893, and the platform extended instead. Business grew after 1877 when Avonmouth Docks were opened. In 1892 the idea of extending the Port and Pier Railway to Canons Marsh was abandoned. The Royal Edward Dock was opened by King Edward VII in 1908. On 5 January 1907 the *WDP* noted that:

anyone waiting for a train at Hotwells station situated at the foot of mighty cliffs, may notice the tunnel is single track. Until some ingenious person discovers a practical way of connecting the Hotwells piece of line with the city,

5.1. Port and Pier Railway Clifton Station/Hotwells. The flat area where the station was, is still to be seen looking from the parapet of the Suspension Bridge. The entrance to tunnel one is to the left of the image. It was a single track railway with the train above the station. It can be seen that it would have been hard to extend to the right (south to the city). (Author collection)

5.2. 1899. An early electric tram judging by the early fender, and its number. The driver and conductor looking very resplendent. No route number indicated, just the route and no boards along the spine. (Peter Davey)

its value for goods traffic will be almost nil. At the Cumberland Basin, the Great Western Railway Company have a regular network. Short as the separation of these lines from Hotwells are, the gap is fatal. The station is little more than a dead end. Even the tramcars only get so far now and then, and a traveller who is very early or cuts the time too finely finds he has to walk for the last two hundred yards. Some have allowed themselves to contemplate when from the fine goods depot at the back of the Cathedral, it will be possible to get by rail to the river's mouth towns, while many more have anticipated when electric cars will continue their journey from Clifton Rocks lift up Bridge Valley Road, past the Zoo, to the top of Blackboy Hill.

The Hotwell station was closed in 1922 to build the Portway.

5.2 Trams

From the 1880s a horse tramway linked Hotwells railway station with the city centre at St Augustines Parade. Horse trams were not numbered.

In 1887, Bristol's first regular horse-bus service (see plate 1.13) started operation from St Augustines Parade and the Suspension Bridge via Victoria Rooms, Regent Street and the Mall. The Clifton terminus at the Suspension Bridge was close to the high-level station of Clifton Rocks Railway. There were no buses on Sunday, so Clifton was not well-served by public transport. It cost 3d to get from the Tramways Centre to Suspension Bridge, so was more expensive than Hotwells which was 1d.

Plans were afoot to use electric trams, which needed extra infrastucture. On 5 January 1894, the Sanitary Authority met to agree to doubling the tram lines where practicable on the Hotwells Line due to the very

5.3. Electric tram in George Newnes' time. Probably rock falls on the right necessitated the fence (in most tram postcards.). The lamp posts are still gas. There is now a change of street furniture from plate 4.5, due to the addition of tram posts in 1889 to hold the wires. The sign says 'Electric Cars stop here'. (Author collection)

5.4. Tram in George White's time. More trams than passengers. There are large posts for the electric arc street lights introduced to Clifton between 1898 and 1906. Note the Colonnade on the right-hand side. The tram is route number 9. (Peter Davey)

large traffic considerably augmented by the development of the passenger steamer traffic and the opening of the Clifton Rocks Railway. In December 1896, controversial plans to use electricity for all lines were adopted. In December 1900, electric trams replaced horse trams between Brislington, Tramways Centre and Hotwells.

Tram Route 9 went past the Railway and often the same tram numbers used the same route each time. Tram 163 (plate 5.2) was the first electric tram here, in 1899, followed by no. 114 in 1900. Other trams seen on postcards by the Railway on the route: Brislington; Tramway Centre; Hotwells are 203, 209, 210, 213, 221, 226 (route 9) and 231 up to 1934. A note on rolling stock[3] reported that nos. 203-232 were a disaster and 4 had to be rebuilt. In 1906, a motor bus to Clifton replaced the horse bus.

A quirk in the Tramways Act of 1870 entitled Bristol Corporation to take over the city's tramway system at book value in 1915, or any seventh year thereafter. Because of this, the main delivery of trams entering service in 1900 remained virtually unaltered until the system closed down. The corporation considered they should run both trams and buses. This battle placed a huge strain on George White who died a year later. Taxi cabs and Clifton Rocks Railway (as a commercial speculation it was described as a failure) escaped this because it was classified as a light railway. The Corporation offered £670,000 for the system but White considered the price to be £2 million. Enthusiasm for municipalisation began to wane, negotiations came to nothing and the system stayed in private hands.[4]

On 8 January 1920, there were complaints that the Sunday tram service, dropped during the war, had not been resumed. This clearly affected Clifton Rocks Railway. There were calls for municipalisation of the trams rather than looking after shareholders. Permission to build the Portway was granted. Part would have tramway tracks in the centre, part with tramway tracks on the side, which would not be accessible by road vehicles. This would cost less to lay and maintain, and enable cars to go faster. The tramway track would have to find a route through the railway tunnel at Bridge Valley Road which led to the question as to who would work the line to Avonmouth. The Portway cost £729,800 to build.

In January 1922, it was proposed to use rail-less electric buses (trolley buses) along the Portway, and the tram lines were never extended beyond Hotwells station.

In November 1925, residents succesfully raised a petition that extended motor omnibus services from Durdham Down to the Clifton Rocks Railway, by way of Upper Belgrave Road, the Promenade, Gloucester Row and Sion Hill – marring one of the most beautiful spots in the city. And yet 141 buses in one day had carried only 319 passengers (2.26 passengers per bus). Fifty nine had carried only one or no passengers. It was pointed out that if the watch committee was to stop the motor-omnibus service from Eastville to Clifton it would inconvenience not just residents living along alongside the Downs, but many more people besides.

There are several images of trams at Clifton Rocks Railway with different fleet numbers. The earliest electric cars were numbered 86-97 and 116-118 and used in Kingswood and Staple Hill. The vast majority of the electric tram fleet arrived in 1900-01 to replace the horse cars. The new stock was numbered 1-85, 98-115, 119-124 and 162-232 and became regarded as Bristol's standard tram.[1]

In 1938, the replacement of Bristol trams by buses began. The last tram to Hotwells was 7 May 1938. One wonders why it went on for so long when the Clifton Rocks Railway closed four years previously, and the landing stage would only have been used at three hours before to one hour after high tides. The lines were never extended beyond the Hotwell station to Shirehampton, even though the Portway was built wide enough to accommodate trams.

5.3 Tram operating times and fares

In 1893 the Rocks Railway opened at 8:30am and ran until 9:00pm. On Sundays the first car was at 2:30pm.[2] This is clearly when horse trams were running.

In 1897, the first trams departed from Joint Station (Temple Meads) at 6:40am, 7:50am (Sundays 2:00pm until 9:50pm) and then every few minutes until 10:50pm to Hotwells (Sundays 2:10pm until 10:00pm). It took seven minutes to get to the Joint Station to Tramways Centre (i.e. St Augustines Parade) and cost 1d.

The first trams started from the Tramways Centre at 6:50am, 8:00am to Hotwells and then every few minutes until 11:00pm. It took 16 minutes to go from the Tramways Centre to Hotwells. Not all trams went as far as Clifton Rocks Railway, and no return tickets via Clifton Rocks Railway were issued on Public Holidays. Tickets had to be bought separately, and booked.

FARES either way.
Joint Station and Hotwells 2d.
Hotwells and Tramways Centre 1d.
Tramways Centre & Clifton, via CRR 2d.
BTCC timetable 1897

There were also special trams to and from the Channel pleasure boats when the boats departed and arrived any hour between 6:00am and midnight, costing 1d.

In 1901, there were electric trams throughout. The first car at Joint Station left at 7:50am and the last at 10:50pm, with a service every few minutes and increased cars during the busiest hours. The first car on Sundays was at 2:20pm and the last at 9:50pm. Cars ran to Hotwells Station only when trains arrived and departed; they stopped at the Rocks Railway at other times. The first tram left Hotwells at 8:17am and the last at 11:17pm on weekdays; on Sundays the first at 2:46pm and the last at 10:17pm. Fares were unchanged with a through fare of 2d.

From 1926-34 the Rocks Railway opened at 8:45am and ran until 9:15pm; Sundays 2:30pm to 9:15pm. In 1926 the electric tram route 8 and 9 ran from Hotwells – Tramways Centre – Temple Meads Station – Arnos Vale – Depot – to Brislington with a weekday start of 5:25am to 10:50pm; Sundays 2:10am to 10:15pm.

5.4 Paddle steamers

In 1858 landing stages were built at Hotwells by the Rocks Railway for passengers of the Cardiff and Newport steamers.

The Bristol Channel was to develop into one of the major areas for paddle steamer operation in the UK after Peter and Alec Campbell, sons of the well-known Firth of Clyde steamboat owner Captain Bob Campbell, sold their business to the Caledonian Steam Packet Company which, from 1889, based operations at Bristol. In 1887, Campbell's paddler *Waverley* had been chartered for use on the channel and so in 1888 was brought down from Scotland. P & A Campbell's White Funnel fleet dominated coastal cruising and ferry operations for the next 80 years. The main cruise destination from Bristol was Ilfracombe with services from Bristol and Cardiff. One could also get on the steamer in St Augustines Parade and Cumberland Basin which were non-tidal docking points.

In 1892 Captain Alec Campbell stated that in the four years his firm had operated in Bristol they had carried 250,000 passengers. A good portent for George Newnes.

On 27 May 1895 the Docks Committee tried to collect poll tax from the passengers from Cardiff on the Ravenswood steamer at the Hotwells' pontoon – which caused absolute chaos and a huge protest. Campbell's of Bristol and Robertsons steamers of Cardiff later refused to collect the tax.

On 3 March 1896, it was agreed to provide improved landing facilities for the public. By March 1913 the pontoon was slipping into the river. Cracks filled with cement had little effect and more openings became visible. Subsidence occurred in the road. The work to repair the bank did not begin until April 1914.

Between 1914 and 1918, the fleet of 13 Campbell steamers (operating from Bristol, Cardiff, Newport, Barry, Brighton and Eastbourne) were requisitioned for mine sweeping duties. West Country trips returned in 1919.

The Bristol Channel district guide of 1925 advises passengers arriving at Temple Meads to take a tram to Victoria Street and from there to Hotwells. Trams ran every few minutes and took 20-30 minutes to get to Hotwells – a distance of two and a half miles – no faster than today (the timetable of 1901 shows 23 minutes). A Bristol Tramways taxi would cost 4s 0d. In 1925 there were 11 steamers operating (two having been lost during the war).

The 1950s saw a significant decline in the excursion trade due to a decade of bad summers, coach travel, foreign travel and motor cars. To this day the *Waverley* (built 1946 to replace the original *Waverley* of 1887) is always popular when it visits Bristol. The old pontoons by the Railway have virtually collapsed.

5.5 Rownham Ferry

The Rownham ferry ran across across the River Avon in Bristol. A ferry was operating by the twelfth century but ceased in 1932, replaced by bridges. It crossed the river at the southern end of the Avon Gorge, between the New Inn in Bower Ashton and the Rownham Tavern in Hotwells.

When the Portishead Railway was built in 1866, the ferry became popular with users of Clifton Bridge railway station.

The ferry was moved in 1873 due to the expansion of Cumberland Basin, with new slipways built on both sides of the river. At low tide, the crossing was a bridge of boats rather than a ferry (see plate 5.6). The toll was 1d – a price reflecting the awkward nature of the tidal crossing and the manpower required to keep slipways free of tidal mud. The slipways are still occasionally visible at low tide.

5.5 Summary

The Port and Pier Railway closed in 1922, replaced by the Portway; the trams struggled on until 1938; the Rownham ferry closed in 1932. All we have now is a busy main road which is impossible to cross, with a very small pavement outside the bottom station. The pontoon for the paddle steamers outside the Railway

5.5. *Westward Ho* outside the Railway. Five trams, lots of people – it must have been a Sunday afternoon. (Ed Scammell collection)

5.6. The ferry is just south of the Railway. (BMAG)

5.7. Tram running past a closed Railway in 1938. (Author collection)

is derelict, and there are not that many passenger boats using the Gorge – partly because it has the second highest tide range in the world. There are few residents living in St Vincent's Parade and the Colonnade. There is no room for a bus stop, and the road is not wide enough for a footbridge – hardly encouraging to revive a twice-bankrupt Railway. And when war broke out in 1939, the conversion of the for multiple uses. would make reinstatement ever more difficult. You will read about this in the ensuing chapters.

six

Maintenance

I have found details of maintenance from several sources: Bristol Records Office, the National Archives at Kew and, unexpectedly, from Robert Woolley who wanted me to have his grandfather Alfred Woolley's day books (we also shared a love of vintage cars). He was concerned that his grandchildren would not appreciate the historic value of the books and wanted the information to reach a wider audience. When I received the books, the first thing I did was to analyse how Alfred Woolley's work had changed over a ten-year period at the turn of the nineteenth century – a time of amazing engineering change. I had suspected the Railway did not allocate enough money for maintenance when it became a company, so to have the actual expenditure records for the period post-receivership, in 1908, was invaluable.

The Railway employed its own engineer (at one stage it was the station manager's son Stanley Pearce who lived in the Colonnade) to do the standard maintenance, and who had a special uniform with gold braid. Mr Woolley would have been called out to do specialist engineering jobs.

After Bristol Tramways and Carriage Company (BTCC) bought the Railway, they provided their own engineers (G. Edwards) and architects (H.A. Penney) to resolve any problems.

6.1 Woolley Brothers Engineers

The records of Mr Alfred Woolley are the only evidence I have found of maintenance in George Newnes' time, but are very detailed. Alfred's day books: December 1899-April 1901 (job nos. 1034-2067), and June 1904-February 1908 (job nos. 6336-7506), detail engineering work he carried out in Bristol including for the Railway and provide a fascinating insight into the kind of work an engineer did at the end of the century, how many men he employed and how much he paid them. The books give a job description, the price, the components with prices, and the hours of each man for each job. The job is lined

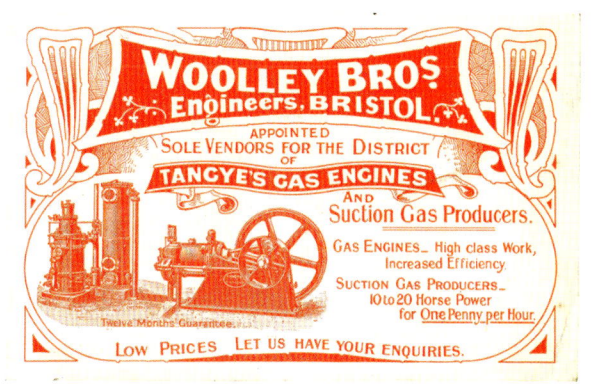

6.1. Woolley Bros., calling card.

through when paid. Sometimes there are diagrams. The books mark the advent of their business in Bristol, since they only started trading as the Woolley Brothers in 1899. Alfred served a seven-year apprenticeship under James Watt, the Bristol Docks engineer, before starting the engineering company, one specialism being gas engines. *Wrights Directory* of 1899 lists them under engineers, machinists and millwrights. Alfred's 1892 certificate of competence of steam engineering is wondrous to behold.[1]

It is fantastic that when I looked at the income and expenditure records for the Railway from July 1908 to October 1914, produced by A.A. Yeatman, the secretary, that Mr Woolley was still in regular work, and that the Crossley engines needed regular attention. Unfortunately, there is no detail about the job itself.

There is an estimate for a job in 24 January 1901 – the first job – where Alfred gives Mr Pearce commission of £7 10 0d to get more jobs. It clearly had an effect, since more jobs ensued. It does seem bad that all the axle shafts and wheels for four passenger cars had to be replaced after only eight years. The specification of contract in the day book was as follows:

Gentlemen,
In answer to your esteemed inquiry we have much pleasure submitting an estimate for work required as specified

by your superintendent Mr Pearce. *Namely*:-

Eight complete turned axle shafts with new steel car wheels firmed up on treads and pressed into position on same

New special made tap brasses, fitted for eight axle lining brackets – each bearing fitted with a 1" adjusting [?] and lubricating arrangement as instructed

To properly adjust and machine [?] pump eccentric straps and fit thereto

Eight new eccentric shim[?] Fitted to axle shaft in two halves – the joints shafed and securely fixed together by double [?]. The main dimensions of the above parts to be taken from your tracing.

Let the whole to be accurately made and fitted in proper working order the four cars now in use. To the entire satisfaction for the sum of £130 0 0d. Commission promised to Mr Pearce £7 10 0d.

...

Esteemed order will oblige and receive our prompt and best effort
AE Woolley

Materials
16 car wheels best steel and fitted on axle shaft
cast stub wheels 1½wt @28/- per wheel £33 12 0
   shafts eight 3" diam x 4'6" long each=36ft @3/10 per ft £6-18-0
mild steel bar at £4-10-0
eight new tap brasses to fit casting frame=weight of brass about 7lb each with pattern   £2-16-0
eight new eccentrics – cast iron to spigot turned for side adjustment split-pattern 2/6
castings 12lb each 1 rolls [?] 14/-, 2/6   £1-0-0
fitting eight axles and shrinking and fixing wheels, allow say 2 days timing on each or 2 weeks £5-10-0
fitting 8 brasses and machining same. One week £2-0-0
cutting keyways in shafts 16 @ 1/6 each £1-4 0
Fitting keys 10s-0d
?eight long 1" screw about 5" long @ and fitted. Cost about 2/- each 16s-0d
Fitting and turning eccentrics and turning screws for same £4-0-0
Fitting in shafts and all complete on cars £3-10-0
Fixing eccentric shafts 16-0d
Russell Tube nuts and sockets 4s-3d

We also shall be pleased to supply and fix in proper working order:

Main driving shaft 2" diam about 42 ft in length with necessary hangers- bearings in ? line collars- fast and line pulleys. Turn elm shaft 19" from end for eccentricities and fit other parts left black.
Also
One another shaft 1 2/4" diam with necessary couplings and bearings fixed complete – the whole lined up and left running to your satisfaction for the sum of £37-5-0

The bill sent also included 1s 9d for tram fares.

6.2 Woolley's table of railway repairs 1904-1908

There seems to have been just as many jobs for the Crossley engine as for making the brake blocks. The safe operation of the engine and pump required an adequate back pressure to be assured even in the event of a leak from the mains. A flow gauge provided information as to whether the mains pressure was adequate or not at any particular time. Although the engines were efficient, the confined space in which they were housed made for difficult working conditions. It was interesting to see that an exhaust pipe was fitted in 1906. There were 26 to 30 brake blocks ordered at a time. There would have been pads needed for the gripper brakes on the track. Ash seems to have been used at least once for the blocks. Hard woods provided better friction. Leather washers were used for gripper brakes. Alfred was also doing work for the Hotel and supplying sundries for the Railway engineer.

6.3 Regular maintenance

There are several Law Accident Insurance Society Examination Reports and other letters in the Bristol Records Office[2] from 1913 detailing what was tested and overhauled.

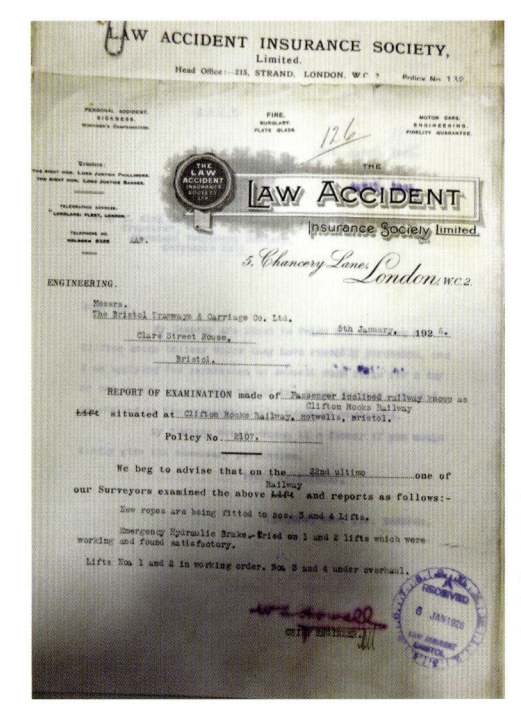

6.2. Ropes and brakes insurance examination. (BRO 39735/8)

Table 6.1.
Mr Woolley's repairs. Note that brake parts are made every year. It is surprising that a new axle, made in 1901, needed boring out and bushing seven years later in 1908 – so must have had heavy wear. It is interesting that the hydraulic lift in the Hotel needed its wire ropes replacing every three years. The jobs Mr Woolley did for the Railway were very diverse. See reference 'Work of an Engineer'.

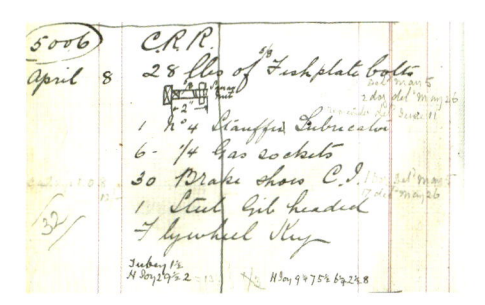

6.3. 8 April 1905: 28lb of fishplate bolts 5/8in (diameter) 2in long, with a square nut. Needed two men spending 53 hours over 10 days.

6.4. 8 April 1905: 1 no. 4 stauffer lubricator. Photo shows one found on a cable wheel.

6.5. 12 April 1905: Six large buffer springs. These can be seen each side of each track just below the cable wheels.

1904		£17 5s 1d
11 Aug	1 gallon engine oil (own tin)	2s 0d
16 Aug	Repairs to undergear of passenger car- taking apart gear wedges and keys	8s 6d
29 Aug	29 brake blocks (1 hr)	
2 Sep	1 cast iron clip and repairs to wheel of car	19s 0d
9 Sep	Spa and Pump Room: to work in connection with renewing 3 suspending ropes and one activating rope to passenger hydraulic lift	£10 17s 6d
27 Sep	Shutting new middle in connecting rod	2s 0d
10 Oct	26 brake blocks	5s 5d
10 Nov	Making new end on plunger rescrewing and new collar turned to fit 1 dozen new springs to sample wire 1/10 1/2 Nov 28th two ¼" gas socket for Mr Pearce	10s 0d
7 Dec	Shutting new end on spindle for gas engine and turning same	2s 6d
9 Dec	3 dozen ½ X 1 5/8 set screws special diam screwed to heads 3/8 thick	7s 9d
16 Dec	3 dozen special small bolts and nuts for pitch chain (for brake)	7s 6d
23 Dec	Fit end pins To brake plunger to fit block	9-0d
29 Dec	Wrought iron bracket	
1905		£12 3s6d
25 Jan	Repairs to bottom plate of passenger car tanks including bolts, plates etc	£3 5s 6d
27 Feb	44 1" 3" pipe screwed both ends, 1 3" bend	18s 6d
8 Mar	Cleaning threads out of flange of gas engine	1s 6d
8 April	28lb of fishplate bolts 5/8" dia. 2" long with square nut. Needed 2 men, 53 hours in 10 days 1 x no.4 stauffer lubricator (photo shows one found on a cable wheel) 6 ¼ gas sockets; 30 brake shoes; 1 stud gib headed; 7 flywheel keys	
12 Apr	Six large buffer springs. These can be seen each side of each track just below the cable wheels.	4s 0d each
26 May	Galvanised bolts 3 dozen	
11 Jul	4 plates, 2 dozen brass screws required 4 men for 24 hours. Cost 1d tram fare.	17s 6d 8s 6d
4 Aug	2 dozen 1" X 3/8" bolts	3s 0d
10 Aug	5 dozen ¾ X ¼ bolts	4s 0d
17 Aug	Cutting off ends of brake block holders and fitting new stud ends for reline slipper block	8s 6d
20 Sep	1 dozen wire rope slings as before	£2 6s 0d
24 Nov	Screwing collar on buffer (see plate 6.4) required 1 man for 1 1/4 hours.	4s 6d
4 Dec	Shaping and drilling cast iron pipe	6s 6d
1906		£7 9s9d
2 Jan	1 small tooth wheel for adjusting connecting rod brasses for gas engine	5s 0d
17 Jan	Shutting new end on spindle	
29 Jan	30 brake block castings as pattern also 6 taken by Mr Pearce 29 Jan 20 ash block to pattern	
1 Mar	One plate 4 ft4 X 12" X 14 gauge	
7 Mar	2 buffer springs 2 ¾" rubber insertions	
16 Mar	Making bent 4" exhaust pipe and flange 12 ft long (to delay the production of a vacuum) for gas engine	£2 5s 9d
7 Apr	3 12" hack saw blades	
25 Apr	Shaping off ends, smelting on plate with ample part Fitting new stick studs and slipper block	7s 6d
2 June	Shaping off worn bracket and casting on stud plate. Also making two for glands to suit plunger blocks sent for brakes	4s 9d
21 Jun	1 doz laces (smith & co)	
10 Jul	Shaping out cast steel plate and fitting and making up piece Shaping out holder for brake blocks and fitting on new stud head Two 14" bastard flat files	5s 0d 6 8s 0d
25 Jul	2 cast iron plungers for brake pump require 3 men for 22.5 hours. Castings required.	19s 0d
18 Aug	Making two metal nuts special. 1 ¼" long 1 1/16 diam	6s 9d
30 Aug	Old slipper blocks	5s 6d
11 Sep	Welding new end on tie rod and making good thread and nut for same	1s 9d
22 Oct	Drilling ½" tap hole in small crosshead	9d
13 Dec	One length of ¼" steam pipe	
13 Dec	Repairs to tank in base car (cancelled)	
1907		£11 19s 1d
25 Jan	50 brake blocks as per pattern sent	17s 10d
18 Mar	One new vice screw	6s 0d
24 May	2 dozen bolts and nuts 5/16 X 1 1 pair clips with bolts 7/16 X 1 1/2	
8 July	Alterations and repair to slipper block	
25 Sept	Repairs to chain and wheel gear for brake New end shaft	1s 0d 12s 9d
8 Oct	Repairs to large sluice valve. Fitting new retaining pieces on valve	8s 6d
8 Oct	One plate 7" wide X 18" long cut down 2 ½ on end and ½" thick Also 8 taper packing pieces for bearings Lent ¼" gas taps	5s 6d
28 Oct	Clifton Spa and Pumproom. Supplying and fitting wire rope (Rubber, trains, rope, flax for lift)	£7 2s 6d
8 Nov	Making 4 springs Drilled fitted with bolts 2 ¼ X 1 ¼ Also about 5ft 1" belt (old)	
16 Dec	½ dozen bush hack saw blades	
30 Dec	1 doz left laces	
1908		
30 Jan	New axle boring out and bushing Wheel drums shrinking on Turning journals facing Eccentrics	£2 13s 6d

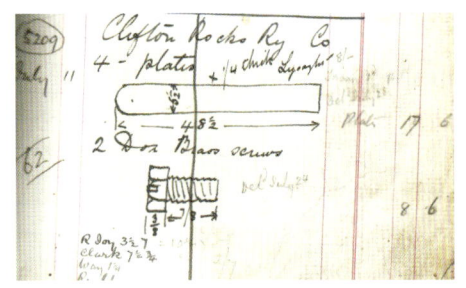

6.6. 11 July 1905: 4 plates, 2 dozen brass screws required 4 men for 24 hours. Cost 1d tram fare.

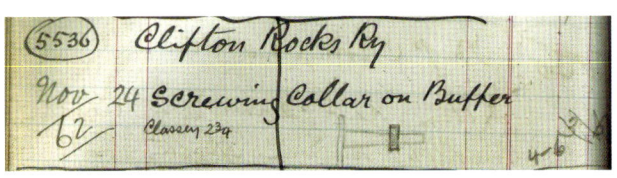

6.7. 24 November 1905: Screwing collar on buffer required 1 man for 1¼ hours.

6.8. 25 July 1906: Cast-iron plungers for brake pump require 3 men for 22.5 hours. Castings required.

date	Insurance check by lift engineer	Car 1,2	Car 3,4
31 July 1919	The condition of the Railway was in good order and that new wire rope had been fitted in April 1919. The permanent stops, sheaves and pins, head gear, runners, locks and gates, safety gear were also confirmed to be in order and no repairs were necessary.	New rope	New rope
17 January 1924	The wire ropes of number 3 and 4 worn		
05 January 1926	new ropes were being fitted to number 3 and 4 lifts and the emergency hydraulic brake had been tried on number 1 and 2 lifts which were working and found satisfactory. Number 1 and 2 lift were in working order.	Running	Not running, New ropes
13 May 1927	ropes worn. 1 and 2 out of commission but no defects. 3 and 4 working, emergency gear tested and found to be in good order. Railway found to be in satisfactory condition and in receipt of careful attention.	Not running, worn ropes	Worn ropes
19 October 1928	3 and 4 working satisfactorily; emergency gear tested and found in order. 1 and 2 examined whilst out of commission, no defects observed. Ropes of 3,4 worn but good for service.	Not running	Running, worn ropes
08 February 1929	1 and 2 working; emergency gear tested and found in order. 3 and 4 examined whilst out of commission, no defects observed. Ropes becoming worn but appear safe.	Running, Worn ropes	Not running
31 July 1929	3 and 4 working satisfactorily; emergency gear tested and found in order. 1 and 2 examined whilst out of commission, no defects observed. Ropes of 3,4 worn but good for service. New ropes recently fitted to 1 and 2.	Not running, New ropes	Running, worn ropes
23 December 1929	Numbers 1 and 2 in commission and working satisfactorily, emergency gear tested and found in order. Ropes slightly worn. 3 and 4 worn but safe. Sections generally in order as seen at rest.	Running, slightly worn ropes	Not running, worn ropes
13 May 1930	Numbers 1 and 2 out of commission; ropes worn but serviceable, no defects requiring attention being observed. Sections 3 and 4 examined at work, emergency gear tested and in order. Ropes worn, but safe. Sections generally in satisfactory condition	Not running, ropes worn	Not running, ropes worn
10 December 1931	One of the safety wedges on car 4 is inoperative. This is receiving attention forthwith, otherwise the cars and haulage mechanism appear to be in satisfactory condition.		
05 March 1932	safety wedge car 1 inoperative, receiving attention, otherwise the cars and haulage mechanism appear to be in satisfactory condition		
30 October 1933	Satisfactory		
26 February 1934	wires worn but otherwise satisfactory	worn ropes	worn ropes
22 June 1934	The ropes slightly worn but fit for further service	worn ropes	worn ropes

Table 6.2. Insurance checks by lift engineer, 1919-1934.

6.3.1 Insurance

One can see in the table left (6.2) that there are insurance checks three times a year, that ropes were replaced about every three years, and that only two cars were operating (saving manpower and maintenance). Mr Woolley would have done the work on the brakes, Cradocks on the wire ropes.

Table 6.3. Bankrupt maintenance expenditure year on year, 1909-1912.

item	product	July-Dec 1908	1909	1910	1911	1912	total
A Pole	Printers and Stationers			£0.77	£0.81	£1.34	£2.92
A Williamson				£0.75			£0.75
Colhurst and Harding	Paint and varnish		£6.40	£23.29	£7.93	£6.27	£43.90
Colthurst and Harding	cotton waste	£1.95					£1.95
Crossley	Gas engine		£0.23	£1.60	£0.78	£1.41	£4.02
Crossley	piston rings	£0.68	£0.80				£1.48
D Cradock	cable				£17.38	£21.51	£38.89
Demature				£5.32		£1.12	£6.44
Dermaline	Hand cream?		£1.48		£1.04		£2.52
Diamond Lubricating	grease				£1.30		£1.30
Durie	uniform	£3.50	£14.33	£20.20	£14.43	£3.50	£55.95
Ernest Day	Blacksmith			£0.19			£0.19
George George	blaster		£2.25				£2.25
H Pook	Saddlers and Harness Makers		£0.38	£0.35			£0.73
Frederick P Prentice	Ironmonger			£4.85			£4.85
Herhelmer Leat				£7.40	£2.25		£9.65
I Merrett (Frank?)	Ironmonger and Gas Fitter			£1.17			£1.17
Leadbeater and Scott	steel				£2.21	£3.56	£5.78
punching tickets		£1.45					£1.45
Rouch and Penny	electrical				£2.20		£2.20
ruffley and willis			£4.13				£4.13
S Cook			£1.28				£1.28
W Goldman	Glass Merchants				£0.80		£0.80
William Thomas	Ironmonger				£0.94		£0.94
Werthumen Leat			£2.25				£2.25
Woolley	engineer		£11.03	£22.75	£14.20	£109.13	£157.11
Total		£7.58	£42.30	£79.71	£61.47	£146.50	£337.56

6.4 CRR year bankrupt maintenance expenditure

Maintenance costs did not go down each year despite the Railway being in receivership, though it is noticeable that in the first six months only £7.58 was spent.

It was rewarding to see (in table 6.3) George George being paid for services, and to find out which companies were providing services – though frustrating not to know the details.

Several of the companies were based in Clifton or nearby: Ernest Day from Rownham Yard, Hotwell Road; Frederick P. Prentice, 5 Boyce's Avenue and 42 Regent Street Clifton; H. Pook & Son, Saddlers and Harness Makers, 13-14 Victoria Street Clifton (now Princess Victoria Street) who would have probably made the leather washers needed for the gripper brakes; William Thomas, Regent Street Clifton; and Mr Woolley, St Georges Road.

Crossley were renowned for providing good service to their customers.

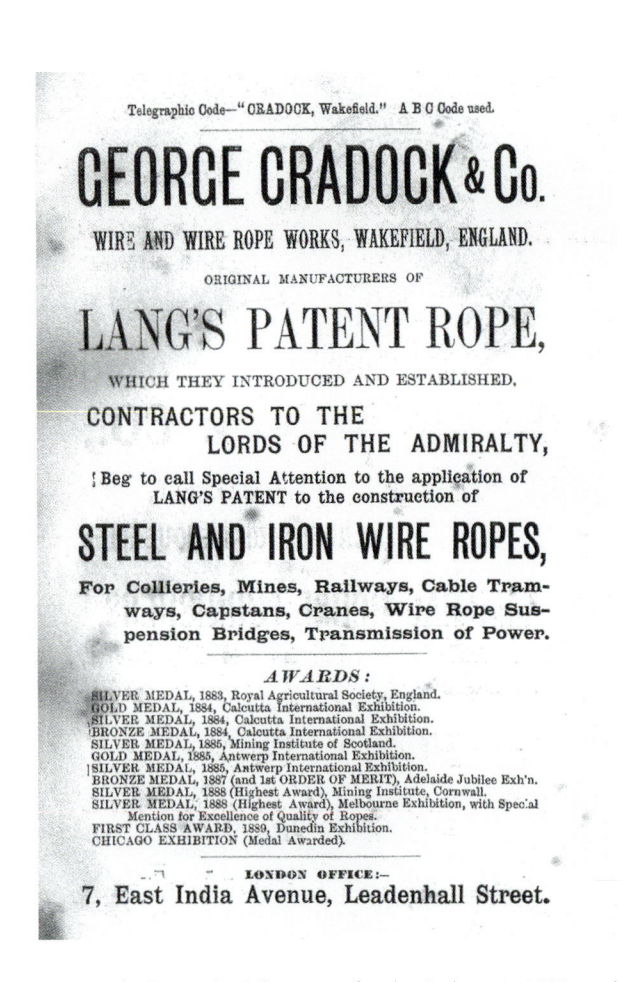

6.9. Cradock supplied the ropes for the Railway in 1893, and thereafter. (Official Description)

6.10. Inspector Wall in the centre, identified by the hat and moustache (see chapters 4 and 8). The female conductor has a different style of hat to a male driver (see plate 5.2). (Peter Swan)

6.5 Maintenance after BTCC takeover

On 18 April 1913, a letter was sent to the treasurer of the Merchant Venturers from Geo Edwards of BTCC asking for permission to deposit rails for a day or two in Princes Lane at the top of the Railway which the applicant understood was their property. This was to enable the old rails to be relaid.

On 28 January 1920, Mr Price the proprietor of the Spa had just bought the hotel and was in the process of turning the ballroom into a cinema. The ballroom had been flooded by an overflow of water from the Rocks Railway water tank (the overflow pipe can be seen in the corridor leading from the top station turnstile room). He wrote to the BTCC secretary. It was agreed to send a cheque for £66 16s 0d to Mr Price – £30 less than Mr Hayes' estimate.

In 1925, repairs and maintenance expenses totalled £293, cleaning and lubrication £9. In 1926, repairs and maintenance expenses totalled £303, cleaning and lubrication £11.

On 12 September 1926, a letter from Mr Price the manager was:

calling attention to the condition of the top station guttering in Princes Lane. It is overgrown with grass and so filled up with dirt that the gutter did not fulfill its purpose at all.

On 23 January 1928, a letter was sent from BTCC Inspector J. Wall to the BTCC engineer George T.C. Edwards that owing to recent rain, wet had come in through the roof into the engine room, more than on other occasions, and there was a danger of it dropping on to switches on the starting panel which drove the motor pump. Mr Wall suggested that some protection be put in place to prevent this. A follow-up letter was sent by the BTCC architect Mr H.A. Penney, which stated that the roof of the engine room consisted of a brick arch which further on becomes the tunnel for the Railway itself.

When the Railway was taken over by BTCC, spots of water were percolating through the arch. The centre portion of the engine room was protected by corrugated iron bolted within an inch or two of the arch (it is not possible to view behind it) and the water conducted to the sides where it was carried away by rain-water chuting.

Originally we had thought the corrugated iron and gutters (shown in plate 6.11) had been installed during the war. We were surprised it had been painted, unlike the corrugated tin down the BBC steps.

Where no protection was provided, considerable quantities of water came in, especially in one place over the wooden staircase (moved during WWII but was on the north side), where it ran down the stairs and into the public area. It was noted that in the old stone steps proceeding from Hotwell Road over the Railway, a new fissure or crack some two or three inches wide had appeared. There had been abnormally wet weather. The course of water would continue until something was done about it. The owners of the land and the steps were the Merchant Venturers according to the district engineer for the corporation. The stone work forming the boundary wall to the steps was in very bad condition and quite a number of topping stones were loose and liable to fall on persons using the Railway.

On 5 April 1928, a letter was written to the Merchant Venturers telling them about the dangerous masonry and fissure and surface water. On 24 May, the Merchants considered there was no liability to prevent surface water from coming onto the roof. The Railway's response, on 6 June 1928, was that the present condition of the steps was such as to allow more surface water than would otherwise find its way into the Railway and requested repairs to the steps. On 29 June, the Merchants still did not want to admit liability but had arranged for some of the fissures to be stopped up.

On 10 February 1932, the weighing machine (installed in 1929) outside the lower station had been tampered with. The lock had been forced, the leather pocket cut and the money taken. Police were called. It was also damaged in July and September and, owing to the heavy cost of repair and small returns, it was decided not to continue with the contract.

On 17 April 1935 the *Western Daily Press* reported that boys had broken into the railway and caused £5 of damage. Nine boys appeared before the Bristol Juvenile Court. BTCC explained that the railway had not been in use since the previous year but was inspected each month and the entrance kept locked.

6.11. Corrugated iron and gutters on the first floor of the bottom station (the engine room). (Author)

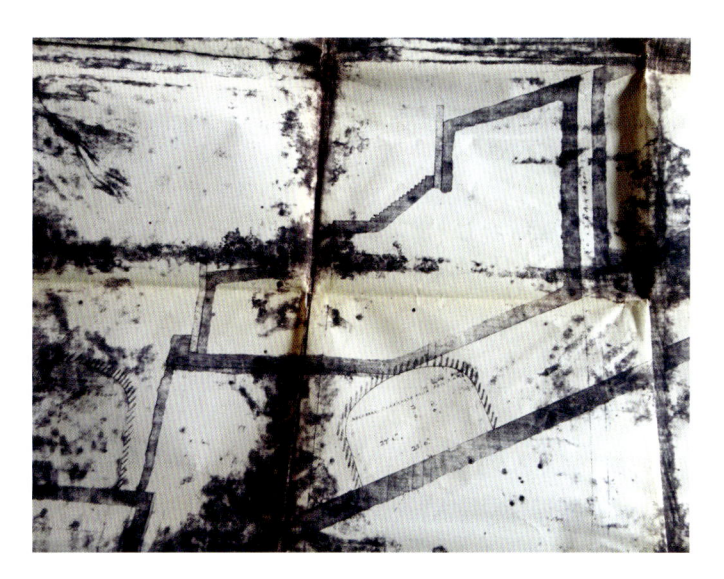

6.12. The 1891 plan by Marks shows the steps above the location of where the tunnel stops being at an angle. This is where the marl (mudstone) was found that caused the rock fall. (BRO)

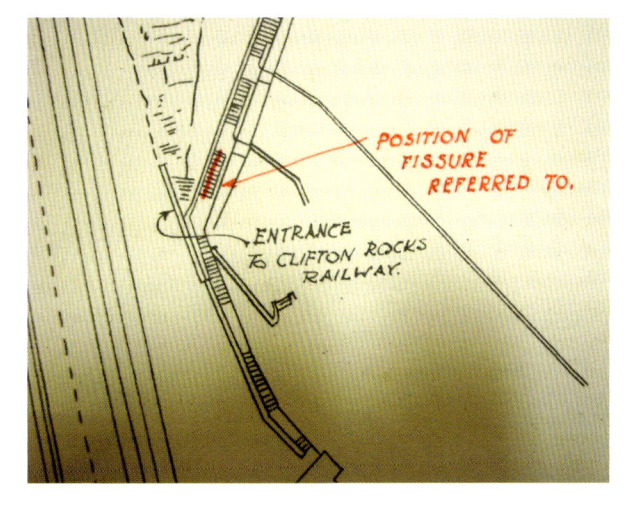

6.13. The 1928 map marks the fissure by the steps where the 1891 rock fall occurred. (BRO 42054.G.Drawer 4/18832)

Someone had broken in and damaged some glass so they boarded the site. They were broken in again and considerable damage was done. There were broken window panes in the cars and the driver's cabin as well as other broken glass. Cushions had been thrown out of the cars into the tunnel. The boys were aged between eight and fifteen. They had been seen by a police officer, and they each admitted having been there, but only two admitted doing any damage. The rest accused each other. BTCC agreed with the police that if the boys had fallen down the incline they could have been killed and remained in there for some time before being discovered. The chairman of the bench advised the parents to be more strict in their supervision. Three of the older boys had been to court before and two were still on probation. One was ordered to a remand home for one week. The rest were bound over for 12 months. The parents of each were required to pay 5 shillings costs. It is a shame that they did not pay to repair the damage.

seven

Railway Artefacts

We found a great many railway artefacts in the rubble on the railway lines from a section sealed during WWII and used to house items from the top station (the most rewarding source), and also in the bottom waiting room and water tank. People have also returned items that they 'rescued' exploring the site when it was derelict and unloved.

We cut a hole in the wall to access the top station artefacts. Plates 7.1, 7.2, 7.4 show artefacts from the top station having lain undisturbed for 65 years in the first chamber (which was used as a bomb absorption chamber during the war, since it was only 15ft under Princes Lane. It was easier to put the items here than take them off-site). There were heaps of railings, gears, valves, turnstiles all mixed up together. Many artefacts and the rubble have been left in here to retain the character of the site. Plate 14.18 in chapter 14 shows the rail section where we got most of the rest of the artefacts before we cleared it.

7.1 Table of Railway artefacts found

category	Found section	type	manufacturer	qty	Height inches	Depth inches	Length inches
water tank	bottom ground	sight tube		2		1 3/8 OD, 1 1/8 ID	18.4
water tank	bottom ground	cable		1	120		
water tank	bottom ground	chain		1	29	.5	
water tank	bottom ground	lead coil		1	0.3	1.2	
water tank	bottom ground	pulley		1	0.5	3	3
water tank	bottom ground	pulley		1	2	2.5	2.5
	exposed rail	bracket 2 hole		1	0.5	0.5	15
rails	exposed rail	greaser	Stauffere, London	1			2 7/8
accessory	exposed rail	mole wrench		1			11.5
rails	exposed rail	sleeper (4 in situ under whole track)		1		3 1/2	64.5
	exposed rail	tube		1	7.2	1.7	1.7
top station decorative	exposed rail	decorative ironwork		2	4.7	2.7	
top station decorative	exposed rail	decorative ironwork		2	5	3	
top station decorative	exposed rail	marble		1	0.75	2	7 1/2
top station decorative	exposed rail	pillar top		1	7	4	2
top station light	exposed rail	prism		61	2.5	2	3.2
top station decorative	exposed rail	railing knob		2	6	4	4

Location	Area	Item	Maker	Qty			
top station	exposed rail	speed limit sign		1			
top station floor	exposed rail	tile		3	0.75	6	6
top station floor	exposed rail	tiles	Adamantine, North Wales	4	0.5	6	3
top station accessory	exposed rail	whistle	acme	1	2.1	0.8	
top station	Sion Hill tunnel	curved bracket		1	11.5		
top station	Sion Hill tunnel	grating		4			
top station	Sion Hill tunnel	grating		1	9.5	0.5	16
top station	Sion Hill tunnel	plate glass		31		0.25	

Location	Area	Item	Maker	Qty			
top station	Sion Hill tunnel	sight tube		2		0.2	
top station	Sion Hill tunnel	decorative ironwork		1	5	3	
top station	Sion Hill tunnel	gas burner		1	1	3.7	
top station	Sion Hill tunnel	sheet glass		3		0.15	
top station	Sion Hill tunnel	tubular gaslight fitting		2		2.5	
Car?	top	large hinge		1			
top station	top	peak cap		1	1.5	9	9.5
water reservoir	top reservoir	ladder 5 rung hooked top		1	52	1.5	12
top station	turnstile	6" internal pipe bend		3		9	
top station	turnstile	bearing support		1			
top station	turnstile	bracket		1			6
rail	turnstile	cable pulley and bracket		3			
rail	turnstile	cable pulley on u support (pulley 3.25" by 6")		1	9	3	16
rail	turnstile	cable pulley support (pulley 6" by 7.75")		1			17.75
car	turnstile	cog with square hole		1		8	
top station	turnstile	conductors wheel		1	15	1.5	15
	turnstile	curved plate attached to electric conductor		1			
rail	turnstile	cylindrical compensator		2	3	6	
top station	turnstile	grill		1	36	8	
top station	turnstile	kettle		1			
	turnstile	I shape bracket with hole		1			18
	turnstile	long bracket		1			48
	turnstile	long bracket	Youngs, Birmingham	1			48
	turnstile	long pulley		1			
	turnstile	pipe gaskets (10.5 ext, 6 int)		1			
	turnstile	rod with levers (2 14" rod, 2 25" lever arm, 42" rod)		1			32.5
car	turnstile	governor drive rod some with missing pinion and levers		4			6
engineer	turnstile	tripod (lifting equip/ hydraulic)	Young, Birmingham	1			50

top station	turnstile	Turnstile barrier	W.T. Ellison and co, Manchester	3	41	7	43
top station	turnstile	turnstile	W.T. Ellison and co, Manchester	3	42	42	21
top station	turnstile	turnstile cover		1			
top station	turnstile	Turnstile counter cover	W.T. Ellison and co, Manchester	1			
top station	Top station	entrance scissor gate		1	7.5		
top station	turnstile	car entrance scissor gate		2	5.25		
top station	turnstile	drawer		1			
top station	turnstile	gas light	Nico	1	23	13	13
top station	turnstile	lamp plus ring		1	2.5	9	9
top station	turnstile	partition		2	178		60
top station	turnstile	partition		5	178	60	
top station	turnstile	railing	Hart and Son, London and Birmingham	1			
top station	turnstile	railing curved		1	41		
top station	turnstile	railing s curve	Peard and co, Birmingham	1	40		27
top station	turnstile	railing straight		1	43		46
top station	turnstile	railing support		1			
top station	turnstile	valve	L Blakeborough and sons, Brighouse	4	30	9	12
top station	turnstile	valve pipe		3			

7.2 Crossley engines

Sadly there is no evidence left of the Crossley engines except for stud marks on the floor, and the article by Dugald Clerk (plate 7.6). It intrigues us how they got the engines upstairs, and how they removed them during the war. However a friend of mine has many stationary engines so I have included a photograph of a similar engine to show the dimensions (plate 7.5). He organised a Crossley party for me, which meant that his friends turned up with their engines. He told me that the two engines had consecutive numbers. He also told me that each engine underwent a day of testing before being released from the factory. The engines were dispatched in component parts (flywheel, crank etc) in crates.

Town gas was cheaper than benzine (petrol). Clerk's reference compares this engine with an Crossley Otto 6hp engine of 1880 to see how they had developed. The gas consumed was 25.9 cubic feet per brake horse power so did not give as good an economy as the earlier engine (34 cubic feet). Exhaust gases were better discharged. Gas consumption per indicated horsepower was 21.19 hp on Clifton gas (was 19.5 hp per hour), brake horsepower was 15.75. Compression pressure was 48 lbs per sq inch up from 31. It was also quieter than earlier engines since it used different kinds of gearing and was belted to the bed against a strong faced flange. It was lit by hot tube ignition. It is interesting that Mr Woolley in chapter 6 adds an exhaust pipe to try to reduce the back pressure. To move a large volume of water up to the height of the Avon Gorge would involve quite a substantial pump. It may well have been a Tangye or Worthington Pump.

7.1. The turnstile section before anything was removed. (Author)

7.3. Finding part of the braking system. More artefacts can be seen alongside the cable by the wall. See plates 14.18, 15.9 for the rail section before anything was moved. (Author)

7.2. Masses of railings mixed up with cogs, levers and valves. (Author)

7.4. Note the gas light by the turnstile. (Author)

7.5. Derrick Hardwick and friend starting a 12hp Crossley engine. It is a similar size to the 9hp, and it looks similar too. There is a knack of starting it: if your arm is in the wrong place you will break it. (Author)

102

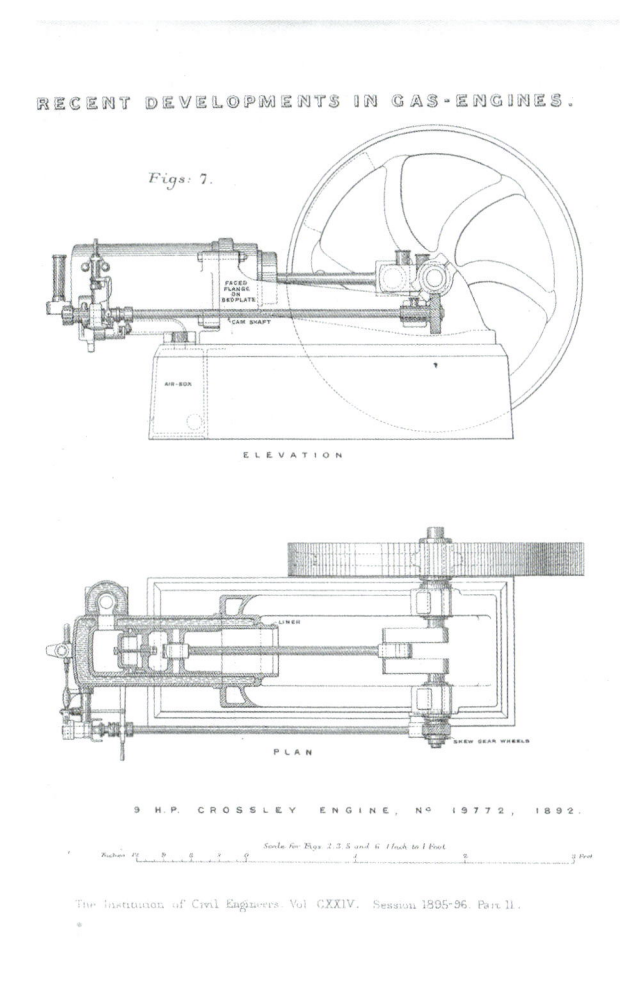

7.6. 1896 drawing of one of the 9hp engines of Clifton Rocks Railway by Clerk[1] (business partner of Croydon Marks), when it was tested for power and gas consumption. (*Recent Developments in Gas Engines*)

7.3 Cable fittings

We were lucky to find accessories for the cables still in place. Plates 7.7 to 7.9 show linkages and cable rollers.

7.9. Cable roller. (Author)
7.8. Cable roller in position – so worn it is broken in the middle. (Author)

7.7. Various levers and cable fittings. The round item on the right is a compensating weight. (Author)

7.10. Governor drive bevel gear. (Author)

7.11. Interlinking straight cut gears of governor drive. (Author)

7.12. These Blakeborough valves are the kind of valves that can be see at Lynton – but only the top is visible there. (Author)

7.13. Valve gaskets made from wood and hessian. (Author)

7.4 Brakes

Parts of the governor drive brake (there are four in various states of completeness) are shown in plates 7.10 and 7.11. All were found in the turnstile section and would have been removed when the cars were winched to the bottom when the tunnel was converted for war time use.

7.5 Top station valve

There was one Blakeborough valve per car to put the water into the car tank at the top station. These can be seen on the top station photographs encased in wooden cabinets. Blakeborough & Sons Ltd were a large engineering company established by Joseph Blakeborough in 1866 and whose core business had been the design and manufacture of industrial water valves. Most of this work was specialised and of bespoke design for the control of liquid flow.

7.6 Clack valve

A clack valve is a valve used in pumps, having a flap or a hinge which lifts up to let the fluid pass, but prevents the fluid from returning by falling back over the aperture. It was so called because of the 'clacking' sound it made.

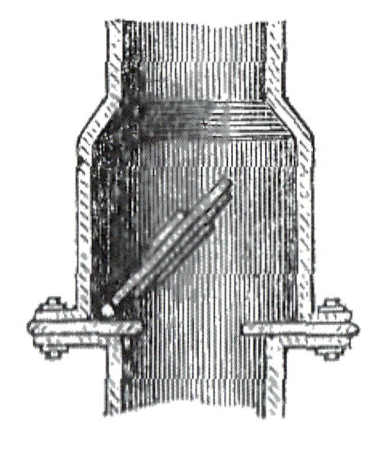

7.14. 1908 illustration of a clack valve showing the flap to let the fluid pass. (Chambers[2])

7.15. This clack valve was lifted from the bottom water reservoir. The filter is to the left, the pipe leading up from the reservoir to the first floor pump is to the right. The water pipe had been cut, presumably when the equipment was taken out during the war.

7.16. Clack valve thimble filter – since the water was reused there would not have been as much sediment as say, Lynmouth Cliff Railway, which used river water. (Author)

7.7 Pulleys

7.17. Porcelain pulley. (Author)

7.18. Pulley for wire that went up the southern side wall of the tunnel. See also plate 3.32. (Author)

The pulleys in plates 7.17 and 7.18 were found in the reservoir of the bottom station and were used for the telegraph system.

7.8 Telegraphic depth indicator

There was a depth indicator at the top station to repeat the level of the water tank. This is in the custody of M Shed. It was manufactured by the Telegraph Works, Silvertown, London.

We also found some glass sight tubes to check the

7.19. Depth indicator hands point to 21ft (maximum 23ft), and 12in. (Author)

105

water level in the reservoirs.

7.20. Top left, side view (depth indicator on right-hand side).

7.21. Above, pendulum and terminals.

7.22. Left, other side view. (All photographs the author)

7.9 Turnstiles

We are lucky enough to have three Ellison rush-preventative turnstiles from the top station, one of which has been put back into the top station in its exact location, the others are left where they were found. The one we removed was clockwise but it was the only one with a counter – it is located on the site of an exit turnstile which was anti-clockwise. We believe the turnstiles are from George White's period (plate 4.33) since the George Newnes' turnstiles (plate 4.7) appear to be a different shape. The accessories shown in plates 7.25 to 7.27 are lovely.

From the mid-1890s until the 1980s, the vast majority of turnstiles fitted at sports grounds, swimming pools, zoos, mines, public toilets etc were manufactured in Manchester by W.T. Ellison or W.H. Bailey.

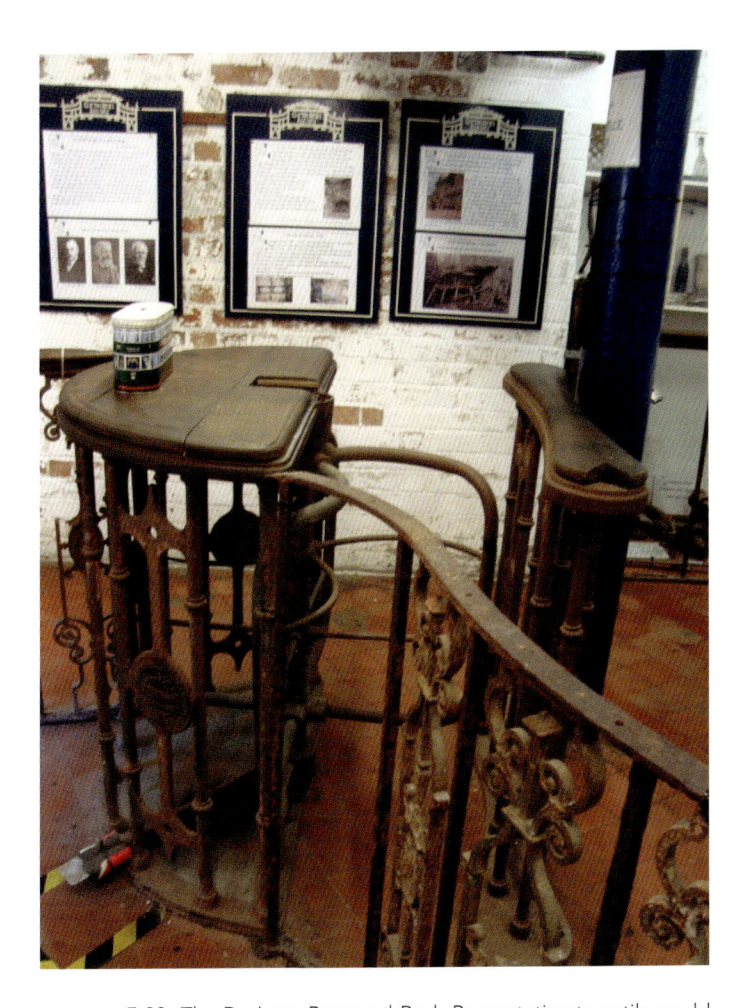

7.23. The De Luce Patented Rush-Preventative turnstile could process 4,000 entrants per hour, or 3,000 if change had to be sorted.[3] We were lucky enough to find two different counter cover plates. The wood has been treated with linseed oil only. They are in good condition considering where they have been stored since 1940.

7.24. In between the wood joints of the turnstile top we found some original Clifton Rocks Railway tickets. (Author)

7.25. The tamper-resistant counter still works. (Author)

7.26. W.T. Ellison, Manchester, manufacturer of the turnstile. (Author)

7.27. This sturdy gun-metal plate was locked over the counter to avoid fraud in the tally of numbers and receipts. There was a hasp under the curved part of the plate. This fitted into a lock on the right-hand side of the counter container shown in plate 7.25. (Author)

107

7.10 Ornamental ironwork inside

The ironwork inside has not been painted so the original colours can then be seen far more clearly, and it maintains the character of the site. This was the advice given to us by conservationists, and a good example of conservation versus restoration, showing where they would have been painted green and gold. Similarly the woodwork in the bottom station is left as it is.

7.29. A nice curve.

7.30. The decoration is not symmetrical. The components have been tied to the frame since they can not be welded. (Author)

7.31. The original green and gold paintwork is still visible. (Author)

7.32. The railings were made by Hart and Son of London and Birmingham. (Author)

7.33. Wooden knob from one of the partitions where one got onto the cars – see plate 4.7. (Author)

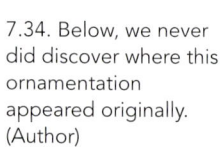

7.34. Below, we never did discover where this ornamentation appeared originally. (Author)

7.11 Grating

This would have prevented passengers getting wet feet when entering the car.

7.35. This grating was found in the top station. It is 9 1/2in wide and 16in long so must have run at right angles to the rails, over the chutes. It is 1/2in thick, and was damaged when removed during the war. (Author)

7.36. Gate to get into a car. This gate is 7ft 6in high – the height of a car. This is still in the open position so needs taking apart to make it operable again. There would have been four gates, one per car. Several people remember going through a gate to get into the car. (Author)

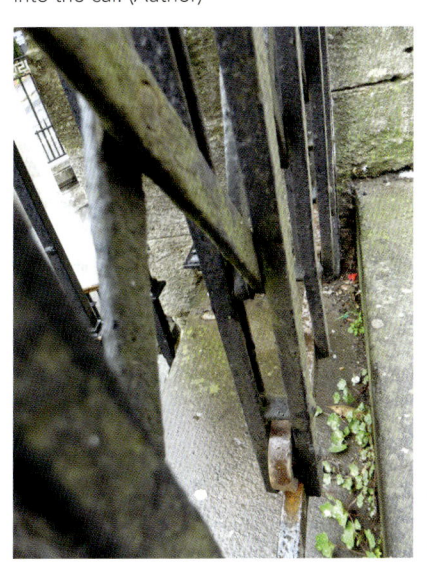

7.39. The bottom of the gate has rollers. The rail splits into two, again to enable the gates to swing open. (Author)

7.12 Folding scissor gates

Folding scissor gates have a combination of slider mechanisms and parallelogram mechanisms, to enable the gate sitting on rollers or wheels, to fold. This kind of gate was commonly used in mines, lifts etc, when one needs the gate to take up as little space as possible when open. Most tend to be simpler than here since the rail is embedded in ground; here, they are not. The bottom runs on big rollers, the top of the gate has round rods. When fully open, the top roller rails drop since they are then not supported by the gate, and the bottom rails are lifted. One then moves the gates through 90 degrees on hinges, so that visitors do not trip over them or bang their heads, unlike when only one side is half open. It is quite fiddly fully opening or shutting the gate since the rails must be in the right position, otherwise you get banged on the head. Heaven forbid taking one apart since there are so many bits. It is far better to keep oiling them.

7.37. Gate to get into the top station, 5ft 3in high. One has to slide the gate to open it, and woe betide the hapless operator who shuts the gate too quickly and traps a finger. (Author)

7.38. The top rail is supported by rods. The separate rails and the hinges on the posts enable the gates to swing open. (Author)

7.13 Partition

A partition separates the passenger from a person selling tickets, or to stop them going into a prohibited area.

7.40. This is the decorative leaded green and clear glass partition one passed when boarding a car (seen in position in plate 4.33). It stopped the passenger from going onto the tracks, and had glass to make the tunnel lighter. The partitions have been left in the turnstile section of the tunnel since they are very fragile. (Author)

7.14 Pottery

It has been very nice, and unexpected, to have found bits of decorative pottery dating from the time of the Railway.

7.41-2. China with Bristol's arms found in Sion Hill tunnel. (Author)

7.15 Beer bottle

In the old tunnel we discovered under Sion Hill leading from the railway, which had been bricked up in the wartime, we found a green glass bottle of the Sunrise-brand from Ashton Gate Brewery from the 1920s. It is complete with stopper and presumably still contains beer. Had this bottle been sitting on the floor in a pile of rubbish for eight years before the Sion Hill tunnel was blocked up? It could not have been placed there after September 1940.

7.43. Bottle of Sunrise beer, still covered in wet mud from the Sion Hill tunnel, and complete with stopper. (Author)

7.16 Plaster and marble

7.44. Plasterwork: side view and, 7.45. top view. (Author)

7.46 Marble side view and 7.47. top view. We are not sure where this decorative work came from originally, but both plaster and marble were found in the turnstile area of the tunnel. (Author)

7.17 Whistle

In 1884 Joseph Hudson of Birmingham, supported by his son, revolutionised the world of whistles by making the world's first reliable 'pea' whistle – the Acme Thunderer. By 1927 they were the world's largest whistle maker and the Acme Thunderer and its variations became the world's best-selling whistle and are still going strong today.[4]

The Acme Thunderer whistle which we found was made between 1930 and 1945 (dated by the Acme company). This style of whistle was used mainly on the railways as a guard's whistle. Acme whistles were also used during the war by ARP officers, army, navy and air force and for civil defence.

7.48. Acme Thunderer found on the exposed rail section of the top station. Since there was no access to this section after the first-aid post was built, it must have been from when the site was still a railway. (Author)

7.18 Ornamental ironwork outside

The beautiful railings were made by Gardiners of Bristol which was founded by Zacharia Cartwright in 1860. When he died, the firm was taken over by Emmanuel Chilcott who took on partner Alfred Gardiner in 1860. The firm expanded to making windows, ironwork, church furnishings, lighting, and ironmongery in Willway Street, St Phillips. They moved to their current site in the Soap and Candle works on Broad Plain in 1958. They are the only railings in Clifton with a manufacturer's plate. Gardiners still exists today as Gardiner Haskins department store on Broad Plain, Bristol.

The railings have survived along with those belonging to houses with basements, because they prevented people from falling from a height – a particular risk during night-time black-outs when street lights were turned off.

But many beautiful railings were lost. The tragedy was that, faced with an oversupply, rather than halt the collection which had turned out to be a unifying effort for the country and of great propaganda value, the government allowed it to continue. The ironwork collected was stockpiled away from public view in depots, quarries, railway sidings. After the war, even when raw materials were still in short supply, the widely held view is that the government did not want to reveal that the sacrifice of so much highly valued ironwork had been in vain, and so it was quietly disposed of, or even buried in landfill or at sea.

7.49. Rosette. We got some more rosettes cast at the Bristol Foundry, and bought some more balls to replace missing ones. There are small rosettes as well as large ones. We now have two small rosettes (on the posts), four large ones on the railings and two large on the new ironwork over the top station entrance when we replaced the George White ironwork in 2016. (Author)

7.50. Zig-zag decoration under the finials. (Author)

7.51. We found a railhead in the rubble, and did not know where they were fitted to begin with. We got some more cast at the Bristol Foundry. There are now eight on the railings and five on the new ironwork over the top station entrance. (Author)

7.52. Gardiners of Bristol sign on the railings, discovered when they were being prepared for painting. I have not seen any other railings in Clifton with a manufacturer's plate. (Author)

7.19 Gas and oil lighting

The regenerative gas light (a typical Victorian gas light) was found in the section of the tunnel where the turnstiles were retrieved (see plate 7.4). The lamp bracket was returned to us by someone moving house and thought we might like it back. We identified where it had been taken from. We found a rim later in the bottom water reservoir. It was lovely that the light fitted the bracket. There are knobs on the side to adjust gas and air. The gas and air flow through separate pipes to supply the burner, and they are heated by the waste products of combustion as they pass through a series of flues. This improves the illuminating power of gas.[5] The yellow flame, which is the best for illumination, only uses a fraction of the gas. The left-over, unburned gas goes up the large vertical tubes to be reused. A glass shade would have been present with small holes to draw in oxygen/air.

7.53. Regenerative gas light when it had just been found in the blast absorption area of the tunnel before the barrage balloon section. It is sitting looking very sorry for itself on the floor of the top station. There is no rim on the base. (Author)

7.55. Gas mantle. (Author)

7.56. Probably a stand for a small oil light. (Author)

7.54. The light after it had been firmed up by two bits of exhaust pipe inside, where the gas tubes had been, and the top lid riveted back on. The back enamel is original. It has not been painted. The photograph was taken after we found the rim. (Author)

7.20 Electrical fittings

7.57. The toggle mechanism provides 'snap-action'. The design was patented in 1916 so these elegant industrial switches must have been installed by George White. The covers were probably brass since the switches are porcelain. The South Western Electricity Historical Society found in their archives some similar switches in a catalogue. They are called Tumbler Switches and the knob is called the 'Dolly'. These are fixed in metal conduit boxes. The metal conduit would be earthed at source so that any leakage would immediately trip the fuses. Nowadays sensitive trip switches do the job more safely. (Author)

7.58. 25-amp fuses. (Author)

7.59. A page from the Sunco Catalogue 1920. The 'Lundberg' shows the type of brass cover that would have been on the switch in plate 7.57. (Peter Lamb)

7.21 More things brought back

We have not worked out precisely where this gatepost (right, plate 7.60) was originally located, but was probably where the first-aid post wall is now. The owner was moving house and wanted to know if we wanted it back. We found the wooden top for it in the turnstile section.

7.60. Gatepost. (Author)

7.22 What we would like back
We have an amnesty. We are grateful for visitors in the past 'rescuing' items from the Railway and returning them. A photograph is better than nothing. In October 2017 an 'S' was returned.

7.61. This Bristol Tramways sign is in an architect's garden in Somerset. He is not ready to return it. Note the ornamentation on the left of the sign. It may be the same as in plate 7.34.

eight

Railway Memories

It is surprising how many people are still alive who remember travelling on the Railway. This chapter gives a valuable insight as to what they remember, and helps to bring the story to life. Some remember the trip and some were frightened; some remember the slot machines; and others give a fascinating insight as to what they did on a Sunday afternoon (and how much walking they did). There is disagreement about the colour of the cars. I have been in contact with about 70 people whose contribution has been invaluble.

Eileen Amesbury
Born 1919 in Chestel Street, later Bishopsworth Road. There were four in her family. They travelled often on Railway when she was six or seven years old. It was easier than walking up the zig-zag to get to the Zoo, or the Observatory, or the Downs. There were normally two cars running.

Mr Bartlett
Born in 1921 in Brunswick Nursing Home, Brunswick Square. He had a brother (five years younger) who lived with grandparents who were retired publicans. He lived in Montpelier, near Picton Street and the Prince of Wales and round the corner from the Old England. He travelled on the Railway from age five or six on Sundays. In the summertime they always had a ride on a tram on a Sunday to visit a pub. At that time, pubs on Sundays were only open 12.00 midday-2:00pm and 7:00pm-10:00pm (the only other places open on a Sunday were the off-licence and tobacconists/sweetshops). They would get off the tram at Victoria Rooms and walk through Clifton and eventually would get to the Railway top station. They would ride down to Hotwell Road and walk along to the basin where the boats came in, ending up at Dowry Square. They would then go to the pub where his father knew the publican. His grandparents had a drink and he would have lemonade. The pub was near Dowry Square (not the Rose of Denmark). They would then catch a tram to the Centre, change and then get the tram to Cheltenham Road. The tram in Hotwell Road would stop further on, by the Suspension Bridge, and would then turn round to return. The railway was very busy; there was always a queue. He did not remember how much they paid but remembered buying tickets at the top. They would go to the left, and then onto a step to get into car. He remembered the ride as a 'fantastic adventure'. He would sit amazed going down. He remembers blue and yellow cars like the trams. The railway was in good condition and the drivers were very smart and tidy.

Ann Beaver
She remembers a penny-in-a-slot machine that was a chicken laying chocolate eggs, in 1913.

John Bidgood
John was born 1925 and remembers riding the Railway at the age of eight or nine. As a child he lived near Trenchard Street, behind Park Street. To get to the Railway would entail a tram ride from the Centre, or a walk.

David Bickerton
David was part of the civil engineering team that performed remedial work on the lower station-front in 1955. The remedial work was necessary as the façade was physically moving out into the Portway. This movement was due to the rock structure of the gorge, which consists of two layers of rock with a layer of clay in between. The instability was checked by installing rock anchors at an angle to anchor one level of rock to the other. The project was co-ordinated by Professor Skempton of Imperial College London, an expert in such matters. During the work, David walked up and down the tunnel to explore. That was 1955 and he is sure that the BBC had long since vacated the premises, leaving no signs of equipment. During the drilling that was necessary, David said that

he located two air vents in the roof of the tunnel, that it would have been possible to climb into. There may well be others too.

Rose Bird
Rose Bird was born Canada in 1916 but moved to Bristol at age eight or nine where she lived in a house in Eastville. She had three brothers and three sisters. The house was run down and her father had no job. One sister worked in Wills, the other in a chemist shop. She wanted to be a hairdresser but you got no wages for six months, so she worked as a door opener. She had no education. At age 14, she worked in Brookes Dye works as a cleaner. There was a good community spirit. Old ladies were helped. She was very happy. They kept chickens and vegetables. It cost 1d for all sorts of vegetables for stew. As a treat they would have fruit-cake on Saturday. She travelled on the Railway between the age nine and eleven. They would walk from Eastville to Hotwell. It cost 1d to go up and they would return via the zig-zag. They would travel at holiday times since it was a treat to go on the railway. They would go to the Downs. There was no queue, mostly children. Two cars were running.

Bill Bonner
Born in 1924, she lived in Clifton all her life. She travelled on the Railway when she was eight in 1932, when she lived in Clifton Park Road. She went with her mother and friends to Campbell's Steamer and back if they had spare cash. It was exciting. She thought the car was brown. It was always a real treat to travel on it.

Colin Boyce
Born in 1928, he lived in 6 Oakfield Place, Hotwells. He travelled on the Railway when five or six. It was a rare treat for being a good boy. He went up the railway and down by the zig-zag. There were only two cars running.

Brian Bradley
Born in 1925 he lived at Little Caroline Place. He travelled on the Railway from eight years old, once a year as a treat, to get to the University bonfire on the Downs. The Railway was not really busy. They could only afford to go up and returned by the zig-zag. The bottom station was more decorative than the top station. As a kid it was very exciting. It was lovely not to have to walk up the zig-zag. He thought the cars had a dark blue body.

Betty Britton
'I can remember how the railway worked. I can remember a lady opposite with a hat with cherries in it. Dad told me to ask the lady for a cherry. I was five and it was very dim in the coach and when we got out the sun seemed so bright.'

William Broadhurst
Born in 1917, he travelled many times on the Railway as a child. He lived at St Nicholas Street in the centre as his father was the licensee of the Radnor Hotel. He travelled on the railway aged about seven, his sister being four years older. They had a regular route on a Sunday afternoon. They went on a tram from the tramway centre to the Downs then the Observatory and then the Railway which was busy with a fair number of passengers. The carriage was full. Two cars running. There was a water pump in the top station. His father preferred horses.

John Buckley
John remembers going up the Railway when he was six. He remembers the gates of the car opening at the top to let him out. He lived by Campbell's Steamer office in Hotwells Road on the corner of Merchants Road.

RG Burgoyne
In 1920s and 1930s, he lived in Bushy Park, Totterdown. Two or three of his older sisters took him on a tram to the Centre and another to Hotwells. After taking the Railway, they would have a few enjoyable hours on the Downs and then return home via the Railway and the tram. He remembers watching the chink of light from the bottom of the tunnel getting bigger and bigger as the train climbed the tracks through the rocks.

Keith Cornish
'I have a vivid childhood memory of riding up the railway, when I would have been three years old. We boarded the carriage at the bottom, which was well lit, although I did not realise until later, that the tunnel itself was in darkness. My father was trying his best to put me at my ease and was telling me to look out for the carriage coming down, which I did. As we passed each other, it appeared to be going faster than us, which of course was caused by the two closing speeds. After it passed us by, it could be seen disappearing into the gloom below. This has remained in my memory for over 70 years and I have thought about it many times.'

Muriel Cox
Born in 1919, she travelled on the Railway as a child. She lived on the hall floor of 8 Gloucester Row. She travelled down to Hotwell Road on Sundays but not often; usually when her mother was cooking dinner and 'we had to get out of the house'. There were two cars, which met in middle. It was rather fun rattly and gentle, not too fast and not too slow, and very dark. There were facing seats. One car whooshed up and the other down. There were always quite a few people using it. 'We crossed the road (easy then), picked up the ferry from landing stage, and went up the river to the horseshoe bend.'

Don Cullen
Born 1924. As a child in the 1930s, he travelled up the Railway with his mother and brother. They lived in Horfield and would take the tram back to Zetland Road. The bottom station was not lit very well and he remembers it flickering yellow. He went through a turnstile and straight into the car, which was very classy with windows. Only two cars were running. He remembers brass rails, reddy-brown wood and flickering yellow oil lamps. He was able to sit down in the car. The top station was very luxurious. It was very noisy with the sound of rushing water. It was dark and frightening but he was not very old (about eight). A real treat he will always remember.

Trevor Dean
Born 1924. 'I am 81 but like to think I can remember much from a young boy's perspective. Although unfortunately I have no memorabilia or pictures, my memories still seem very clear. At the time I lived at Zetland Road junction with my parents. Occasionally an aunt who lived in Cornwallis Crescent, Clifton would have Sunday lunch with us and would in the summer take me back to Clifton for a trip on the Rocks Railway. We entered via the top gates in Clifton and having walked down a flight of steps I remember there was a wooden partition with large glass windows right across the bore of the sloping tunnel so one could see to the bottom very well. I recall there were four tracks for the four lifts but I only saw two lifts working each time we were there. Being inquisitive, I asked how this railway worked, and my aunt said by the weight of water in the tank of the top car which I saw was attached to the other car via a steel cable so that when the brakes were released on both cars the top one was heavy enough to pull the other car at the bottom up the rails, even if the top car had no passengers, and the car at the bottom was full of people. We boarded the top car, and the first thing I can remember was the very large bright copper pipe which I estimate must have been about a foot across by six inches. It was oval in shape which coupled extremely neatly to the fixed permanent water supply pipe of similar shape and also of bright copper. When the driver was ready to go he would sound a bell on his car. I think it was a similar type of bell to the tramcar bells, which the driver could operate by stamping on a steel button. I noticed that if the Driver at the lower end of the system was ready, he would reply by sounding his bell in a similar manner and the cars would then move as soon as the brakes were released from both the cars. The journey was very smooth and almost silent. When we arrived at the end of the journey the passenger sliding doors on the car were opened and one could hear the sound of rushing water as the now lower tank on the lower car prepared for its ascending trip. Later in the afternoon we returned to the top once more which was just as thrilling.'

Mona Duguid
Born 1915. 'I was about nine or ten when we lived in Redland. Having relatives in Cardiff we made frequent visits. We would get a tram to Blackboy Hill where the tram terminated. We would walk across the downs to the Clifton Hotel, get on the rocks railway down to Hotwells. I used to be fascinated to see the tram on the way up passing and for a moment would be side by side. I used to look out of the window and see water running down the rock face. When we reached the bottom we went on to the Hotwell Road, and cross over to the Campbell's Steamer (*Waverley* or *Ravenswood*). There would be a black man with a little boy and they would sing, and his father would play the concertina and entertain the people when the steamer arrived. Pennies would be dropped into his hat. On our return we would go back up the railway, back over the Downs and catch the tram. When I was 5 or 6 I remember every Sunday morning my father would take me for a walk and look at all the work being done on the Portway.'

Phyllis Farmer
'I used the Railway between 1930 and 1935 when I lived in West Mall and mother took me to visit my grand parents in Knowle. Rather dark and eerie as we passed grey shining walls, and used to wave at passengers in the ascending car when we passed half way. When we reached the bottom we boarded a tram which took us to the Tramway Centre, where we changed to another tram which took us along Baldwin

Street, Redcliffe Street to the three lamps Totterdown and Wells Road.'

Molly Foss
Born in 1918, she once travelled on the Railway, aged five, with her grandfather. Two cars were running and they got out by the hotel. She said 'it was exciting'.

Ralph Fryer
He remembers riding uphill because mum would only pay to go up not down. The family lived in Southville and he travelled on the Railway from the age of two every week on week days or weekends. Normally there were two cars running. He can remember the turnstiles of the bottom station, where it was dark and cave-like. The cars were brown and yellow. The railway was in excellent condition.

Judith Gauchi
The Railway was a special treat, a tram ride from Temple Meads to Hotwells. It was very dark, with only a light in the carriage. When one carriage started up, the other started down.

Jean Gillingham
Born in 1922, Jean grew up in 1 Victoria Terrace. As a child she could remember being startled by the lights at Clifton Rocks Railway. She was frightened by the banging of the metal gates as they closed behind her. The treat was to walk up to the top station, take the car down and to return up the zig-zag. She remembered a well-dressed beggar who always stood at the bottom station. There was a man who sold lemonade and sweets outside at the bottom. It was considered an amazing treat to be allowed to buy a sweet.

Norman Goldsworthy
He travelled on the Railway in 1932 at the age of ten. His father took him when he was small to see his grandparents in Brislington.

Christine Hamm
'My grandfather (Rosewell) was a brakesman on the Railway for many years. He died young, way before I was born. My Mum told me about it and I have the 1901 & 1911 census, which say Brakesman. I also have a bad copy of an article in the newspaper when he died from 6 May 1924. It said six staff members were bearers at the funeral at Arnos Vale. I can't quite read how long he worked there, it could be 13 or 17 years. The article says he was an engineer. In 1901 he was in lodgings with another Brakesman. [Details of 1901 census RG 13/2368, page 23, Part of Bristol West, Part of Christ Church Clifton. Both lived in no. 5 Gloucester Street, Clifton.]

Mary Hingston
Lived in Dowry Parade. Would go up Rocks Railway. Whoosh of water from the tank scared her.

Harold George (grandson of George George)
Harold George was born in August 1914 in Point Villa (a very old house), Lower Bridge Road – the same house as his grandfather, George George, who is described in detail in chapter two. Harold was 95 when interviewed (Harold died in 2015 at the age of 99). Point Villa is shown on the 1840 tithe map but not on the 1828 map. His father's grandfather had been born there. It was just by the bend after the Bridge Valley Road traffic lights. He was the last of the family that had lived there. His father John died in WWI in 1914 when he was four (there may be a memorial to him in Bristol Cathedral). His mother had a telegram to say his father was killed. His grandfather George George, who was a famous blaster, brought him up. He never knew his mother's family. Her father was in the police force. Harold never knew his maternal grandmother. Nine children lived in Point Villa. He had two step-brothers and two step-sisters. Between the two tunnels of Bridge Valley Road was a cottage that belonged to a family who lived at the bottom of their garden. He used to climb over the garden wall and watch the trains leave Tunnel 1 and enter Tunnel 2. This was the only other house nearby. He remembered the platform outside the tunnel. He has a picture of where he lived being demolished to build the Portway (see plate 2.13). He remembers when the family moved, they pushed furniture up to the bottom of Bridge Valley Road on bogies. They moved to California House on Oldland Common, Bristol when Point Villa was demolished. George took his gateposts with him. He went to Hotwells School of Hope Chapel, and then to National School. His mother had a brother. She lived with her in-laws in one big bedroom. There were 4/5 beds in there. His grandmother was well known for keeping pigs because she kept them well. They grazed along the land from the house to the railway. His grandfather would kill them for the local butcher in Hotwells (Alfred George). They would take the pigs in their high trap pulled by Doris the pony. They even took the pigs to Berrow, Weston-super-Mare (it would have taken some time). They would also take the pony and trap laden with vegetables up Brandon Hill, which made the pony puff a bit. His grandmother kept

chickens – they hatched in her apron and would cause some merriment if you were talking to her. The garden was to the left of the house. He played round the platform and climbed the fence when sailing boats went by. His mother would go to the platform by port and pier tunnel and get spring water. He remembers boats being stranded, and a sheep falling down the cliff. The tram terminated beyond the Suspension Bridge and when the train was due he had to run to the Port and Pier station to warn his grandparents to get refreshments ready. He remembers travelling on the Clifton Rocks Railway – it was cheaper (a halfpenny) to go down than up (a penny). When he got to the bottom he used to go back up to the top by the zig-zag. The tram stopped at Clifton Rocks Railway, so he would run down to the Port and Pier station platform. The drivers at the top and bottom communicated that they were ready by a call sign. He also travelled on the Aberystwyth Railway which his grandfather had also worked on. He started work at Robinsons the printers, then Wax Paper, Fishponds. He also worked for Parnalls at Yate. He got 1 guinea wages, below the working wage, but he did not belong to a union. He wanted to avoid having to go the Home and Colonial to get cash. It took three hours to queue at the Labour Exchange. In 1936 he got a job in GEC Birmingham and worked there for 40 years, so he was housed by George for 30 years. He only had enough in his pocket to buy a single ticket 2s 9d and a tram fare. He came back to Bristol in 1976 after retirement. There was a house for sale on California estate, and his grandfather had moved to California house. His memories keep him going – he had had a fantastic life. Harold used to read the paper (*Daily Express*) to George. He had happy memories.

Ray George (grandson of George George)
Born in 1932, and so 18 years younger than his brother Harold, Ray lived at California House. Ray's father died and his wife married his father's brother. Uncle John (Harold's father) may have a memorial in Bristol Cathedral since he died in WWI. The family broke up after 1944 and he had no further contact with George George (who died in 1950 when Ray was only 11). Harold explained his difficulties to Ray when his father died when he was four. He had a sister Frances. Harold did not make it easy for Uncle Will (Ray's father). Harold gave Ray an Aberystwyth guide book which had been signed by Croydon Marks (who was engineer of Aberystwyth Railway) and given to George George in August 1896. Ray gave me the book for the Railway Archives. There is a photograph of construction of the Aberystwyth Railway showing George George in characteristic bowler hat and watch chain. Croydon Marks may be in the photo too. This is shown in the chapter two. Ray never heard George George talk about his work at all. He had amazing blasting skills. His father or a charge hand at the quarry must have promoted his skills.

Shyll Slade (great-niece of George George)
Born in Hotwells in 1921, she travelled on the Railway in 1929 when aged seven or eight. She used to visit her grandmother in Clifton in Exeter Place. She was not scared of the Railway. It was very quick. There were only two cars running. Her grandmother's brother was George George. The trams used to run to just beyond the Suspension Bridge and turn round so they could get to the station. The bottom station was very compact, lit, plain and functional with turnstiles. There was very little space. She used to love watching the car come down; it was a very similar shape to those at Bridgnorth. It was cream and brown – the livery of Bristol Tramways. The glass was polished and well cared for and it was very decorative. Sir Stanley White made sure it was run properly. She remembers all the wrought iron in the top station which was far more ornate.

Ruby Hengrove
She lived in Ashton and would walk to Hotwell and take the Railway up to Clifton on Sundays.

R Howell
'My grandparents had a restaurant in Colston Street, and my family lived in Sea Mills from 1923 to 1927. We would take the train to visit them, picnic somewhere in the Avon Gorge, and take the Railway up to the Downs. Mother and father, my uncle and aunt and our cousins, then living in Sion Lane, often used the Railway. It could not have cost more than a penny each for none of my folks ever had much money.'

Reginald Incledon
Born in 1917, he has an 1893 opening day medallion. He lived at Chessel Street, Bedminster as a child. They were a large family. They would walk to the Railway and get the train up. It was very expensive to take the whole family. They would go to Nightingale Valley to play, the rock-slide, Observatory and cave. He can remember peering through the glass at the Railway carriages and described them as gleaming.

Lilian Jenkins
Born in 1920, she lived in St Werburghs. There were five children in the family. She travelled on the Railway when seven years old, a couple of times a year on bank holidays. It was a treat. Her spinster aunt would take her. There was always a queue, even with four cars running [presumably because it was a bank holiday]. She had an ice-cream at the bottom station.

Lorna Leach
Her grandfather was from Bristol and very proud of it. His aunt lived at St Vincent's Parade Hotwells, the last house and the one nearest the Railway, and would offer teas in her garden for visitors to the Railway.

Cecilia McCarthy
Born 1926. 'I must have been about six or seven years old when my father took my brother and myself on the railway. We probably got there by tram, but I can remember travelling to the top. I remember the dark, and the proximity of the tunnel walls to the car windows. My grandfather who was a bricklayer, helped to build the railway, but would never travel on it as he thought it unsafe.'

Heather McOmie
Born in 1923, Heather was eight when she went on a school trip from Penarth by Campbell Steamer (it would have taken about 1¾ hours). They got off at the landing stage by the Railway and went up in the Railway. They then walked to the Zoo. She remembers it was a lovely sunny day, but the railway was very gloomy and eerie. They sat facing each other.

Doreen Mallett
Born in 1926, she remembers seeing weeds growing from the tunnel walls. She travelled on the Railway to visit her aunt in Clifton. She, her brother, and mother used to watch for the car to come down. She did not remember it being in a tunnel. It was a real treat to ride on. She also recounts the story of the fender on the car being a great temptation to cling onto as it ascended so her brother was warned not to. There was a boy who lost both arms doing that on a tram in the Hotwell Road. He became a good artist by holding the brush between his teeth, but he contracted poison in his gums which poisoned him.

Jon Miell
'I remember being taken up from Hotwells at the age of four in 1934. It was very dark and very scary for a young lad.'

Pam Millard
Born in 1920. When she was a child, her treat on a bank holiday was to visit Clifton. The family would walk from St Werburghs to Clifton. They would go down on the railway for half a penny since it was cheaper than going up. They would finish the day by walking home along the Portway.

Iris Mitchell
Born 1921, she lived in Victoria Square. Her father worked for the Railway and they had a privilege ticket. She can remember penny slot-machines at the bottom and brown and yellow cars.

Ann Osbourne
Born 1910. Ann, at 95 years old, remembers riding the Railway in the early 1920s. Her only memory of the ride was that it was very dark. She lived in Eastville at the time, but her family took occasional trips to the Clifton Downs and Zoo.

Brian Owens
He went on the Railway from age 5 or 6 until he went to school. They would catch the tram to Henleaze, then to Zoo and walk to the Railway. He remembers the ticket office at the top, and turnstiles. He remembers lots of dark stained wood, and uniformed officers. He could hear water. He can remember the ticket man and a man in the car. His father had a motor car so he would be picked up from Dowry Parade.

David Pearce
Born in 1928, he first rode on the railway in 1934 when five or six – and remembers going through the turnstiles to board it. He remembers two cars running and water flying in and out.

Vera Price
Born in 1904 at Ashton Gate, she was a cypher officer in the Foreign Office in WWII. She remembers travelling on the railway from about 10 years old on Bank Holidays for a treat, and enjoyed it very much. She was usually going to the Zoo.

Mr Roach
Born in 1927, at 83 years Mr Roach remembered riding on the Clifton Rocks Railway as a boy. His father and mother would take him and his younger sister for a walk after Sunday evening service at Hope Chapel. They would go to the Portway and sometimes were treated to a ride on the Railway to Clifton. He

remembers it was darkish after the sunshine outside and seats that ran lengthways down the carriage, not traverse as on the trams. 'It was exciting even if it was slow, and it was scary with the rushing noise of the water being transferred from one carriage to the other. When we got to the top we made our way to the Promenade where large numbers of people all dressed in their Sunday best were walking along. I remember my dad telling me to keep my back up. At the end of the prom there was a Wall's ice cream tricycle with the logo "stop me and buy one". My sister and I shared a 3d snowfruit which was a prism-shaped cardboard tube containing orange-flavoured frozen water – the forerunner of the modern ice lolly. We then took a steady walk to Ambra Vale East where we lived at the time. No-one had cars, so walking was the order of the day.'

Joan Shapland
Joan was six when taken by her father for a ride. She heard water running out at the bottom when they got out and remembers shining dark brown seats and brass handrails.

Joyce Smith
Born in 1929, she was three or four years old when she rode on the Railway. The family lived in Bedminster and for a Sunday outing they would take the tram or walk to Hotwells, go up to Clifton in the car, and walk back down the zig-zag. There were only two cars running. She remembers the tunnel being dark, the bright lights of the other car passing in the tunnel. She could see the water gushing from the car into the water tank before going up at the bottom.

Stan Snook
Born April 1914 in Dowry Parade, Hotwells, his father had previously lived in a railway cottage in Oldfield Road. His father had worked on the Railway as a bricklayer. 12 bricklayers were employed. They would go on family outings to Leigh Woods, Nightingale Valley. They were lucky because their father was regularly employed as a Corporation bricklayer, so was 'well off'. He would often use the zig-zag, but since he lived nearer the Joy Hill end he normally went up Granby hill to get to the Spa. The Railway was used by clients of the Hotel. His father once took him on the Railway for a novelty. When the Port & Pier Railway was extended he recalls going from Clifton Down to Severn Beach.

James Steeds
Born in 1928. James, his father, grandfather decided to travel on the Railway on its last day He was six years old. The car was crowded with people standing.

Elise Symonds
Born in 1921, she lived in Sea Mills and travelled on the Railway occasionally. It was, she said, very exciting.

Herbert George Wall
Peter Swan remembers Herbert George Wall (born 1871, died 1952). He was an Inspector on the trams, Hotwells to Brislington – and it is just possible that he and the other man were asked to stand in the entrance of the railway in 1912 by the photographer, to make the shot more interesting (see plate 4.34).

Peter Webley
'My grandfather, George Webley, worked on the Clifton Rocks Railway as a Railway Brakesman at the time of his marriage in 1906. He was born on the 21 September 1876, one of two twin boys. His father, my great-grandfather, was a butcher, confectioner, publican and cab proprietor and lived at the Clyde House now the Clyde Arms on the corner of Hampton Road and Redland Park where my grandfather was born. Later my grandfather lived at 20 West Mall, Clifton at the time of his marriage and then Caledonia Place, which was just round the corner from the Clifton Rocks Railway so he did not have to walk very far to work.'

D. Williams
'I would have been at the age of nine or ten, having been with my family on an early morning cruise on the steamer. Opposite was the Clifton Rocks Railway. On entering was a deep well of water on the left-hand side and a rather loud noise of sh, sh, sh – no doubt the pumping of the water to the upper car. The rear end of the waiting car towered majestically above oneself, so tall. In front were the very steep steps leading way up to the top of the tunnel. We entered the car to find it limited but adequate in accommodation. I thought I would be able to get my bike in there. There was some sort of electrical signal between the attendants of each coupled car. My father told me they could go when the tank on the car above was full of water. The attendant said "off we go then". Slowly we ascended the steep incline.'

Robert Willis
In 1932 with his parents, he would take the tram to

the Centre with their gramophone and a record. They would come on the two-penny steamer from St Augustines Parade to Hotwells stop – just by the Pump House. The steamers were not really steamers – they were launches with a canvas top. They would walk to the bottom station and use the Railway to get up to the Downs for a picnic. This was a special Sunday treat. He could not remember if the Railway was noisy or not. The Railway was busy on a Sunday because it was a big day out and was a popular way to get to the Downs. Most people had no cars, only motor bikes. Going on the railway was regarded as a special treat as part of the day out. He remembers being confused because he thought the tram floor should be on a slope since it was going up hill. It was only 2d to go up. He would lay the cloth out for the picnic for his mother on the Downs. They would play *Blaze Away* (the only record that they had). They might buy a windmill, or a bird on a stick. They would then go to the top of Blackboy Hill where there was a space like Speakers Corner where his father would make a speech. They would then get the tram back to Eastville. He remembers the Colston Street signal because trams could not get up Park Street, so they had to stop the traffic to enable the trams do a big sweeping turn to go along Park Row. If it was wet it was too hard for electric trams to get up Park Street. Two horses were needed to get up Park Street when horse drawn. The stables were where the Council House is today.

Robert Woolley
Robert was born in 1935. His grandfather, Alfred Woolley (died 1956 or 1957 aged 88), may have installed the Crossley engines in Clifton Rocks Railway. He formed Woolley Brothers at 61 St Georges Road, College Green. Cabot House opposite the library. They lived in Lime Kiln Lane. His father was James Woolley (1902-88), also the Railway engineer (1911-22), and worked there till 1934 doing bits and pieces. Robert gave us Alfred's ledger book 1900-2, and 1902-3 detailing engineering work carried out in Bristol for our archives. They kept the hours for each job. The invoice was lined when paid so it acted as In book and Out book, purchase ledger and sales ledger. The Woolleys were called to sort out the more difficult repair jobs in the Railway. Flat four engines were hard to clean. One had to lift out the pistons to decoke them. They made gas from sawdust. The engines moved so slowly that one could put one's hand on the floor and feel as they went back and forth to test temperature and bearings. They were very talented, diverse engineers having several patents to their name.

In L shed (the storage area for M Shed) there is still his machine to test paper strength, and one to make paper bags. They worked for City Engineer as a subcontractor. We were also given photos of Alfred's workshop which had overhead shafts and a 3ft 6in Broadbent lathe. Robert served his apprenticeship to his grandfather using slotters and shapers.[1]

Unknown caller
His wife's cousin, a Mr Maggs, had been a conductor on the Railway.

8.1. Alfred Woolley. (Robert Woolley)

nine

Pump Room and Spa

Development from 1880

As the nineteenth century drew to a close there was a last attempt to revitalise Clifton as a spa town. As described in chapter 1, Clifton Spa Company, who just wanted to build a spa, had failed to raise the capital in 1888. Other people, such as George White, had wanted to build a lift from Hotwells to Clifton but not to build a spa. In the end the driving force was publisher George Newnes, who was not a local man, but had just built Lynton and Lynmouth Railway and could see the huge potential of Clifton. In October 1890, the Merchant Venturers allowed Newnes to build a cliff railway provided he built a Spa; provided no alteration was made to the exterior elevations of the buildings in Princes Buildings; that the Spa be used for no other purposes than permitted at the Bath Pump Room; and no alcohol. The Spa grounds would not be thrown open to those travelling by the Railway, or to the public indiscriminately, so that Spa visitors were of a very select character. There is much reference in this section to the Merchant Venturers, who were acting as planning control.

The result was the Clifton Rocks Railway hidden away in a 500ft tunnel (the Merchant Venturers had refused to allow a line with a view) with two pairs of cars, built in the garden of 15 Princes Buildings, with two discreet entrances (one at the junction of Sion Lane, and one from the Pump Room), and a Pump Room with an incredibly ornate façade down Princes Lane, but a plain façade along Princes Buildings. When we do our trips and Open Days for visitors they are always astonished to find this beautiful building down a private lane.

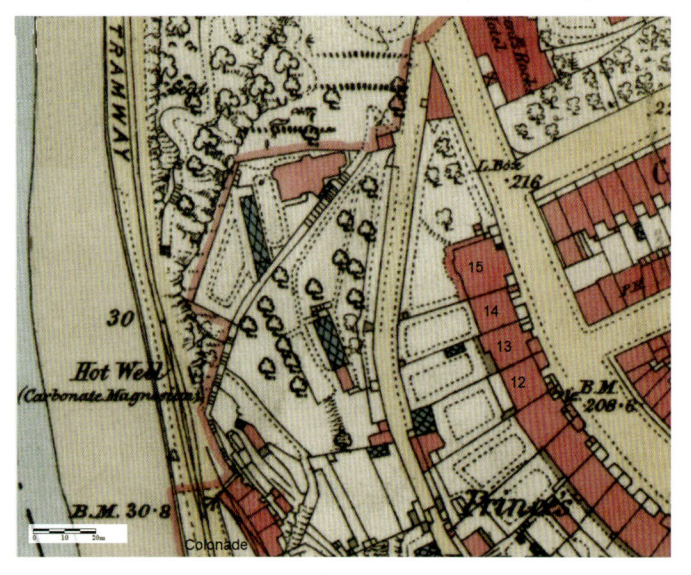

The Pump Room (later known as the ballroom) and Hydro proved a more enduring success though it had its problems as you will see.

This book is about the Railway so I will not discuss the architecture in detail and instead describe how George Newnes built up the enterprise. It will include the development of the Hydro: since each stage had their problems, nothing ever opened on time, and how the Railway was affected. It is also interesting to see how he celebrated each opening since he had been so generous and flamboyant when the Railway was opened – but then that had been more controversial; this time there were no silver inkwells and expensive clocks.

9.1 Problems with the plans

On 27 February 1891, new plans to reduce the height of the new buildings were submitted on behalf of Newnes, for the Pump Room and of the Railway top station – this would avoid the necessity of obtaining the assent of the owners of Caledonia Place to alter the covenants. The plans were approved by the Merchant Venturers and the following press release refers to the Pump Room:

9.1. This 1880 map shows the location of no. 15 Princes Buildings and gardens before Clifton Rocks Railway and Pump Room were built, and the Colonnade (sometimes spelt Colonade) at the side of the Gorge. It shows the steep terraces and glass houses between Princes Lane and the Colonnade. Sion Spring House is shown as St Vincent's Rocks Hotel. Note the zig-zag path (shown on 1828 map in chapter 1), an early footpath between Clifton and Hotwells is shown north of Tuffleigh House. There is a stable at the top station of Clifton Rocks Railway. (knowyourplace)

9.2 This 1900 map shows the Railway (opened 1893) and the Pump Room (opened 1894) in the garden of no. 15 (lowered so the view was not obscured). The Turkish and Russian Baths (opened 1898) were under the terrace. The Grand Spa Hotel and Hydro opened in 1898 after the purchase of more buildings, including nos. 12-14 Princes Buildings. The chimney (indicated by a circle) is for the laundry and boiler room. Note the Hotel was called the Grand Spa Hotel until the 1980s (now the Avon Gorge Hotel). You can see why residents were concerned about losing their view from Caledonia Place if the Pump Room had been too tall. (knowyourplace)

The plans for the spa are of a very elaborate character and displays of wealth of design, and making provision for the most extensive accommodation for visitors. It was hoped by the designers that they would be able to erect a building that would be an ornament and a credit to Clifton: but it appears that one of the inhabitants of Caledonia Place objected to the construction of the edifice on the ground that it would contravene the covenants on which he had purchased the property and would interfere with his vested rights of light and air. Mr Newnes made endeavours to come to an arrangement with the objecting proprietor, but without avail, and now alternative plans have been submitted to the Merchant Venturers Society, who are the owners of the site, and these have been accepted. The elevation which the lower scheme offers to the front is not of a striking character [see plate 9.5], seeing that a height of five feet only, precluded any attempts at ornamentation, but if the elevation to the front on Zion Hill is not of an attractive character, nothing but praise can be bestowed upon the elevation to the river and the private road, while the interior is of a highly enriched style of ornamentation.

On 28 February 1891 the *Western Daily Press* reported:

The indefatigable energy and perseverance of the promoter of the above scheme have brought to a successful issue by the official passing of the whole of the plans for the work. At a committee Meeting at the Society of Merchant Venturers, held yesterday, a letter from Messrs Osborne. Ward, Vassall, and Parr was read offering on behalf of their client Mr George Newnes, MP, to at once commence the work in accordance with one or the other of two schemes prepared by the able engineers.

We can congratulate Mr Newnes upon the ability which his engineers have displayed in meeting every possible difficulty and providing alike for the comfort and pleasure of the residents and visitors of Clifton. It may be a matter of some surprise to many to learn that the scheme which the society have sanctioned is not the one which the engineers and the promoter recommend, neither is it the one which will give to Clifton a structure of such external elegance and attraction which all would most heartily approve. It appears that there is a covenant existing in the leases of a few of the gentlemen holding property to the immediate neighbourhood of the spa site, which prevents any building from being erected above the height of a wall now actually standing on the site. The common-sense reader will probably conclude that this covenant could be at once waived if sufficient inducement were offered to those having the right to release this society from the obligations of permitting a building to be erected higher than this Wall. The proverbial generosity of Mr Newnes is sufficient assurance that this has been done, and we understand that no pains, in addition to actual offers have been spared to meet the wishes of all interested and to obtain the consents for carrying the more elaborate scheme into effect. Messrs Marks and Munro at last felt that the only course open to them was to prepare alternative plans showing one building as they would greatly desire to build it and the other with the pump-room sunk to a level which would give a height not in excess of the

9.3. Clifton Grand Spa and Hotel insignia above the main entrance of the Avon Gorge Hotel. (Author)

9.4. The entrance from Princes Buildings incorporates the 1694 entrance from the Old Hotwell House. There is a massive marble staircase behind the black door which runs down the side of the Hotel. This entrance is so plain compared with the back, which always catches visitors by surprise. (Author)

9.5. The limited height of only 5ft precluded any attempts at ornamentation. (Author)

existing wall. How far they have succeeded may be judged from the fact that both these plans have received the entire approval of the Merchants' Committee, who have given their consent to the work being at once commenced and carried into effect.

The high level or upper scheme is, however, one that would have presented the finest architectural feature in the district: and it is a matter of extreme regret that this plan is likely to be frustrated. This building gives a grand pump room, lounge, reading-room, smoking-room and other accommodation, of a nature likely to have formed a veritable palace of ease and recreation, not only to the visitors to the spa, but to the residents of Clifton. It has been a reproach to Clifton for many years that nothing in the way of an assembling hall and general lounge has been provided for the visitors desirous of benefiting by the salubrious climate, which the locality has, and always must, continue to enjoy on account of the elevated position. Yet now, when a bold scheme is formulated and actually brought to the point of inauguration, it may be marred at the last moment. The time-honoured expression that nothing is impossible to the engineer has; once again been fulfilled by the works which Messrs Marks and Munro have already done for the enterprising promoter Mr George Newnes and we unreservedly wish the venture every success.

On 26 March 1891, Colonel Rawlins of 1 Caledonia Place had a letter put in the *Bristol Mercury* through his solicitor, to say that he had not acted improperly but fairly, having due regard both to his own interests and Mr Newnes. Mr Newnes had appeared, intending as a private speculation to create a Pump Room some 30ft high which would seriously interfere with the special advantages and apparently without the slightest intention of considering him or his comfort or pleasure. He had bought his house for occupation owing to the quietness of the situation, and the open and pleasant view, and without the knowledge of the Spa. He would rather sell his house at a price based on the premises by Mr Newnes' advisers rather than be an obstacle to the scheme. He had received £150 on behalf of Mr Newnes for the release of his rights of his view.

On 30 September 1892, The Merchant Venturers confirm that the plans for Newnes' Spa buildings, with the elevation being kept down to the level of the existing wall, were approved.

9.2 Opulence, lots of marble, no expense spared

The first report of pushing forward with the Pump Room is on 1 December 1892 in the *Western Daily*

Press. Surprisingly little is reported on progress, compared with the Railway. The plans had been drawn up by Munro and Son, and Croydon Marks. The contract had been let out to C.A. Hayes again. Whereas all the Railway publicity was about safety, this publicity is about opulence for wealthy clients and advertising Newnes' standing, and being the best in the country:

> Nothing has been spared to make the Clifton Pump Room worthy of its position and surroundings. The principal reception room will be about 100' long and 55' wide. It will be richly decorated with choice marbles and marble mosaic. This portion of the work has been entrusted to Messrs Arthur Lee and Bros [chief importer of foreign marbles based in Canons Marsh, Bristol] who have promised to leave nothing undone which their skill and technical knowledge can suggest. The roof will be supported by twenty huge polished columns and twenty four pilasters. These will be of Cippolin marble, one of the most beautiful of coloured stones, which has been recently discovered at Saillon in Switzerland. On one side there will be semi-circular recess, in which from an ornamental fountain the mineral water from the hot spring will be drawn. These will face the principal entrance from Sion Hill [Princes Lane]. The level from the roadway [Princes Buildings] will be reached by a massive marble staircase.

On 7 February 1893, when the first Railway journey was attempted, it was reported that:

> Messrs Lee had been busy at their works with the vast quantity of marble necessary for this elaborate building.[1]

On 16 February 1893, the *Bath Chronicle* stated:

> It is a pity that from difficulties unforeseen [and unknown], that there has been great delay in commencing the erection of a splendid new Spa which is in close proximity to the top of the Zig Zag and a portion of the same scheme, a great quantity of both foreign and English marble will be used in the building, and has for some time in preparation. No expense or effort will be spared to make this Spa worthy of beautiful Clifton; and resuscitate the attraction to invalids of the medicinal springs of Hotwells.

On 13 March 1893, *Bristol Mercury* reported Newnes' hopes to complete the Spa in August or September (actually completed in 1894):

> In connection with the Spa, the Pump Room which will be one of the finest and largest rooms of its kind in England, is now in the course of erection. Supplied with the leading newspapers and magazines, open to the use of visitors coming to take the Spa waters, and the Pump Room will be conducted on the usual plan in vogue at all pump rooms.

On 15 March 1893, the *Clifton Chronicle* stated:

> Meanwhile the second part of Mr Newnes' scheme, the Spa and Pump Room, is to be pushed forward rapidly. Rumour has recently been again busy declaring that the Spa was a thing of air merely, are repeated denials of such statements notwithstanding. By the end of next summer or in early autumn Mr Newnes expects to have just below Sion Hill one of the finest Pump Rooms in the Kingdom. Here visitors can quaff the healing waters of the Sion Spring which once made Clifton a noted watering place, and here they will be able to listen twice a day to a first class orchestra concert, attend an occasional ball or special concert, or in the other rooms of the Spa amuse themselves in various ways. [Nice to see what facilities will be provided]. It offers a bright prospect for Clifton, and our readers, with ourselves, will wish Mr Newnes success in a generous and enterprising undertaking.

On 27 October 1893, Newnes' solicitors notified the Merchant Venturers that the Railway had been open for six months, and that the Spa Room was being roofed, so it was resolved that the deeds be executed.

On 24 November 1893, Newnes asked permission to use no. 14 Princes Buildings as a Hydropathic Establishment, and to connect the gardens of the Spa room with the garden on the other side of Princes Lane by an underground passage. The Merchants agreed. We are not sure if they built the passage.

On 3 January 1894, 'Attention should be called to the completion of the Clifton Spa and Pump Room and the Clifton Rocks Railway'. This is confusing since it was not opened until July. Note that the Clifton Spa is often the general name for the spa facilities, so at this stage includes facilities for those who wish to be in residence and read the papers – the hotel had been converted from two houses. Sometimes the word Clifton Hydro instead of Clifton Spa is used to describe the hotel, but later on refers to the Turkish Baths. It is not always clear what bit of the Spa is being referred to, but the Railway is not part of the Spa so is always referred to separately and is a different company.

9.3 Opening preparations

On 23 July 1894 it was reported in the *Birmingham Daily Post* that elaborate preparations were being made for the celebration of the 62nd annual conference of the British Medical Association (on 1 August 1894 – only the second time it had come to Bristol, and lasting for five days). It commenced with a general meeting in the Victoria Rooms. One of the 'most important functions of the following day will be the opening of the Clifton Spa and Pump Room'. The *Dundee Evening Telegraph* described the event as opening a palatial Spa and Pump Room and that between 30 and 40 representatives of the London press were to be brought down in a special saloon. The fact that so much press came from London would have delighted Newnes. To combine the opening with a prestigious conference was a very good way of increasing the advertising of the Spa so would have been done deliberately – especially with so many doctors to comment on the healthy aspects of Clifton. There were several long articles extolling Clifton's virtues:

> The balmy air of Clifton has long been famed for Sims Reeves [foremost English operatic, oratorio and ballad tenor vocalist of the time], in the zenith of his fame, made this his headquarters whenever he was touring in the west because of its beneficial effect upon his singularly sensitive larynx. Medical visitors will be able to appreciate to the full the boon to the public health of the 60 acres of Clifton and Durdham Downs.
>
> In the last century, baths and spas were all the fashion, and were believed to cure all the ills that flesh is heir to. The long peace reopened the continent to English travellers and the spas fell into disrepute as undeserved as their previous renown was excessive.
>
> Bristol have too greatly underestimated the immense possibilities of Clifton. There are very few suburbs of a great city so richly endowed by nature. The Downs are incomparable, and they form the centre of a remarkably picturesque and otherwise interesting district. The Avon Gorge looks like a bit of scenery plucked from some famous Scotch panorama. Clifton itself, perched on the heights, has all the appearance of some stately classic town; and indeed, the prospect from whatever point it is viewed, is one of either impressive grandeur, or of sweet pastoral simplicity. Mr Newnes has, no doubt, like a skilful general, calculated upon all these things whilst maturing his plans for the creation of the Pump Room. He has detected charms and possibilities where those accustomed to the magnificent panorama would see nothing but a familiar and pretty scene.

Another report states that Newnes,

> whose manifold energies demand varied outlets, has taken Clifton Spa under his fostering care, the assembled doctors will have an opportunity of seeing what he has done towards proving a Pump Room on the higher level, with water nearly as hot as that of Bath.

A letter by 'an old inhabitant' on 26 July, 1894 who is looking forward to the next stage of development:

> I suppose that now we may congratulate Mr Newnes on success of his Hotwells and Clifton enterprises. The Rocks Railway has been working well for a long time, and has proved a success, not only financially, but is very great public convenience. The Spa, or Pump Room is now almost complete, and it is one of the most beautiful buildings, both as to design and workmanship, inside and out, anywhere in this neighbourhood. Mr Newnes still has in store other designs for Clifton, namely Turkish and other baths, to be constructed on the most scientific principles. We may hope that the new Clifton Pump Room will, by its healing and health giving waters be as popular and efficacious as the old Hotwells. Clifton is well known as one of the most beautiful health resorts in Britain; and with this long neglected, but special advantage, again brought to the front, it can not fail to be more and more popular.

So it is almost complete, being due to open five days later. It clearly took six months to fit out.

9.4 The Pump Room

The front elevation of the Pump Room faces the Leigh Woods and the far famed Suspension Bridge which spans the Avon Gorge, and is extremely elegant. The design is of the Grecian order of architecture, and the material is Bath stone, with a deep moulded plinth of rubbed and sanded blue Pennant stone. The six huge windows form a considerable portion of the front and with rounded heads and cathedral glass they are extremely handsome. Around the heads are really ornate tracery panels with central shields and moulded dentilled cornice; the frieze is broken with triglyph guttae and bells, with well-defined, sunk, moulded and carved patera [oval features] breaking the intervening section; the entablature is supported by fluted pilasters.[2]

The line of the parapet, which is beneath the sight line from Sion Hill [to keep Colonel Rawlins happy] is cleverly broken by grotesque heads of the early Georgian period, above the line of the pilasters which intersect the windows.

In the centre of the Pump Room front is a bold and handsome portico some 20ft by 12ft, supported by six massive clustered Doric columns [stout columns in the Greek style], and beneath the portico and over the main entrance is a noble looking canopy [now removed].
The site and the carriage drive was cut out of solid and hard rock.

9.6. Grotesque head. (Author)

9.7. Triglyph, with bells in frieze above, pilaster to sides, shields, animals below. (Author)

9.8. Princes Lane Entrance of the 100ft long and 55ft wide Pump Room, 18ft below Princes Buildings road surface, and dug into the rock. The drive is steep, and not easy for horse and carriage. (Fine Art Photographs of Bristol, Clifton & District)

9.9. Portico now removed showing the entrance detail better. The six lofty arched windows have been removed. (Author)

9.10-9.11. Caryatides supporting the pediment, adorned by buddleia and (out of picture) plastic drainpipes. (Author)

9.12. Ornate architectural detail above doorway to the ballroom. (Author)

9.13. The insignia of Newnes is to be seen above the central porch. (Author)

It is not generally known that if Mr Newnes could have carried out the design according to his wish there would be three fine dome lights rising above the beautiful panelled ceiling. The centre one of these domes would have been 10 feet in diameter, the others six feet. With the lantern domes if frontagers had allowed it, the general effect would if possible have been still more graceful.

Clearly Mr Newnes chose to publicise the fact that he had not been allowed to have all his own way. This is presumably because it would be too visible from Caledonia Place.

There was another entrance to the Pump Room though the subway from the Railway. One of the conditions that the Merchants set in November 1890 was that the:

> grounds of the existing house were for the recreation of those who may be staying within the new Spa building and not be thrown open to those travelling by the lift, or to the public indiscriminately, so that a Spa of a very select character will be formed.

Possibly the people on the south most pair of cars were intended to go to the Pump Room if four cars were running. The subway is as original, and is a curved brick arch corridor adjacent to the top water reservoir – see plate 4.8.

9.14. North end of an 1893 Pump Room plan showing the corridor from the Railway top station to the back door. The brick subway is at the top of the plan. At the end of the subway, one turns right into a corridor by the side of the Pump Room. Access to the music gallery was by spiral staircase. This is where current visitors to the Railway exit and explains why there is a rounded wall. You can also see two blocked off exits to the Pump Room shown by steps. There is a reclining room at each end of the corridor adjacent to the Pump Room. The room by the back door also held a toilet and sink. The cistern of a toilet can still be seen there. The 'WAITING ROOM' label for the lift (the Railway) is confusing as this is the location of the water tank. Symbols for the Pump Room columns can be seen. (Building Plan 28/70/b BRO)

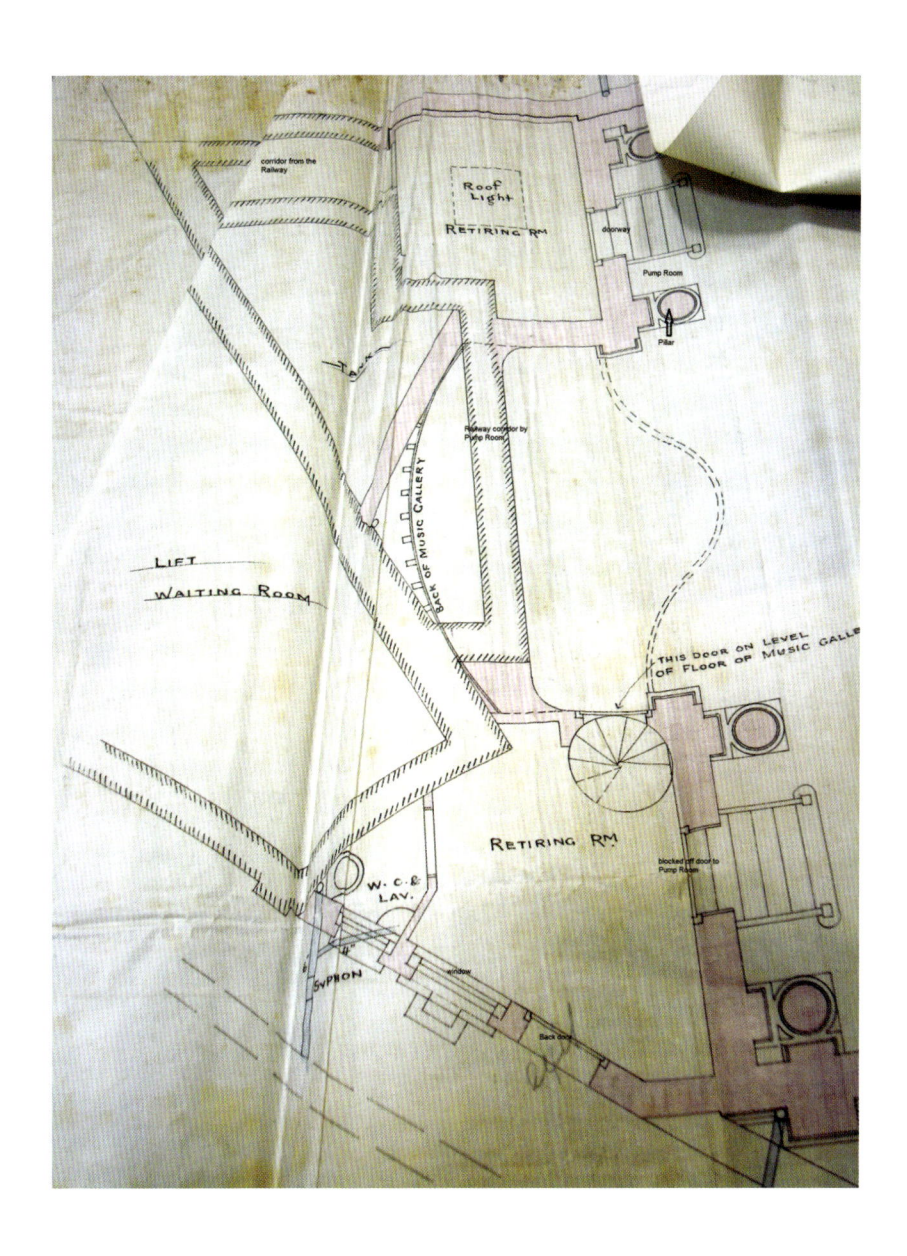

9.15. Right, the low solid ceiling is under the music gallery. To the left of the wall is the Pump Room, but closed off. (Author)

9.16. Far right, view of low solid ceiling from the back door. To the right of the wall where the photographer stands is the Pump Room. Where the circular spiral staircase once was can clearly be seen. There is a blocked doorway to the side. (Author)

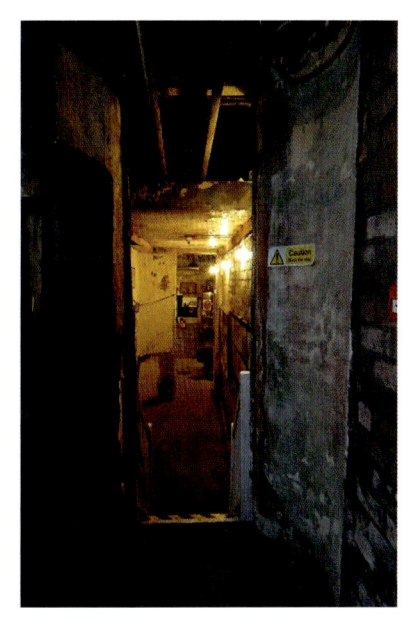

9.17. Note the music gallery at the end, the fountain on the right, the beautiful ceiling. The marble staircase is behind. The portico entrance is to the left, opposite the fountain. (Fine Art Photographs of Bristol, Clifton & District)

9.18. Scrolls in the crown of the arches are carved in foliage. Marble figures above the recesses are still in position. (Author)

The walls of the long apartment will have arched recesses, supported by columns of Swiss marble, a beautiful specimen in white cream and green [Cipollino], the veins of which will run vertically The effect of these marble columns with moulded and rich architrave and the pilasters of the same marble, will be exceedingly chaste and beautiful.

The marble pillars, with Corinthian heads, were painted black in the 1970s, when it was a night club. There is oak panelling. The fountain was in a recess opposite the entrance, made out of statuary marble with a raised fluted basin resting on a short Ionic column with leaf spandrel underneath. The mosaic recess tiles are still there (this is where the band played later when the fountain was removed). Above the alcove were the words *Aqua Bibe*. The elevated music gallery had wrought iron balusters, rich panels, shields and musical instruments. The whole of the ceiling of the hall was divided into panels, with architrave frieze, and cornice, the latter being 5ft in girth, and the roof being divided into 21

9.19. The sketch[3] shows the fountain more clearly. (Royal Album of Bristol and Clifton)

bays. Most of the plasterwork ceiling survives.

> The oiled oak floor was carefully grooved and tongued and specially laid for dancing. [This was famous in its day but has gone.]

9.5 Grand opening

On 1 August 1894 the Pump Room was opened by the Mayoress of Bristol. The event was celebrated by a luncheon (described as refined and sumptuous) at which the enterprise of Mr Newnes and the beauty of the building were warmly eulogised and 200 members of the Medical Association attended.

> Residents in Bristol and Clifton generally will watch with interest the career of an enterprise which owes its inception to a stranger. We use the word in a limited sense, only that it implies the absence of 'residential qualification'. To the Rocks Railway, Mr Newnes has now added another monument to his own perseverance, and as another tangible expression of his confidence in the capability of Clifton to develop into a flourishing social centre, the magnificently appointed marble halls. Of the material effect and elegance of this addition to the amenities of Clifton there have already appeared copious descriptions, and there is no need here to descant further on that phase of the enterprise, though there is a temptation to indulge in an architectural paean in which the note of praise would not be deemed extravagant.

> Mr Newnes entertained the guests in the elegant chamber which will, it is hoped, add to the superb natural beauties of the neighbourhood those social attractions which Clifton has so conspicuously lacked.
>
> The Mayor and Mayoress of Bristol with other representative citizens, were invited to meet the guests, and during the luncheon the Spa fountain was started by the Mayoress (Mrs Symes) and by Mrs Newnes, both of whom pressed down a tastefully designed silver dinner gong, and made an electrical connection with the valve of the fountain. They then tested the waters. A deputation representing the inhabitants of Clifton, waited upon Mr Newnes, and presented him with an illuminated address [with a drawing of the entrance to the Spa, surmounted with the city arms and a picture of the Suspension Bridge; also in bound form with a list of signatories], and expressed their heartiest appreciation of his endeavours to bring back to the district the past success which resulted from the throngs of visitors who attended the old spa.

They hoped that soon they would see that:

> those objectionable tickets, 'this house to let' and 'apartments to let' would soon have vanished. Their success meant the success of Mr Newnes, and for him they wished all success and extended the right hand of fellowship in his undertaking.

They said the signatories would have been much more numerous but that less than a week had been devoted to it. All Cliftonians were unanimous in their thanks and good wishes.

> The President [of the British Medical Association] expressed his conviction that the waters of Clifton Spa were as efficacious now as in the past, and that the fashion which had set in by the journeying of invalids to foreign spas should henceforth be altered now so that so magnificent a spa in so lovely a situation had been provided in this country. Might he commend those waters to the members of the British Medical Association and express the hope that they would be of the greatest utility to rich and poor alike (applause). Mr Newnes gave the health of the The British Medical Association. He characterised the medical profession as the noblest of all the professions, and said 1200 doctors now assembled in Bristol would be able to go back to their practices in various parts of the country, and when all other resources had been exhausted they would be able to cure their patients by sending them to the Clifton Spa.
> Mr Newnes in responding to the toast proposed by the Mayor said that the scheme for completing the spa was still getting forward, and that all kinds of baths for the treatment of invalids in the latest scientific manner were now being added. The building was pronounced by all as being extremely handsome, and the highest compliments were paid to Mr Croydon Marks and Messrs Munro and Son, the engineer and architects for the ability displayed in designing and carrying out the scheme.
> [Mr Newnes also said that they had] got the identical spring which was so famous for its healing powers years ago (applause). The spring used to come out at the bottom of the hill, but the reason why it was stopped at that point was in consequence of the necessity of widening the river Avon, in order to make way for the huge vessels which carried the commerce of Bristol up and down the river. The machinery which was laid down for the Rocks Railway would be utilised for pumping the water from that spring to the fountain in the Pump Room.

We do not have full details of how this happened because the Crossley engines at the bottom station were used to supply the power for the pumps to pump the water from the bottom water tank back to the top water tank – thus reusing the water. The pipework was taken away from the fountain adjacent to the north of the railway during the war. There is reassurance that another bore hole was sunk to the rear of no. 4 the Colonnade (see later) on the boundary of Newnes' land, 150 feet from the river to avoid contaminated water. It was a prolific spring and pumped underground so pipework was needed.

> To the right of the Pump Room will be found the subway, a bright and cheerful passage constructed of white glazed bricks and well lighted by electricity. This is the entrance to the Spa gardens. These gardens are very extensive, and consist of four wide terraces, reaching from the Spa to the back of the Colonnade adjoining Hotwell Road. Amongst the many attractions of the gardens will be numerous summer houses, kiosks, a tennis lawn, an extensive conservatory, an open air tea-house with cooking room attached, and last but not least in such delightful surroundings, a band stand. The gardens will be electrically lighted, and will be utilised for soirées and open air conversaziones. No doubt that many a delightful soirée will be arranged in connection with the Clifton Spa and Hydro.

Details of the proposed Turkish Baths were also given for the doctors' benefit, despite the fact it was not completed until 1898. In the evening there was a concert at the Grand Spa (the hotel) with opera singers and orchestra, and people were invited to inspect the Pump Room. In the interval Croydon Marks spoke on behalf of George Newnes and gave a programme of pleasures in store:

> He said it was intended to charge daily visitors 2d per day for the use of the room and to drink as much water as they liked free of cost. In the room there would always be the daily and weekly papers, and they hoped this Spa would be made a sort of lounge. Arrangements were to be made for season tickets to be issued at a cheap rate, and promenade concerts held on Thursday evenings, the admission to be sixpence to each concert, or else by season ticket at a reduction. On Bank Holiday promenade concerts would be held in the afternoon and evening (applause).

On 7 August 1894, it was reported that the Pump Room was filled with a large and appreciative audience – 600 visitors passed the threshold in the afternoon. At an interval, Marks on behalf of Newnes thanked them for their patronage and said the management motto would be 'Everything pure, nothing low, no trash (applause)'. Soon they hoped to have lounges in the recesses and space allocated for promenading during concerts. The music he could promise them would be of a high enough order to satisfy any critic.

On 30 August 1894, the Spa applied for a music and singing licence (but not alcohol). The Spa had been built at great cost, and would no doubt prove of benefit to Bristol and Clifton. It was intended to give high-class concerts there. The Chairman of the licensing board said that he had visited the building, which was provided with every requisite for the safety of the public. The application was granted.

9.6 Use of the Pump Room

Marks certainly tried hard to promote the use of the Pump Room. On 30 August 1894, it was reported that people were being turned away from the concerts due to popularity. On 16 October 1894, new features were added. A vocal and instrumental concert was given instead of an orchestral concert on Thursday; a concert of grand opera by a London company every Saturday; a lecture on 'Interviews and Interviewing' by Mr Harry How of the *Strand* magazine 'with sixty oxy-hydrogen slides specially prepared and the new duplex dissolving system will be operated by Mr Bromhead in the Pump Room'. This sounds like new technology. (Mr Bromhead had a well-known photographic studio in Whiteladies Road, Clifton from 1884. He carried all his photographic equipment on a specially adapted tricycle.)

> A reserve reading room has been furnished to enable readers to attend concerts; and a Spa band is being formed. We understand that Mr Croydon Marks has in course of formation a Ladies Club, with a suite of rooms at the Spa for the use of members; and daily band concerts will be instituted at an early date. It is quite evident that the residents and visitors are being catered for in a manner which should be exceedingly popular.

However, numbers on a Saturday night in November were disappointing, and there were other problems:

> A large sewer leading from Clifton lift upper station and the Spa, in the private lane at the back of Prince's Buildings, burst, and did considerable damage to the residences of Mr Colman [Henry G Colman is listed at 13 Princes Buildings, no occupation given] and Mr R Moore. Since then the Sanitary Authority have been repairing the damage, but the rainfall increased the pressure of water to such an extent that the walls of Mr Colman's stable gave way, and as they came in the horse and Mr Colman's coachman had a very lucky escape. The torrents of water coming down the terraced gardens washed through Mr Colman's house and did much damage.

On 17 November 1894, *Clifton Spa Monthly*, the latest addition to local literature, reported:

> The Saturday entertainments given the Clifton Grand Spa has induced the management to engage artistes with high credentials, and the experiment has certainly justified the trying. Despite this, the following week there was not an exceptionally large attendance at the weekly concert at the Clifton Spa Saturday night, but those who had put themselves the trouble being present had the pleasure of listening to an enjoyable entertainment.

On 24 November 1894:

> Mr Newnes having asked permission to use no. 14, Princes Buildings as a Hydropathic Establishment and to connect the gardens of the Spa room with the garden of the other side of Prince's Lane by an underground passage. His requests were acceded to by the Merchant Venturers.
> [On 21 December 1894] Mr Newnes' plans for Baths, a Smoke consuming apparatus etc., were approved by the Merchant Venturers.

On 14 January 1895:

> [The *WDP* reports] Mr Croydon Marks in the management of the Clifton Spa is determined that he will, if possible, gratify concert goers, and as a partiality has been shown for military bands, he has arranged for an early appearance of the bands of the Grenadier Guards and the Scots Guards.

On 28 January 1895:

> [The *WDP* reports] amidst of the snow which was whirling through the streets of Clifton, when Mr Gould stepped upon the platform at the Clifton Spa to give his lecture he saw before him only small number of hardy enthusiasts, who were dotted about in a desert of chairs.

In March 1895, leave was given to Newnes by the Merchant Venturers, to alter the elevation of no. 13. Princes Buildings, and to make an addition to the Spa room as a retiring room.

In 1895, the Merchants granted tunnel and water easement to nos. 14 and 15 Princes Buildings (for the Spa room), and lease of land in 15 Princes Buildings and land at Hotwells for the Railway.

On 2 July 1895, for the first time, the Railway announced it would coincide with the tram service until the close of concerts. Up to then, the Spa had avoided mentioning the Railway: remember that one of the conditions that the Merchants Venturers had placed upon the Railway was that passengers should not mingle with people using the Spa gardens. However, it seems the concerts were not as popular as predicted, perhaps because of price: 2s for reserved seats, second seats 1s, and unreserved seats 6d. On 27 January 1896, it was noted that 'audiences were by no means so large as the quality of the performances merited' both in the afternoon and evening. On 3 February 1896, the last concert was held before extensive alterations and additions to the Hydro by the new Clifton Grand Spa and Hydro Company.

On 14 April 1898, Pump Room concerts began once more with a performance by the Bristol Artillery Band. There were evening concerts every day and on also on Saturday afternoons. The Spa Company arranged a series of monthly tickets, and visitors arriving at Hotwells and buying their sixpenny ticket at the offices of the Railway would be conveyed to the Spa for free, with programme and seat at the concert.

On 21 October 1898, the Merchant Venturers reported that 'An application of the Spa Company for leave to place gates at the top of Princes Lane was resolved to be entertained if the Company obtained the assent of all persons having rights there.' Were too many undesirable people trying to get to the Hotel gardens?

On 5 December 1898, entertainments included literary recitals (there is a complaint that the size of the room makes it hard to hear), mandolin recitals, Crimean and Indian Mutiny veterans' concert (the 1000-strong audience taxing the accommodation of the large room), annual Ladies' Night of the Bristol Gleemen (90 members) on Thursdays.

On 20 December 1898, it was announced that the Pump Room was now available for 'balls, banquets, concerts, receptions, at Homes, etc. A new dancing floor is under course of construction. With 'Electric Lighting and the Finest Ballroom in Clifton. Supper served in the Dining Room of the Hotel. Manager Mr F. Edens.'

9.7 Turkish bath and Hydro

The biggest Turkish bath in Bristol was at Brunel House, College Green. This was the first and largest of Bartholomew's establishments, opened in 1859 and survived until at least 1955:

> The Hydro is not a Nursing Home but an Hotel for Visitors and those who wish for Rest and Special Attention. Turkish Baths or Treatment optional. For the maintenance of health and prevention and cure of disease.

The second Bartholomew baths opened in Bath in 1880 and he went on to open three more.[4] Newnes was trying to emulate the success of such baths and went on to provide vapour, plunge, Turkish and Russian Baths – no wonder he needed smoke consuming apparatus.

9.8 Quality of Hotwell spring water

On 9 August 1894, it was reported in the *Bristol Mercury* that:

> the old discussion had arisen with regard to the qualities of the Hotwell water, which was freely attacked by the advocates of Bath in the last century. There are three springs, one of which Mr Hughes has lately restored, and which is possibly the oldest, and another which comes from beneath the bed of the river and is conveyed to the foot of the rock near the Suspension Bridge. As this does not bear an unquestioned reputation, Mr Newnes would not draw his supply from it, but purchased the Colonnade property in order to acquire the third spring, which is situated in the rear of these houses, and in all probability is that from which the old Hotwells Spa was supplied 100 years ago.
>
> The spring here is no less than 150 feet from the river and its depth was 60 feet. It is sunk in solid rock but in order to make absolutely sure, Mr Newnes deepened it by sinking through a very hard stratum of close-bedded limestone so tools were repeatedly broken. At the foot of the boring the water gushes up with great force and in great volume, and was carried in a pipe from the bottom of the well to the Spa. The City Analyst was happy with the absolute purity of the water.

On 14 September 1898, a letter to the editor from Charles Jones of Leigh Road South, Clifton, reassured readers about the elaborate means taken to preserve the Spa water in a perfectly pure state:

> The water is pumped underground and not contaminated by river water. It contains several important salts considered valuable for certain chronic diseases, such as chlorides of sodium and magnesium, carbonate and sulphate of magnesia, carbonate of lime. The Clifton water sparkles with carbonic acid gas, quite clear, pleasing to the taste, medium temperature of 23.5 degrees Centigrade and keeps good for a long time in properly corked bottles, advantages in its favour compared with other medicinal waters of Matlock, Tunbridge, Wells, Buxton, Epsom and Harrogate. A covered Spa swimming bath for both sexes would be a valuable addition to our new local institution, which need not interfere in any way with the comfort of the visitors in the Hydro, any more than the explosions and noises caused by the blasting of the rocks [quarrying] and the siren steam whistles.

On 28 September 1912, radium was discovered in the spring water. The water is currently pumped from two deep wells.

> The general opinion is that the water drunk at the Hydro could be used for bathing purposes. This opportunity should be seized of placing Clifton on a level with the well-known spas in the country.

9.9 Proposals

On 2 August 1894 details of the proposed Turkish Baths were given at the opening of the Spa and Pump Room in Clifton, despite the fact it is not completed until 1898 because they have to create a new company in 1896 to fund the scheme. Newnes used the British Medical Association conference to sound out the potential.

> There was no bath that has been pronounced of medicinal value which has been omitted from the scheme.
> The baths will be beautifully arranged, and contain special douche, spray, and rain baths, with convenient dressing and drying rooms. There will also be an extensive cooling room, containing a fountain, and handsomely appointed, and the whole will be laid with mosaic marble flooring of elaborate design; so also will the plunge bath, which being also fitted up with marble steps and kerbing, will present a very handsome appearance. This bath will be 30ft in length by 14ft in depth, with a depth of water varying from three feet to four feet, and the water brought to the requisite temperature by means of steam pipes beneath the floor. There is one item which must not be overlooked at the Clifton Spa, being something of a novelty – we refer to the Russian steam bath.

On 25 January 1895, Newnes' application that the lease of the Colonnade property now held by him, should be converted into a lease for a long term of years, was declined by the Merchants. The 1896 prospectus shows 40 years, whereas other buildings are for 1,000 years. Clearly he needed no. 4, since that was where the spring was located, but I am not sure why he wanted the rest of the terrace. On 29 March 1895:

> Leave was given by the Merchant Venturers to Sir G. Newnes to alter the elevation of no. 13. Princes Buildings and to make an addition to the Spa Room as a retiring room.

On 2 January 1896, some surprise was expressed in the *Western Daily Press* that the baths were not completed, due to 'wanting to purchase the adjacent properties for a fair market value rather than a purchase value enhanced by the success of the Spa'. Representations had been made to Newnes that a larger establishment should be provided. Three adjacent houses with extensive gardens and grounds had been purchased on behalf of a local company proposing to associate themselves with Newnes. They already had nos. 14 and 15, so presumably this is when they bought nos. 12 and 13. The company would be called the Clifton Grand Spa and Hydro. Croydon Marks would act as chief constructor, already being chief engineer and manager of the Spa.

On 18 January 1896, it was proposed to absorb the existent Grand Spa and Pump Room into the Hydropathic establishment:

> [The *Bristol Mercury* reported) For those who are actually invalids and deserve perfect quiet, Tuffleigh House has been set aside. In fact the visitor to the Hydro has the seclusion and the scenery of one of the most rural spots that could be sought. The Hydro will accommodate 150 visitors. I think it is clear how much of its success will minister to the liveliness and prosperity of Clifton. Without speaking unkindly of establishments which are well conducted and do a useful work, the Hydropathic establishment as it is understood in this part of the country is a modest and inexpensive sort of temperance hotel, general placed ten miles from anywhere and only suited to those whose tastes are simple and who are satisfied with idyllic quietude and solitude. But there are at Matlock and other places in the North of England and in Scotland at Straffpeffer, and elsewhere hydros on a far more ambitious scale, where luxurious baths of all kinds and descriptions are provided and where the health seeker can go through the regimen and treatment needful to recuperation.

9.10 Funding

The property, including the mansion, 14 Princes Buildings, the land, buildings, fixtures, fittings and appurtenances, together with the option to purchase the adjoining property, cost £7,200 (lease rents £40 on no. 14 and £60 on no.15). We now know it cost £23,000 to build the Spa so far. Newnes had certainly managed to buy a lot of property to continue the development. He had only developed no. 15 for the Pump Room, and no. 14 for the residence, so it would triple the size of the hotel footprint. It cost £64,000 to develop the Hotel. The Railway had cost £30,000. The original anticipation was that it would £30,000 to build both Railway and Pump Room. Further capital was needed to convert the houses is £21,000.

Compare the 1900 map showing Railway, Hotel and Hydro (plate 9.2) with the 1880 map (plate 9.1) (which has the house numbers of Princes Buildings so you can work out what has happened). The Hydro has been tucked behind the hotel, the drinking terrace added on top. The gardens of nos. 13-15 have been built on. The Hydro plans are in plate 9.23. The tall yellow chimney that one sees today at the back of the Hotel is a laundry chimney and used by the Turkish Baths. The prospectus opened on 9 January 1896, and closed on 23 January, with the object of forming a limited company and raising £40,000.

Clifton Grand Spa and Hydro Limited
Capital £40,000 divided into:
1000 Ordinary shares @ £5 each £5,000
300 Cumulative 6 percent @ £5 each £1,500
Present issue:
4000 Ordinary shares @£5 each £20,000
3000 Preference shares @£5 each £15,000
600 Ordinary shares and 600 preference shares [£6,000] paid up to the vendor Newnes in part satisfaction of the purchase consideration.

Newnes (from Putney Heath) and Croydon Marks (from 55 Chancery Lane) were still directors. Croydon Marks remains director of Clifton Rocks Railway. Job Smith chairman of the Matlock water company, was a director too along with Philip Fussell of West Gloucestershire Water, L.A. Atherley Jones MP (radical British Liberal Party politician for Durham and barrister), James Dole of Redland House (provision merchant, and only local person), and Henry Serpell of Plymouth (biscuit manufacturer).

> The Company has been formed for the purpose of acquiring from Sir George Newnes, the newly-erected Grand Spa and Pump Room at Clifton and other adjoining properties, and to establish in connection there with a first-class Residential Hydropathic Establishment, constructed on the most modern and approved principles; and to provide various attractions and amusements for the Residents of Clifton and for Visitors. Sir George Newnes agrees to accept as consideration for his interest a sum which will simply cover the purchase moneys and his actual outlay upon the properties. He has also agreed to join the Board of Directors.
> The Company is to acquire:
> The Grand Spa and Pump Room, with its furniture and fittings, universally described by the Press as the most imposing Pump Room in the country.
> The Sole right to an unlimited supply of water from the new Well constructed by Sir George Newnes at a great expense under scientific advice by boring through the rock to a great depth, together with the pumping gear and machinery.
> The four houses in the Hotwells Road, known as The Colonnade
> The three large and commodious mansions known as nos 12, 13 and 14 Princes Buildings, and the picturesque house and grounds commanding position on the cliff, and known as Ghyston or Tuffleigh House
> The beautifully terraced garden grounds reaching from the Pump Room down to the Hotwells Road
> The sum to be paid to the vendor in respect of the whole of the above-named property and outlay is £23,000 as follows:
>
> £3,000 in ordinary shares, paid up
> £3,000 in preference shares, paid up
> £7,000 in debentures
> £10,000 in cash
> £23,000
>
> The Mansions in Princes Buildings will be immediately converted into a residential Hydropathic Establishment. The Baths will be constructed and equipped upon plans designed for this undertaking by the late eminent Hydropathist, Dr Hunter of Smedleys, Matlock who personally supervised the whole of the proposed arrangements and who had consented to act as Consulting Physician for this company. They will comprise Turkish, Plunge, Russian, Sitz, Douche, Shallow, Way, Spray, Pack, and other Baths, while special arrangements will be made for Medico-Electric treatment and for Massage. It is believed that a large revenue will be returned from the Baths alone, apart from the Hydro.
> Tuffleigh House will be made available for any special medical cases needing more repose or retirement than could be afforded in the general establishment. The extensive grounds will be provided with Tennis Courts, and tastefully planted with shrubs and flowers along the beautifully terraced walks, whilst the adjacent celebrated Clifton and Durdham Downs, with an area of 650 acres, will afford an unrivalled promenade in pure invigorating air. The plans of the Baths and particulars of the scheme have been submitted to some representative members of the medical profession in Clifton [in 1894], and they have expressed their entire approval of them.
> It is estimated that the further Capital required for the construction of Baths, and the conversion of the houses into a Hydro, together with the cost of furnishing, will amount to about £21,000.
> It is proposed to issue Share Capital to the extent of £35,000 and to issue Debentures at four and a half percent for a sum not exceeding £20,000. £5,000 Debentures have already been subscribed, in addition to the

£7,000 to be issued to the Vendor. To protect the interests of the shareholders it is proposed that further Debentures shall not be issued until the sum of £10,000 has been spent on the conversion of the premises into a Hydropathic Establishment, when £5,000 further Debentures may be issued; but the remaining £3,000 will not be issued until the Baths and Hydropathic Establishment are completed.

The Hydropathic establishment will be conducted on lines similar to Smedley's, well-known Hydropathic Company at Matlock, which has proved so successful and profitable, paying a regular dividend of twelve and a half percent to its shareholders for the last fifteen years. It is confidentially expected that owing to the unique position and charming scenery surrounding the Clifton establishment, a much larger number of permanent guests than is usual in similar institutions will be induced to reside there.

The Buildings when enlarged will accommodate about 150 Guests, which at a moderate tariff, and reckoning only half this number as an average for a period of thirty-four weeks, and one-fourth the number at a reduced tariff for eighteen weeks, may reasonably estimated to yield a revenue of about £3,700 after providing liberally for all working expenses. The Baths will be open to the general public as well as to visitors to the Hydro, and it is estimated that the net revenue from this source may possibly amount to £1,000, making a total estimated revenue per annum from these two sources alone £4,700.

The fixed payments will be

Interest on Debentures say £12,000 at 4 $^{1}/_{2}$%   £540
Dividend on Preference Shares £15,000 at 6   £900
Leaving a balance of £3,260

available for Dividend equivalent to 12½ percent on the Ordinary shares, and leaving £760 for Reserve Fund, Directors fees, etc.

It is probable that an income may be derived from the Sale of Water, as inquiries have already been received from many parts of the country for Water to be forwarded regularly to patients suffering from various diseases, although no advertisement or public notification, except the newspaper reports of the opening ceremony of the Pump Room has been given.

As nearly all the first-class Hydros are in the North, this establishment which will be one of the very best, will have the exceptional advantage of being within reach of the large towns of the South and West of England and of South Wales, as well as being accessible from the great Midland towns. The long distance to the North prevents very many people from taking advantage of the efficient Hydropathic treatment.

The attention of investors is particularly directed to the solid and substantial character of the assets, which consist mainly of property nominally leasehold, but almost entirely held for long terms of years on small ground rents, which make it nearly equal to freehold, while nothing is included for goodwill or any expenditure on the property beyond a fair market price, which is confirmed by the report of the City Surveyor as follows:

The Grand Spa, Pump Room, no. 14 Princes Buildings, the Gardens overlooking the River, Term 1,000 years from 1836. Rent £50 per annum

No. 13 Princess Buildings, Stable and Coach House Term 40 years from 1892. Rent £9 10s per annum, perpetually renewable every 14 years on payment

No. 12 Princess Buildings, Stable and Coach House Term 1,000 years from 1864. Rent £20 per annum

Ghyston or Tuffleigh House, Hothouses, Vineries and Garden. Term 1,000 years from 1874. Rent £15 per annum

Nos 1,2,3,4 The Colonnade with Gardens adjoining. Term 40 years from 1870. Rent £10 per annum

Building Site at Hotwells, adjoining Station of Rocks Railway. Term 75 years from 1895. Rent £15 per annum

I am of the opinion that the value of the same as a going concern, without taking into consideration the Tenants' profits which is believed will accrue from the undertaking, is the sum of £19,400. It should be born in mind that in considering this valuation that several thousand pounds have been spent by the Vendor upon the properties incapable of present valuation apart from the special purposes for which the buildings are intended.

The Vendor undertakes to pay the expenses of and incidental to the formation of the Company up to the date of the first allotment of Shares.

Letters to local newspapers in the next few days, including from the Mayor (who wanted five ordinary and five preference shares) were very positive, and on 28 January (five days after closing the applications) it was confirmed that there was enough interest to warrant the directors to proceed with the conversion of the buildings.

9.11 Spa competition from abroad

A very determined effort was being made to assure the crowds of visitors and patients who annually leave the country for Continental Spas that, within a few hours journey they could obtain equally efficient treatment in Clifton. According to the *Derby Mercury*:

Medical science today recognises that a change of diet, a course of baths, and a residence in a luxurious hydro will dispel care from a stubborn patient's mind and will effect a cure of diseases when drugs and prescriptions alike fail. The Clifton Spa will provide an establishment wherein the comforts of a luxurious residential home may be obtained with the charm of society, and the elaborate suite of baths will leave nothing to be desired by the resident for massage or other treatment upon the lines now considered possible only in the Continental Spas. The extensive gardens, which overlook the River Avon will afford visitors great interest in watching the passage of the vessels in and out of the busy port of Bristol, while the adjacent Downs provide a promenade and riding area, charged with breezes from the Channel.[5]

9.12 Closure of the Pump Room and progression with the Hydro

On the 3 February 1896, there was the first concert in the Pump Room since the formation of the company, and the last for some time as the extensive alterations and additions would begin immediately. Croydon Marks hoped that by Christmas:

> they would have a suite of baths which would not be surpassed in this country. He was going to Continental places himself to inspect similar institutions, so that he might gain a knowledge of what is done on the continent, and in order to ensure that there should be no suggestion that things were better there than they would be at the Clifton Spa.

It was actually opened in March 1898.

9.20. The date carving above the entrance says 1897 but it actually opened March 1898 – only 15 months out from original press release. Newnes' insignia above. The rest of the carving is very ornate with Neptune, dolphins, etc. Harper and Harper (Birmingham) were the architects, not Munro, but the building is still beautiful. (Author)

On 4 June 1896, the minutes of the Merchant Venturers reported that 'The Clifton Spa and Hydro Company having asked for terms for the conversion of the leases of 13 Princes Buildings and of the Colonnade into grants for 1,000 years, and the surveyor reporting that the ground rents should be £60.15s, it was resolved to offer the properties at a rent of £80.' So the prices had increased from the prospectus. Very confusing, since the ground rent for 13 Princes Buildings was £9 10s, and for the Colonnade £40.

On 6 August 1896:

> The rapid progress now being made in the erection of the Clifton Grand Spa and Hydro, and consequent interference with the Pump Room, have caused the directors of the company to decide upon closing the Pump Room. It is expected that the whole building will be thrown open in the Spring of next year.

Concerts did not start again until April 1898. This would cause a massive loss of revenue since it closed in February 1896.

On 25 September 1896, a bricklayer fell off the scaffolding at the back of the premises where alterations were in progress. He died in the Infirmary within a quarter of an hour of arrival. At the coroner's court it was thought that the man overbalanced. Had another rail been added it would have made it safer for those above, but more dangerous to those below. The scaffolding was 70ft high. A verdict of accidental death was returned.

On 27 October 1896, Croydon Marks returned from Budapest, having visited some of the oldest baths in the world, including a Turkish bath 350 years old. 'These and mud baths have been specially studied. It was desirable to embody all the successful features of Continental practice.'

On 16 November 1896, Arthur Scull, sanitary specialist of Redcliffe Street gained the contract for sanitary appliances and interior plumbing.

On 1 December 1896, there was an 'amazing' public auction of building materials removed from Clifton Spa: 100 panelled doors, iron columns, stoves, a two-ton iron boiler and interior of a gentleman's two stall stable.

On 25 March 1897, it was thought that the Grand Spa and Hydro, under the supervision of Croydon Marks, would be 'opened in the middle of June'. Still very optimistic. Provision was being made in the hotel for 100 bedrooms.

On 26 March 1897 the minutes of the Merchant Venturers report announced that: 'The Clifton Spa Company's plans for alterations required by the addition of no. 12 Princes Buildings to their property were approved.'

On 2 June 1897 the company sought a 'Manager and Secretary capable of undertaking the whole of the duties under the Medical Officer; Head Bathman; Head Bathwoman thoroughly accustomed to Turkish, Hydropathic and other baths.'

9.21. One of the doorways is still in situ. (Author)

9.22. The plan shows the location of steam cylinder, the electric bath and mostly pack baths (a warm bath with hand friction, then envelop the patient in a sheet wet in cold water, and confined to the body by several quilts or blankets. Sleep was perpetuated by magnetism[6]). The boilers and furnace rooms do not appear to be near the chimney which is a bit strange. (BRO)

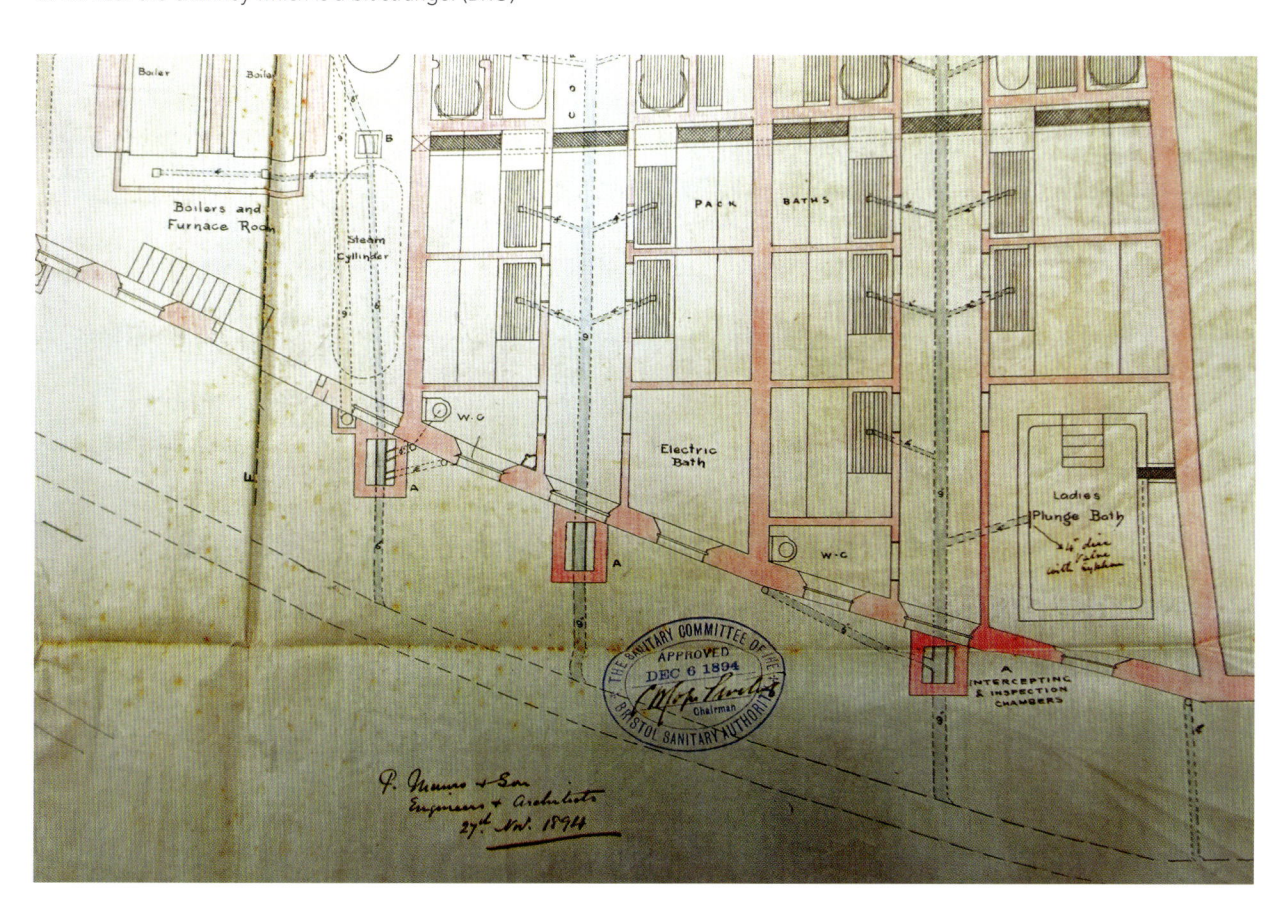

9.23. Bartholomew's Turkish Baths in Brunel House shows patients treated for rheumatism or other disease which requires steam or medicated vapour, the hottest the patient can bear. It is declared the best remedy for eczema when it is located at the extremities.[7]

9.24. The Turkish influence in the Hydro can be seen. Observe the heat bricks. (Mike Edwards)

9.13 Opening the Hydro

Adjoining the Pump Room is the Clifton Hydro, two fine mansions arranged on most approved lines for those electing to be in residence. The buildings contain, in addition to a large number of bedrooms and private suites of apartments, a general dining room and a drawing room.[8]

This was later to be known as the Grand Spa Hotel, and currently the Avon Gorge Hotel. Interestingly enough there was a hydraulic lift (which cost £500 to install) to convey the residents to the Turkish Baths.

On 16 March 1898, it was announced that the Grand Spa and Hydro was to be opened by the Duke and Duchess of Beaufort on 31 March.

Guests to the Hydro would enter from Princes Lane to ensure privacy. The *Bristol Mercury* suggested that:

> Some visitors may wish to make the Hydro their home. The Pump Room has been redecorated and retained as an assembly room for guests, reading papers, club and lounge with occasional entertainment. Hydraulic passenger elevator (dark oak and mirrors) to all floors; electric light and artificial heating provided by controllable ventilating radiators; billiard rooms and smoking rooms.

On 18 March, it was announced by the Duke of Beaufort in a letter to the *Bristol Times and Mirror* that he had not been asked to open the Hydro, and, since it was largely commercial, he felt he could not take part in the opening ceremony. Solicitors' letters followed. It was not a good start.

On 31 March 1898, the interior reconstruction of three large houses in Princes Buildings had been undertaken by a joint-stock company, and Clifton Grand Spa Hydro opened. At the grand opening ceremony, a musical reception was played by the band of the 1st Life Guards. Chairman Newnes and other directors spoke to nearly 700 guests.

> Newnes narrated the history of the undertaking, observing that his engagement with the Merchants Society had now been fulfilled and that the institution would bear comparison with those of some of the most celebrated continental watering places.[9] [Newnes was rather on the defensive.]

> He expressed gratification at the large attendance of the leading people of Bristol and Clifton. He added he did not want to strike a discordant note, nor did his co-directors wish it. He was perhaps a stranger, although he had taken in a certain way the lead in making a connection between Bristol and Clifton by the Rocks Railway. He did not think anyone could say that railway was an injury to anyone. On the whole it was perfectly successful; but there was another clause, as the lawyers would term it, in the scheme, namely that a spa and hydro should be built there – he was not sure whether the word hydro was used, but at any rate a spa – should be built there, otherwise the concession of the Clifton Rocks Railway was not to be made.

> Well he consented to that clause and had carried out the contract, and now they had a hydro as well as a spa. The question was how far this was a public matter. They knew there had been suggestions that it was a commercial

undertaking, but at any rate, if he lived in Bristol and had any interests there he would think it was in the nature of a public undertaking which would be likely to bring further prosperity to the city. He would not labour that point, but his connection with the matter in Bristol and Clifton was for no personal reasons of his own except to carry out the contract which he had absolutely entered into. There had been, he was afraid, a certain amount of nonsense, about this being used for political purposes in some way. There was not the smallest reason for any suspicion of that sort, and he did not know why anyone should try to say that he had done things there for political reasons – there was absolutely no truth whatever in it. Since he had been out of Parliament [in 1895] he had had some 10 invitations to go back, some of them he admitted very hopeless, others promising, and perhaps if he had chosen he might have won one of the seats offered; but he was not in any way politically connected with Bristol and Clifton, and let it not be thought he had the smallest intention of in any way interfering in local politics in the smallest degree. [He became MP for Swansea in 1900.] They had blasted 8200 tons of rock and laid enough miles of piping to go from Bristol to Bath. Nothing they had done was in desire of personal aggrandisement except for a natural desire to get back if possible, interest on ones money, and he had the most sanguine hopes for the future of the hydro.

No expense appeared to have been spared, with mosaic Terreste flooring, heavy Japanese wallpaper, Louis XIV style furniture and Moorish divans.

> It is described as one of the most luxurious of its kind in England, but not ostentatious. Even the Bristol Corporation electricity report singles out the Hydro as one of the more important buildings fitted with electric light. There was a comment that the Hydro was by no means finished. It was also commented that Bath cheerfully maintains and extends its baths on the security of rates, Clifton costs the ratepayers nothing.

On 14 December 1898, we find a description of new electric heat baths – a bizarre therapeutic novelty (invented by Mr A.E. Greville 'who has devoted much time to electrical study'). It used generators consisting of handsome aluminium receptacles of different shapes and sizes, according to the limb or part of the body they are meant to accommodate. It exposed the patient to intense dry heat from electric wires. When in use:

> the patient lies comfortably on a couch with the arm (or other limb) previously bandaged with lint, in the generator the interior of which is then raised to the required degree of heat. The usual temperature is 300 degrees Fahrenheit, which in the course of forty minutes or so produces free perspiration, and generally relieves any pain that may be present.

Such baths were installed at Tuffleigh and offered only when recommended by a qualified practitioner.

9.14 Licensing issues and other news

It is interesting how the Merchants and the licensing magistrates worked together.

One of the original conditions in 1890 of being allowed to build the Railway and Spa was that 'no licence for liquor be obtained' so on 22 February 1901, Merchant Venturers 'agree to refuse application of Sir G. Newnes for waiving of prohibition against applying for a licence for the Clifton Spa premises.' Predictable, since that was an original condition. It is clear though that the numbers of guests have dropped probably owing to their not being able to buy alcohol. Since so much money had been sunk into the enterprise, they wanted to make full use of their facilities (at the expense of the residents – nothing changes.) This was refusal for any kind of licence if it offered a different facility from Bath Spa facilities.

On 20 September 1904, the Hotel asked the licensing magistrates for a renewal of the music licence. The police reminded the licensee that as dramatic entertainments were held there, a stage entertainments play licence was necessary too. The chairman stated that the licensee would render himself liable to a heavy penalty if it happened again. The music licence was granted. New rules were introduced to make all places used for public amusement safe, as far as possible, in the case of fire or panic.

In 1905, the Hotel bought the garage in which I live at 97 Princess Victoria Street for use as a Hotel Garage. I still have the original Hotel garage parking signs.

On 7 March 1905, Mr Vachell of the Hotel applied for a licence to sell wine and spirits for the Clifton Grand Spa and Hydro. The Clifton Down Hotel and St Vincent's Rocks Hotel opposed it. The applicant stated that there were very few hotels offering the accommodation required for a health resort. He hoped that the hotels opposing the application would not have the monopoly. £64,000 had been sunk into the building and equipping the house. It was very difficult to carry on trade without a licence. The company did not want a tap or a bar, except for serving purposes, just facilities for supplying people in the Hotel. The building was rated at £350 a year (the Clifton Down at £840), and £200 was spent on electric lighting. It could accommodate approximately 90 visitors. The opposers said there was no problem ordering from a wine merchant. The chairman said the bench was very reluctant to add to the number of hotels in the immediate neighbourhood, so they were not prepared to grant the application unless one of the other hotels was extinguished. Mr Vachell offered to pay a monopoly value but this was declined. They were told to make another application the following year. It was worth a try.

On 31 March 1905, the minutes of the Merchant Venturers note: 'Resolved that increased Ground rent of £50 a year be asked of the Spa Company in return for leave to use nos. 11-14 Princes Buildings as a licensed hotel.' No. 11 had been last leased for 1,000 years with ground rent £17 5s. No. 12 for £20 per annum. This was expensive. No. 11 has been added to the portfolio (and is now the White Lion Bar).

On 10 April 1906, the licence granted to sell wine and spirits came up for confirmation by the licensing committee. It was accompanied by three conditions: payment of a monopoly value of £2,000, restriction to a term of seven years, and the extinction of the Royal Shades licence (in Waterloo Place). The owners of the Clifton Down Hotel, St Vincent's Rocks Hotel and Royal Shades were also present. Mr Vachell contended £2,000 was too large since it was only granted for seven years. He did not even think they would sell £2,000 of liquour. They did not want a bar, simply a means of supplying those who were staying in the house. The company under the conditions of the lease, would have to pay the Merchant Venturers an extra rent of £50 for the premises if they obtained a licence. He maintained that on figures of the Clifton Down Hotel, the net profit from the licence would not be more than £230 on the Spa. Even if the £2,000 figure was retained he asked that payment should not be required at once. Counsel reminded the justices also, that a licence in the neighbourhood was to be bought by the applicants and extinguished.

Mr Wansborough (of Clifton Down Hotel, established 1865) asked that the licence should be refused altogether. It was a great hardship upon those who had embarked their money in similar businesses in the neighbourhood to have this competition thrust upon them. He was sure the Bench would never have granted a provisional licence for a new house on these terms. Within 530 yards of the Spa there were 12 fully-licensed houses, six beer houses and five grocers' licences. The chairman announced that the justices confirmed the licence on the terms previously imposed.

On 27 April 1906, it was reported in minutes of the Merchant Venturers that the Grand Spa Company had obtained a hotel licence for premises leased from the Society. The Merchant Venturers had also provisionally approved plans for increased dining room accommodation.

In 1907, a lease was also taken out on no. 10 Princes Buildings to Clifton Grand Spa and Hydro Ltd for 40 years with ground rent £9 10s per annum (same terms as in 1891). So 10 had now been added to the portfolio.

By 1908 George Newnes' businesses were failing and this affected his health. By 1910, his fortune had gone, he had retired to live in Hollerday House where he died on 10 June 1910 of diabetes. He would have been aware of the licensing issues.

On 23 February 1912, there was a desire to use 10-14 Princes Buildings as a hostel for a training college by the University of Bristol. This was declined.

On 4 March 1913, the 1906 wine and spirits licence was now practically spent, and it was necessary to seek not a renewal but a new licence. It was reported in the *Western Daily Press* that the Spa was now:

mainly for the benefit of guests rather than shareholders, but they had been able to pay in the way of provisions, wages, salaries etc the sum of from £5,000 to £5,500 a year. Taking an average from 1906 to 1911 the receipts from intoxicating liquors compared with the whole of the general takings were less than one seventeenth (about £300), whilst the average consumed per person per day amounted to only 9d. There was no bar on the premises. The Spa was looked upon as being a great boon to the people of the neighbourhood, and he wished to make a special point of denying the rumour which had somehow or other got abroad that the baths connected with the place were to be discontinued.

Alderman Cope-Proctor said that the terms given seven years ago with regard to this licence were given on much the same information as in the preceding case. It had been clearly shown the amount was excessive, and having in consideration the fact that this was not a hotel pure and simple, but filled a much needed want in Clifton as regards its baths and massage treatment, they were prepared to take a generous view. So long as the Spa was carried on upon that basis, they were prepared to make an annual monopoly charge of £25 per annum. If the baths should at any time be discontinued, these arrangements would at once be reviewed. The terms were made from year to year, and not for seven years.

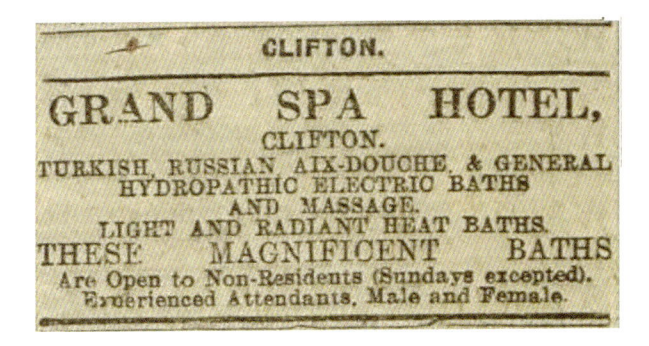

9.25. Above, advertisements after 1917 never mention the baths. This is from May 1917. (*WDP*)

9.26. Left, grand opening of the cinema on 6 September 1920. (*WDP*)

That is a lot better than £2,000.

From 12 to 21 December 1914, several letters were published in the *Western Daily Press* (who reported all licensing applications) criticising the Licensing Bench for refusing a stage entertainments play licence, including one by Muriel Pratt's company which was due to stage a children's play at the Spa. The Clifton Improvement Society stated:

> This association regrets the decision of the Licensing Bench in refusing to grant a temporary dramatic licence to the Clifton Grand Spa, and is of opinion that such action is detrimental to the best interests of Clifton. It has been the policy of the association to foster such improvements by way of healthy and intellectual entertainments for the benefit of residents and visitors.

The application was opposed by two theatres.

On 31 October 1917, Her Royal Highness Princess Louise stayed at the Clifton Grand Spa after opening the Y.W.C.A. Hostel at Clifton, returned to London by the mid-day train, so the great and good visited.

9.15 Sale of site and conversion of Pump Room to cinema

On 16 July 1918, Clifton Grand Spa and Pump Room, no. 14 Prince's Building and gardens overlooking the river were sold to Mr F.J. Price, who was connected to one of the largest film companies in the kingdom. He also bought all the properties including furniture and effects of nos. 10-13 Prince's Building, and Tuffleigh in Princes Lane. (No mention of nos. 1-4 The Colonnade which were bought to gain access to the spring.)

His wife would help run the hotel.

> There is a licence attached to the hotel, and the baths are a feature of the property. The large concert or dancing-room with its spring floor is always in very great demand. The furniture is of a most substantial character, and many improvements have been effected at the hotel including an electric lift [hydraulic] which cost about £500. The hotel has enjoyed a somewhat chequered career, but has of late, under efficient management, had a run of success, and the returns have improved in a vast degree. If it is permissible, it is proposed to add other attractions to the hotel.

A similar licence, for a cinema and for singing and dancing at the proposed Whiteladies Road cinema, was refused on the same day and because the time was inopportune. On that occasion they preferred building houses to a cinema. The wonderful, listed, art deco Whiteladies Road cinema later opened on 29 November 1921. Looking at the plans for the Clifton cinema, the fountain in the Pump Room appears to have vanished, and the first room in the Hydro appears to be a Gentleman's Lounge rather than a boiler room. The back door of the Railway is used by the cinema operator and the public enter from Princes Lane.

On 9 December 1919, an application was made by Mr Price, new proprietor of the Spa, to convert the Pump Room into a cinema theatre. 'It was estimated that there were 15,000 people residing within one mile of the site, and the nearest cinema was nine-tenths of a mile away' (the Victoria Rooms at the junction of Queens Road and Whiteladies Road).

> There was seating accommodation for 650 persons, but Mr Price proposed to reduce the figure to 500. A speciality would be of high class music, and Mr Price who had had a large experience as a cinema manager and a film renter, intended that the entertainment should be of the highest possible character. Counsel emphasised the fact that there was no hall to be built – a matter that was of consequence at a time when so many dwelling houses were needed. There was no opposition, and the application was granted.

On 6 September 1920, the Clifton Spa Cinema presented an attractive programme, open daily from three o'clock. The films were of unusual merit and variety. A capable orchestra rendered a delightful selection of music. This 'most beautiful and luxurious cinema in the country deserves to be well patronised by Cliftonians and others, for the proprietors are displaying commendable enterprise in arranging first

9.27. The projectionist's room was located where the band used to play upstairs. One can still see the sliding doors which were needed in case of fire (a common occurrence with celluloid films). (Author)

9.28. Reference to Clifton Rocks Railway to get to the cinema, 8 September 1920. (*WDP*)

9.29. No mention of the Spa facilities, 14 December 1920. (*WDP*)

class entertainments.' 'With a high standard of merit maintained the Clifton Spa Cinema should be well patronised.' (Despite the existing cinema at the Victoria Rooms.)

On 11 January 1921, 'The pleasure of a visit to the Spa Cinema is greatly enhanced by the extreme comfort of the hall, its heating and lighting, its facilities for refreshments, and the quality of its orchestra, making it one of the most attractive halls in the West.' On 21 January it 'has established its claim for a prominent position in Bristol picture land, and patronage has largely increased in the last few weeks. With the present style of fare continued, future success should be assured.'

On 13 May 1921, the *WDP* reported that the newly-formed Clifton Association was discussing a comprehensive campaign to maintain and improve Clifton, with the object of:

restoring to Clifton the prosperity enjoyed during the latter half of the eighteenth century
Instructing, amusing, and pleasing residents in Clifton
Attracting visitors
Stimulating the circulation of money among all classes
It was further suggested that persons to be catered for should be classified in groups viz:
Literary and scientific
Music and dramatic
Sporting
Educational
Health seekers

Clifton, they declared, deserved a degree of individuality, and was not to be dominated by the city of Bristol. They needed another hotel in Clifton, since the Clifton Down Hotel had been requisitioned by the Ministry of Health during the War. It was now the Pensions Headquarters. They had acquiesced in being overshadowed by the remainder of the city, and yet they had paid about a third of the total rates. It was time, they declared, that the beauties and advantages of Clifton were more widely known. Doctor Bartholomew (the Turkish Bath owner) suggested 'that there was a great possibility in the reviving of the Clifton Spa.'

On 21 June 1921, the cost of admission to the cinema was reduced by half. Ground-floor seats were now 1s, balcony 1s 6d. In competition with the Victoria Rooms and the new Whiteladies Cinema, it was struggling after only 21 months.

On 29 September 1921, Frederick Price surrendered the lease of nos. 10 and 13 Princes Buildings according to the Merchant Venturers records, and took out another lease on 10 Princes Buildings for 40 years.

On 8 November 1921, Sidney Jones, a well-known conductor and composer, was employed (after advertising for a hotel band in September 1921).

On 8 February 1922, the announcement that the Clifton Spa Hotel was to be offered for sale in the summer was the subject of interesting discussions, especially as regards the age of other local licensed premises. The Grand Spa was to be offered together with other properties. The Grand Pump Room which was famous for its twenty massive Saillon Cipollino marble columns supporting the roof, and now converted into the popular Spa Cinema, would be offered for sale at the same time.

9.16 Failure of the cinema and licensing battles
On 7 March 1922, at the annual licensing session, Mr Price applied for a dancing licence for the Spa. He was asked if he had given up the cinema. Mr Price confirmed that the cinema was closed, despite having had good press coverage over the short time it was open. He 'did not intend to use both licences. He thought with a dancing licence he would be in a better position to negotiate. The justices intimated that the application would be granted when the cinema licence was given up.'

In May it is announced that the Hotel would be sold by auction on 6 July 1922 by Mr and Mrs Price, who would retire. It consisted of two dining rooms, lounge, two smoking rooms, card room, writing room, double drawing room, private sitting rooms, billiard room, 70 bedrooms, well-appointed bathrooms, adequate domestic offices. There was electric light, central heating, service and passenger lifts, terraced gardens, Tuffleigh House, the Hotel Garage. Included in the sale was the Grand Pump Room now converted into the Spa Cinema, fitted with seating accommodation for 350. Also included, the leasehold reversion of the

> Clifton Rocks Railway, held under a sub lease for 872 years unexpired, at the yearly rent of £50 payable to the vendors. The whole extends to 1½ acres and held on leases. Total ground rent £205 10s per annum. To be sold as a going concern. Furniture and Equipment taken at valuation.

The site was not sold, the Prices did not retire, but threw themselves into organising dances in the Pump Room. On 2 December 1923, Mr Price applied for an extension of his licence to sell intoxicating liquor at the Clifton Spa and Spa Hydro on four occasions until 12:30am. The application was strongly opposed by the police who challenged his right to sell alcohol at all. The extension was in respect to 4 supper dances to be held between December 1923 to February 1924. It was confirmed that in 1913 a licence had been granted. In law, if a licence was granted to a certain building, that licence applied to every part of it. A condition was there should be no bar or counter sales. In 1918 the Pump Room had fallen into disuse, was converted into a cinema, and the dancing licence cut out. In 1922, he surrendered his cinema licence in order to use the room for dances. He was quite entitled to sell drinks up to 10:00pm to guests at the dances. He realised he could not allow bar or casual sales or serve casual visitors. In February 1924, the licence was renewed under the same conditions. The police opposed the application on the ground the conditions had not been complied with. The dances were publicly advertised and the extensions were for the purpose of supplying drinks to the general public who came in. The drinks were being sold from a buffet on wheels. Conditions of the licence were not broken by a sale to members of a society to whom the licensee had let the room for dances, but was for functions organised by Mr Price and advertised to the public.

On 17 January 1925, Mr Price applied for an extension to the licence for selling intoxicating liquor at the Clifton Spa Hotel until 2.00a.m. on the occasion of a pantomime artists' ball. Mr Price explained that the ball was confined to artists in the two Bristol pantomimes and did not start until 11:00pm. By reason of the evening performance the artists would be unable to obtain refreshments between six and eleven in the evening. The police opposed it on the grounds the extension was too late. He thought 1.00am quite late enough. The chairman of the Bench set the extension to 1:30am.

On 7 April 1925, Mr Price made an application to extend the music licence until 1:30am on various dates in April, and to sell intoxicants until 11:15pm and 2:30am on other dates. The justice stated they were not in sympathy for granting licences for music, singing and dancing for any time after 12:30am. The police stated the conduct of attendees after entertainments had been satisfactory. Mr Price said the bookings had already been made and tickets issued. The Justice announced that the applications in respect of music would be granted, the applicants for intoxicants refused. Next time the Justices would take into consideration the lateness of the proposed hour.

On 15 April 1925, the following week, Mr Price applied for a dance on 28 April until 2:30am. It was for a subscription ball for the Bristol Day and Night Nursery for which the Duchess of Beaufort was president. Tickets had been on sale for weeks. The police objected on the grounds of the previous weeks that 12:30 was sufficiently late. One could not make applications for extensions more than 14 days ahead of the event. Audiences and dancers were well behaved. In view of the facts of the case, the large numbers of tickets sold and the application could not have been made earlier, the application was granted until 2:00am.

On 5 May 1925, concern was shown by magistrates that the number of applications for music and dancing licences ending after 12:30am was increasing as were licences for intoxicants. If they came from the propri-

9.30. The grand ballroom or Pump Room. The fountain has gone to make space for the band to play in, instead of in the orchestra area at the back of the room, which would have been highly modified when it was converted into a projectionist's room. Compare with plate 9.17. (Peter Davey collection)

etor promoting entertainment for personal profit the limit of 12:30am was desirable in the public interest. Raising funds for charitable purpose removed the motive for personal profit: there was no tendency of the number of applications to become excessive.

Mr Price had applied for a Derby Day dance extension from 11:00pm until 1:30am. This extension was granted until 12:30am and, on the following day, until 11:45pm. The Chairman remarked that Mr Price was often before them and hoped the morning's expression of opinion would be noted by him.

On 23 July 1926, plans for extensions of the Grand Spa Hotel premises were in part approved in part deferred for legal advice by the Merchant Venturers. On 24 September 1926, plans for increased dining room accommodation at Clifton Grand Spa Hotel were provisionally approved. Hugh Frossard's well-known band was the resident orchestra from 1927-29.

9.17 Sale of the Hotel and Railway again
On 25 February 1928, Mr Price sold the Grand Spa to the Grand Hotel Company of Broad Street. The *WDP* reported:

> During the ten years Mr Price owned the property he has changed the character of the business from that of a Spa and Hydro to a first class residential and tourist hotel, and the results of his efforts in regard to the Regent ballroom are well known. No doubt a large circle of friends will learn of his departure with regret.

Presumably the dances and concerts continued.

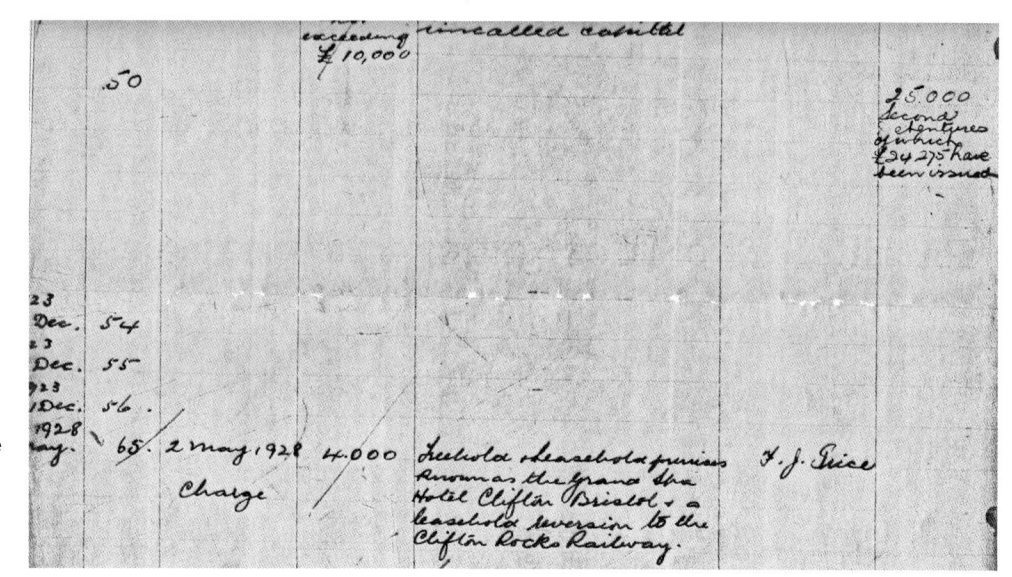

9.31. Mortgage details of the Hotel, including one for Clifton Rocks Railway in May 1928, were given in my garage deeds. (Author)

9.18 War breaks out
In September 1939, the Hotel, Pump Room, Tuffleigh House and my garage were requisitioned by Imperial Airways for use as Barrage Balloon Headquarters.

The Hotel and Pump Room reopened in 1948, by which time the hotel was known for its dances with all the top bandleaders playing there.

9.19 Attempts at revitalising the Spa
Money was spent at each stage, but you will see that the Spa facilities failed despite all the latest gimmicks, and huge amounts thrown at buying up properties, not helped by the cost of leases increasing, and particularly having to pay a premium for getting an intoxicants licence between 1906 and 1913. By 1908, George Newnes' businesses were failing and this affected his health. He would have seen the Railway fail, and the large amounts of money thrown at the Spa. He had tried to jump on the bandwagons of Victorian fashions – cliff railways, spas, Turkish baths, but he had to contend with the Merchant Venturers, Clifton residents and businesses, and the fact that Clifton was a suburb, albeit in a beautiful location, rather than being in the centre of Bristol.

The subsequent cinema failed too, but once the music and dance licences and the intoxicants licence had been resolved, the Pump Room was a very successful location, especially for charity concerts and dances.

Incidentally Bristol's only Turkish baths remaining – at the Hydro Hotel, College Green, which had been open to the public since 1860 – closed down in October 1948.

The current dilapidated state of the Pump Room was due to planning issues in the 1970s and 1980s which you will read about in chapter 14. The Pump Room is still a beautiful building and Grade 2 listed. Unfortunately, it has been on the 'at risk' register for far too long. Many people remember dancing there, myself included. The Hotel still thrives, and has the best view in Bristol.

ten

New Uses for the Tunnel
Wartime Conversion

In the 1930s, Bristol was a highly strategic location in the South West. It was an important administrative centre, it had a huge harbour complex and the associated industries, it had the largest concentration of aircraft and other manufacturing plants, and good transport links by river and rail. It was the seventh largest conurbation in the United Kingdom with 415,000 people residing within the city boundaries.

To begin with it was not believed that German aircraft could get as far as Bristol, and many government offices were relocated to Bristol as a safe area. But once the *Luftwaffe* had taken the French airfields, then Bristol was a good target within reach. It was easy for bombers to reach Bristol from the French coast. Bombers trying to reach Birmingham would fly up the Bristol Channel. If they could not reach Birmingham, they would drop bombs on Bristol on the way home. Bristol ended up being the fifth heaviest bombed city in the country with only London, Liverpool, Birmingham and Plymouth receiving more attention.

This chapter introduces the war-time role of Clifton Rocks Railway: as a barrage balloon workshop and offices, air-raid shelters, and home to the BBC Home Service. Later chapters will deal with operation of the BBC, the shelters and Imperial Airways (later British Overseas Airways Corporation).

10.1 Ownership of Clifton Rocks Railway
Since there was joint ownership of the Railway and a covenant to maintain it, it was difficult to convert for wartime use. It also made life difficult that so many groups wanted to use the tunnel during the war.

In March 1937 Bristol Tramways, as lease-holder, ascertained that it was their responsibility to maintain the tunnel and property. It was not until July 1941 that the surrender of the whole to Bristol Corporation was agreed and eventually enacted on 1 July 1942. By agreement the Merchants and the Hotel (as head lessors) had the benefit of covenants by which BTCC must maintain a railway – the dreaded 'restoration clause'. Much of the correspondence, particularly with the BBC (held in the BBC archives at Caversham)

10.1. The plan shows all the brick structures that have been built on top of the railway lines. Shelter 1 was for Air Ministry staff. Shelters 2 and 3 were for residents to shelter at night. Visitors to the shelters and the barrage balloon section would enter from the top, visitors to the BBC section from the bottom. The many walls demonstrate just how hard it would be to get the Railway running again. One now has to consider which story is the most important for each section.

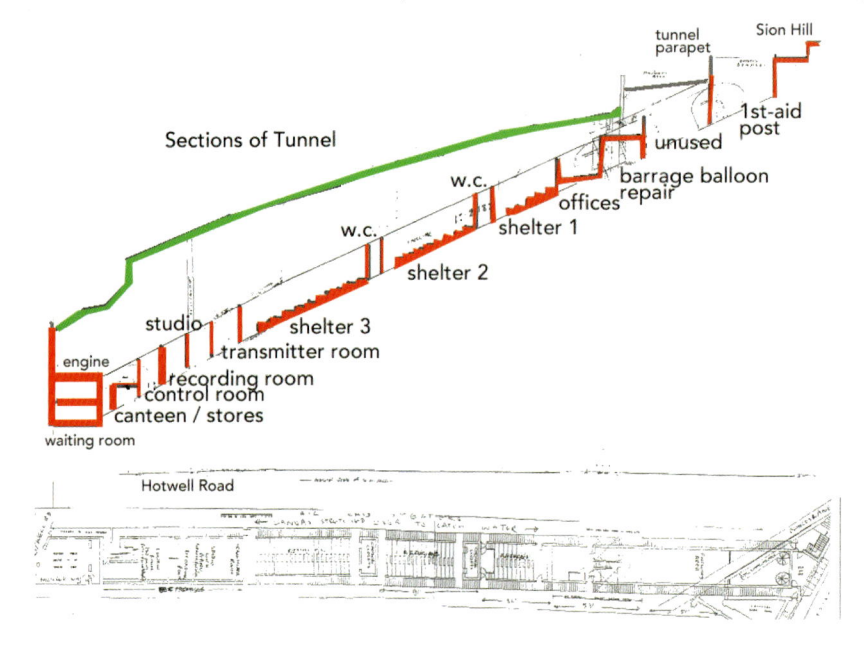

Merchant Venturers	Free holders 1686-	Charge BTCC 5 shillings per year for lease. Have benefit of covenants under which BTCC must maintain a railway	Merchants Hall, Clifton Down
Grand Spa Hotel	Head lessor 1922? -44	Charge BTCC £50 per year ground rent. Have benefit of covenants under which BTCC must maintain a railway	Princes Buildings
BTCC	main leaseholder 1912-42	Must maintain covenant to maintain a railway. Undertaking required to convert back to its former use. Alterations to the tunnel were subject to the approval of the BTCC Surveyor.	(E.G Kingston Secretary, J.F Heaton Director) St Augustines Place, Tramways Centre
Bristol Corporation	Relates to shelter in bottom station 1921-46	Purchase of ground rent agreed for £1500 from Hotel on 1 April 1941. Releases covenant. Corporation to grant a lease for 21 years at a nominal rent and to pay £400 towards it. Must keep bottom station entrance in good repair.	Town Clerk (Josiah Green), The Central Library, College Green.
Ministry of Works	hold sub-lease for 2 large shelters in tunnel.	£100 per year. Contributed £800 to lease granted to Bristol Corporation. Must keep top station entrance in good repair. 1942-46	10 Woodland Rd (Mr Simmonds).
BBC	hold sub-lease for rooms at the bottom of tunnel. 1941-1960	21 year lease. 1s 0d rent to 1950, Contributed £400 29 July 1941 when lease granted to Bristol Corporation. £10 rent after 1950 + water + rates + tax. Must keep bottom station entrance in good repair.	Broadcasting House, London; 7 College Fields, Clifton; Engineering Dept, Whiteladies Rd, Clifton.
BOAC/ Imperial Airways	hold sub-lease for balloon offices and small shelter	Contributed to lease granted to Bristol Corporation.	Grand Spa Hotel.

Table 10.1. Summary of ownership.

deals with how to avoid having to restore the tunnel to a functioning railway. All the structures breached the Merchant Venturers covenant requiring the tunnel to be kept fit for use as a railway tunnel. But in war-time, there is limited time, and one has to weigh up peoples' lives against not being able to use a railway tunnel with a poor transport record for running the railway again. The table above summarises the situation.

10.2 Air-raid Precautions in Bristol

Air-raid Precautions (ARP) wartime regional headquarters were based in 19 and 21 Woodlands Road. The following timeline shows that there was slowness to begin with, and then things moved very quickly from February 1940. The following dates put into context when the ARP started to be prepared, and when bombing occurred.

On 28 February 1938, thousands of volunteers were required, to be prepared for emergency work: 5,000 wardens, 2,000 auxiliary firemen and 1,000 special constable duties. They applied to the Central Police Station, Bridewell. On 2 March 1938, in the *Western Daily Press*, there was a warning that householders in all parts of the city should 'read, mark, learn and inwardly digest' the report of the ARP Committee published in the newspaper, and which would be debated in a week's time. The Government threw great responsibilities upon the local authorities and its citizens to protect life.

On 27 September 1938, in the *Western Daily Press*, the latest ARP news reported that there are 400,000 gas masks available for the public to collect, and shelter trenches were to be constructed in parks to hold up to 50 people (the maximum number to be held in any one shelter, to avert mass slaughter).

On 9 January 1939, gas masks were distributed at Hotwells National School. On 11 November 1940, concern was shown about the policy with respect to free shelters. They could not be provided out of rates. Direction with respect to lavatory accommodation was needed. Also, there was concern about whether to isolate people suffering from various complaints, from healthy people. On 24 December there was a discussion in the *Western Daily Press* about how any type of underground shelter was better than a surface shelter, though only offering better protection if sufficiently deep, and soundly constructed – of which there were few. The preference for this type of equipment had led to a considerable number of people overcrowding in church crypts, basements and one or two other refuges (an oblique reference to the Portway[1] and Clifton Rocks Railway – available from September 1940). It was the policy of the Corporation to aim at dispersal rather than grouping in large numbers. It was not intended to equip public shelters for sleeping purposes. Steps were being made to make shelters drier. They were not meant to be used for occupation throughout the night – just daytime use. Large communal shelters had a number, and a list of households allocated. They were fitted with electric light, sanitation and bunks for sleeping. In every shelter there should be at least two voluntary marshalls.

On 15 March 1941, the *Western Daily Press* reassured its readers that atmospheric conditions were measured between midnight and five or six in the morning, by the health authorities in nocturnal shelters to preserve the city's health and to check for bacteria. There had been no epidemics, but one person I spoke to, who had sheltered in the Portway tunnel, had suffered from impetigo (a skin complaint). He remembered wheals over his body which itched. See chapter 12.5.

10.3 Ministry of Works

In 1938 it had been decided that the country should be split into 12 regions in the event of breakdown of communications with central government. Each regional office held a headquarters comprising a senior regional officer, a treasury officer, regional officers of the ARP, a Ministry of Health general inspector, regional police staff officer, regional fire officer, city engineer (Mr Webb), liaison officers from the ministries (hence BOAC dealing with the Air Ministry and BBC dealing with the Works Ministry) and departments concerned. The war room was located in the cellars of nos. 19 and 21 Woodland Road. Day-to-day business remained in the hands of the local authority. The director of Lands and Accommodation was based at the Office and Works, 10 Woodland Road; the director of Premises and Stores was based at 46 Portland Place.

To implement the ARP, Bristol was divided into six civil defence divisions which included the Clifton division (Hotwells, St Augustines, Clifton, Westbury-on-Trym, Henbury, Southmead, Stoke Bishop, Henleaze, Redland, Cotham and Kingsdown).

10.4 Bombs in Clifton and Hotwells[2]

On 19/20 June 1940, Bristol Aeroplane Company at Filton, as well as the docks at Avonmouth and Southampton, were targeted but bombs fell only on the beach at Portishead. Filton and Avonmouth were the most strategic targets. In July 1940, air raids targeted Filton. *Bristol at War*[3] has photographs of bombings in Clifton. The following dates of bombing have been gleaned from peoples' memories and several books.

24 November 1940 was the start of Bristol blitz in earnest. It had concentrated on the central area, with further damage occurring in Clifton, but greatest destruction took place in the heart of the city from Broad Quay to Old Market. Exceedingly large calibre bombs were reported as having fallen at Eastville, Speedwell, Temple and Totterdown, while for the greater part of the night the city was blazing furiously and many well-known buildings were totally destroyed and others gravely damaged – such as the University Great Hall and St Andrew's Church in Clifton. Harley Place was hit and incendiaries fell on Lansdown Place and Richmond Terrace. A bomb fell on Clifton College Prep School's New Field,[4] narrowly missing two shelters. Originally believed to be an error by the *Luftwaffe*, in later years it became clear that the College, a source of officers and soldiers as well as having a boarding house for Jewish boys, was very much a target.

Further raids were as follows:

2 December 1940, Fishponds, St Pauls, Bedminster, Redland, Shirehampton and Clifton including nos. 30-36 Cornwallis Crescent, Dowry Parade, 5 All Saints' Church in Pembroke Road, Fosseway, back of no. 4 Victoria Square, nos. 3-5 Victoria Square (6 people killed) and Mortimer Road; incendiary at no. 13 Lansdown Place, and Harley Place hit again. Residents of Victoria Square took drinking water in buckets from Oakfield Road. Parachute flares were used to light up the area.

3 December 1940, Clifton College on School House Close, School House garden, and Pre-School Hall set on fire. Squash courts hit and Wiseman's House and Polack's House. Boys were sent home early and most of the school evacuated to Butcombe and Bude. All Saints' church in Pembroke Road gutted.

6 December 1940, Central, Knowle, St George and nos. 7-12, 12a-14, 15-23, Royal York Crescent[1] in Clifton. Incendiary bombs drop on the Suspension Bridge and Tuffleigh House garden.

3 January 1941, Bedminster, St Philip's, Hotwells and

Cotham, 24 York Place in Clifton, Temple Meads Station and the City Docks sustaining damage.

16 March 1941, main attack roughly east to west of a line from Stapleton Road Station, through the city centre to Clifton Down Station. Areas most affected were city centre, Fishponds, Eastville, Whitehall, Easton, StPauls, Montpelier, Kingsdown, Cotham, Redland and Clifton. Holy Trinity Church in Hotwells bombed.

29 March 1941, Floating Harbour.

11 April 1941, The Good Friday Raid on Bristol. In this attack, three high explosive bombs fell in a line from Hill View (off Constitution Hill) hitting Bellevue Crescent and the site of St Peter's Church and Hall at the junction of Jacob's Wells Road and Hotwell Road. Clifton Park Road blocked. Canons Marsh gasometers.

6 May 1941, unexploded bomb near the Portway tunnel.

8 May 1941, incendiary bombs drop on Sion Hill (no. 29 was demolished, and was only redeveloped in 2009), Cordeaux in Regent Street, no. 1 Princess Victoria Street, Wellington Terrace (people killed). All very close to Clifton Rocks Railway.

10 May 1941, bomb damage occurred in the Clifton, Westbury, Sea Mills and St Anne's areas.

2 July 1942, Floating Harbour.

Raids went on until 15 May 1944.

10.5 Imperial Airways and ARP seek use of the tunnel. BBC departments move to Bristol

On 1 September 1939, the variety, religious and music programme departments of the BBC were evacuated from London to Bristol along with listener research and some other administrative departments. Several halls around Bristol were found to help accommodate them. Those most exposed to the bombing were Sir Adrian Boult's Symphony Orchestra involving nearly a hundred personnel. At first Gerald Daly (engineer-in-chief of the BBC in Bristol) had rented the Clifton Spa Hotel and its ballroom to accommodate them, but to their horror, and just before they arrived, Imperial Airways requisitioned the whole hotel. The BBC had no powers of requisition, not being under the aegis of the government.[6] They had great difficulty in finding alternative accommodation for such a large orchestra but eventually the Co-op Wholesale came to their rescue (Gerald Daly joined the Co-op forthwith in gratitude) and rented them their large hall and offices in the Centre.

It is not known exactly when Imperial Airways took over the Spa Hotel for their headquarters, but barrage balloons started to be deployed at Clifton on this date (see chapter 11), and we know that the BBC were not able to rent the Hotel on 1 September 1939.

On 7 September 1939, BTCC wrote to the Merchant Venturers that the ARP officer of the Air Ministry intimating that Clifton Rocks Railway might be requisitioned for the purpose of an air-raid shelter. BTCC thought the matter might be dealt with by agreement. BTCC wished to obtain consent to the removal of the present restriction of use that it could only be used for the purposes of a railway tunnel. The Hotel and Bristol Tramways raised no objection provided their interest was protected. The Merchant Venturers replied on the same day to the BTCC secretary to say that, in view of the state of emergency, the treasurer was prepared to enable the tunnel to be used as an air-raid shelter on condition they undertook to take such steps at the end of the period of emergency to enable the *tunnel to be converted to its former user if desired,* and to indemnify the Society against any claims that might arise against them by reason of variation of use.

On 12 September 1939, BTCC prepared to rent out the tunnel at £100 per year. to Imperial Airways. They were told by the Merchant Venturers that Imperial Airways were pretty certain to go ahead. Imperial Airways wanted to adapt the tunnel as an air-raid shelter for their staff occupying the Spa Hotel. *An undertaking is required to convert back to its former use* at the end of the war. Alterations to the tunnel were to be subject to approval of the BTCC surveyor.

On 15 September 1939, Imperial Airways was asked to return the keys to Bristol Tramways. On 30 October 1939, BTCC wanted to know if ARP still want to proceed. On 31 October, Imperial Airways confirmed to BTCC that they were waiting for a decision from the Office of Works and for engineers to carry out a survey. On 19 February 1940, the Office of Works declared they were willing to pay up to £100 a year in rent. On 26 February 1940, BTCC confirmed they were prepared to rent the railway provided *that the premises restored to their original condition at the end of the tenancy* by the Office of Works. The lessors (Merchant Venturers) required an indemnity for temporary change of use. At present two cars were at the top and two at bottom so these might need rearranging. On 7 March 1940, the Ministry of Works was prepared to accept the conditions to use it as an air-raid shelter. Provision of ARP refuge was urgent. It was preferable that the four cars be placed at bottom of the tunnel at once. On 25 March 1940, the Ministry (for Imperial Airways) arranged an annual tenancy at £100 per annum to convert to an air-raid shelter for staff. This was for the small top shelter (number 1). It had nine ledges to hold 72 people seated. On 29 March 1940, a tenancy was agreed with the Director of Lands at 10 Woodland Road at £100 per annum excluding rates, taxes except property tax, land tax, tithe rent charge. This was for the two large 22-ledge shelters, each to hold 176 seated. Shelter 2 for Clifton residents, and Shelter 3 for Hotwells residents.

On 1 April 1940, there was a BTCC survey report, when part of the Railway was taken over by Office of Works, which gave a statement of condition before work starts:

> *top* is externally stone, frost-weathered; gates work; railings good; ornamental sign fair over Sion Hill entrance but sign with lamp rusty on north entrance; hand rail rotten; asphalt bad; glass windows broken; top internally – turnstiles good but counter missing; dilapidated fuse box and conduit; cabin rusty; glazed screen across tunnel rusty with 10 panes glass broken.
> *bottom* ground floor – sliding doors working; glazed office doors fair; roller shutters rusty; glazed door upstairs OK; timber covers surrounding vertical pipes fair (now missing); ornamental iron queue barrier had 3 broken panels; one gas light and five electrical light fittings fair; portable oak right angle seat fair. Pump house bottom first floor badly chipped concrete floor; wall plaster poor; corrugated iron ceiling lining fair; sink poor; WC (water closet) pan and seat poor. Pumps in fair condition; wells greased and preserved; but electric motors missing.
> *Tunnel* sagging brick lining; four tracks of flat bottom rail very rusty and scaling as were fishplates and rail clips. Timber platform at bottom badly worn; protective barriers poor; hand operated water pump working condition; electric switch and fuse boxes poor.
> *Cars* poor condition – timber bodywork rotting; cushions and seat backs mildewed; 3 panes glass broken.
> *Decorations* of bottom station fair; top station poor.

This gave an insight as to what was present, and what condition everything was in. Clearly much was to be removed – it all looks to be a sorry picture when it had only been abandoned in 1934.

On 2 April 1940, BTTC found it harder to rearrange the cars than expected. They were not charging more than the bare costs. The invoice to the Office of Works was adjusted. On 4 April, the Office of Works was surprised that BTCC were charging them to remove the two cars from the top station. They took the view that this work was necessary before the property could be let for conversion into an air-raid shelter. The cost was more than expected so it was not equitable to ask BTCC to carry the whole of it, but not fair to expect the department should meet the entire

bill. They offered to pay £26-10-0 to cover winch, material and half the labour cost, so BTCC were recompensed accordingly.

On 10 April 1940, the Office of Works thanked BTCC for indemnifying them against any claims that might arise as an air-raid shelter. They also wished to confirm fire insurance would be undertaken by the Government, noting no reference was made during negotiations for the tenancy with BTCC. They imagined that any insurance policy currently refered only to cars and fittings – of no interest to the Ministry. There was reference to the statement of condition stating that the brick lining of tunnel should be confirmed when scaffolding reaches the point where it sags and examination be made. They noted the reservations on the brick lining to BTCC and ask for the Board's surveyor to get in touch. Time was ticking on. Look back at section 10.4 to see when the bombs started to fall.

10.6 Plans for the top station and the tunnel

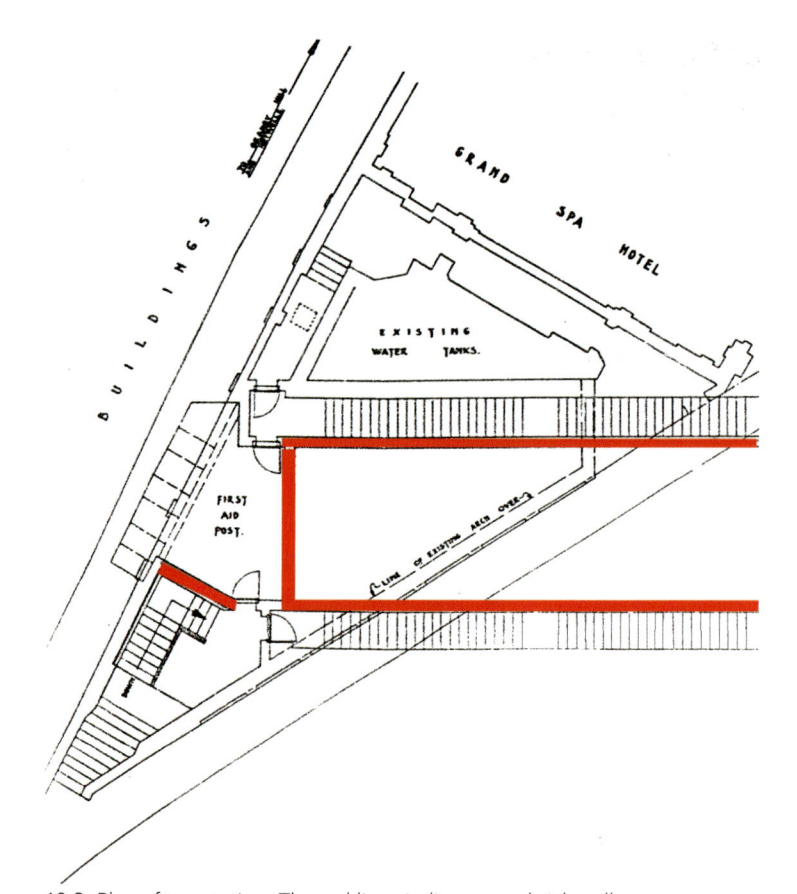

10.2. Plan of top station. The red lines indicate new brick walls. Visitors can no longer enter the top station where they used to when the cars were still running. There is now a short flight of stairs to the right with a new doorway. Steps are built on top of the outside railway tracks so now only six tracks can be seen rather than eight. The west wall of the top station has been built on top of the cable wheels.

10.6. Chemical closets (bins) plan for Shelters 2 and 3. Eight 'bins' for 176 people. Paul Adams mentions them chapter 12. (BRO)

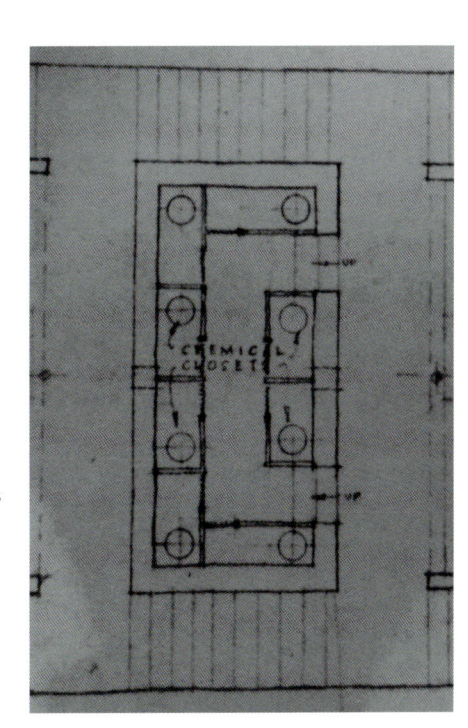

10.3. Above, the first-aid post wall built where you would have originally entered the top station. There are pennant stone steps until you reach the wall, and then concrete steps to the right. The first aid post has a 6in concrete roof, and 14in-thick walls.

10.4. Imperial Airways' first plan. The stairs on the left-hand side are the stairs shown in plate 10.3. The first section is where one can see the rails and is not in the tunnel. The second section is where the turnstiles came out, since you are too close under Princes Lane to have enough protection from bombs, hence the comment that the sections are isolated. The outside edges of the stairs are the tunnel walls, the inside edges are 14in thick walls.
(BRO 42054.G.Drawer 2/08153)

10.5. Reference to Imperial Airways rather than BOAC so plans drawn up before 1 April 1940. Note that originally there were just three shelter areas, and a bigger blast absorption area. The smallest ledge area on the left is for staff, the next for Clifton residents, the lowest for Hotwells residents. There are no offices to begin with to the left of the staff shelter. The two largest shelters have chemical closets, the staff shelter gets two chemical closets later. No plans for the bottom. (BRO 42054.G.Drawer 2/08208)

10.7. Later plan of 1941 to show the BBC taking the bottom quarter of the Railway, and the barrage balloon area being extended. Two water closets added. There was a door to get into office area. Reference to Air Ministry rather than Imperial Airways. (BRO42054.G.Drawer 2/08208)

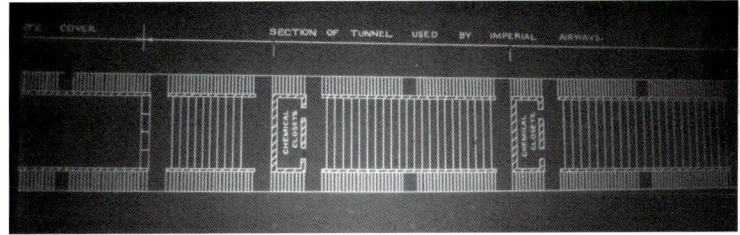

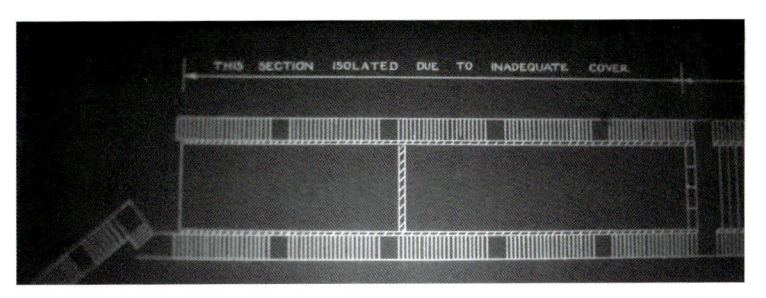

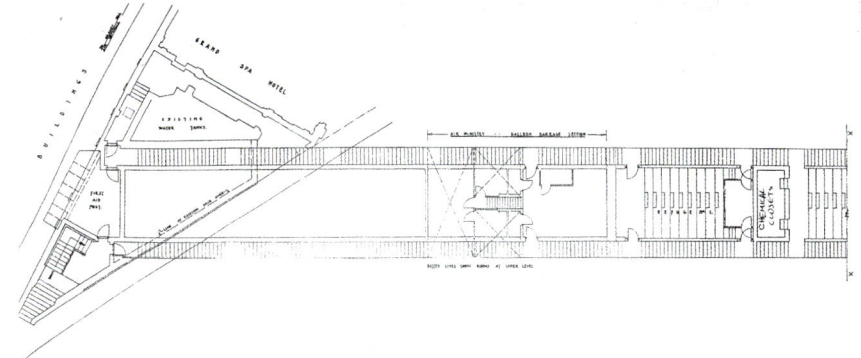

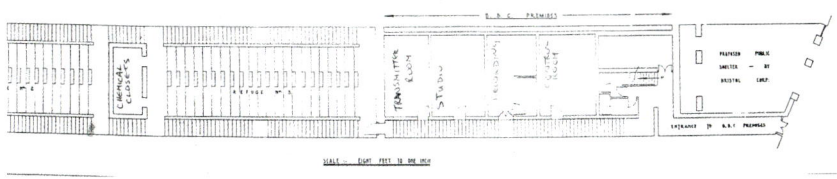

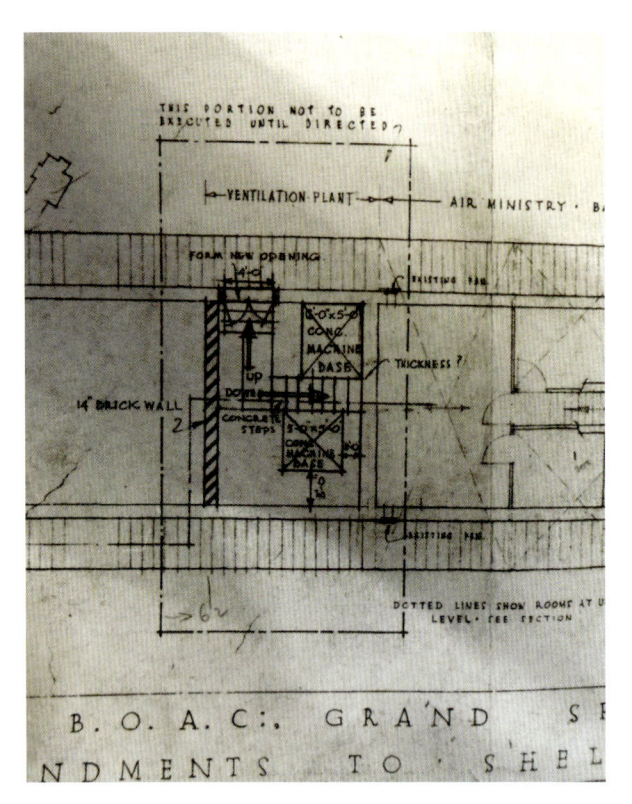

10.8. 30 Aug 1941 BOAC draw up a later plan to add ventilation – never implemented. There is a large hole in each wall down the stairs with a fan. (BRO)

We are lucky that so many plans have survived and are sitting in the Records Office. Air-raid regulations are to be found in chapter 12. Before the BBC moved in, people using Shelter 3 would enter at the bottom and walk through the cars (see chapter 12.4). Look at plate 10.18 from February 1941. It shows the four cars at the bottom, the track about to be cut. The bottom station partition is to be seen in the background. After the BBC moved in, Shelter 3 was only accessed from the top.

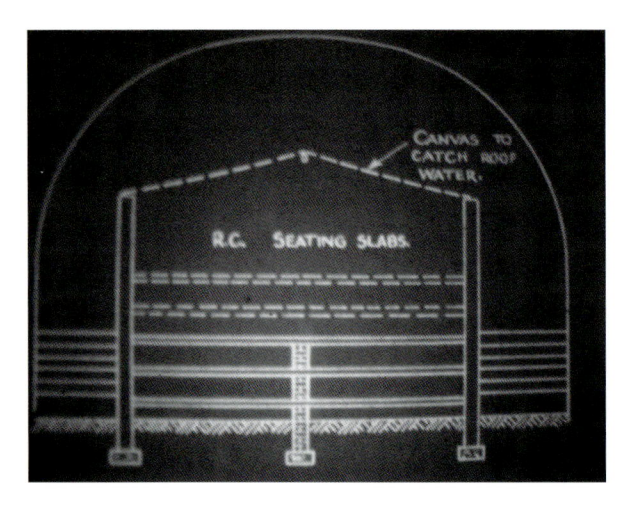

10.9. Canvas over shelters with wooden guttering to catch water caused by condensation or surface water.
(BRO 42054.G.Drawer 2/08153)

10.10 Canvas still in position in the 1960s.

10.11. Trough still in place, with drainpipe to take water away.

10.7 Plans for the bottom part of the tunnel

These are invaluable since they show what was there before.

In May 1940, bomb-proof record rooms were set up in Tunnel One of the Port and Pier Railway (see chapter 5). Tunnel One is 73 yards long and terminates at Bridge Valley Road. Some items were not returned until 1958. On 27 April 1940, the Ministry of Works confirmed Government departments always ask about maintenance and fire insurance with tenancies.

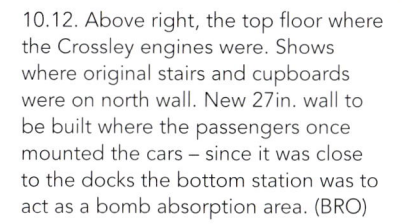

10.12. Above right, the top floor where the Crossley engines were. Shows where original stairs and cupboards were on north wall. New 27in. wall to be built where the passengers once mounted the cars – since it was close to the docks the bottom station was to act as a bomb absorption area. (BRO)

10.13. Bottom station plan ground floor, 10 April 1941, showing where the four cars are stored (where the generator room for the BBC will be built). Bristol Corporation want to turn bottom station into extra shelter for St Vincent's Parade residents. The steel emergency exit door still survives (plate 10.14). Water closets were added. (BRO 42054.G.Drawer 2/08153)

10.14. The steel emergency door measuring 2ft by 2ft and is the only way into the bottom station now, from the BBC generator room. The photograph is taken from the waiting room side. The wall is 2ft thick. Visitors have great fun in getting through by various techniques depending on how tall they are, or how big they are. It is not possible to cut another hole in the wall for security reasons, and to help keep the bottom station dry. It is also an original feature so cannot be enlarged.

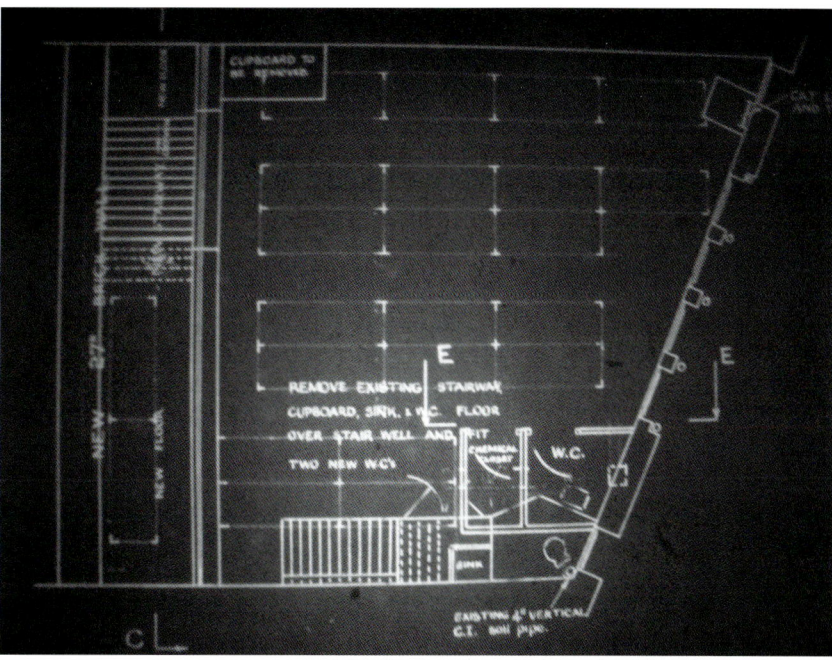

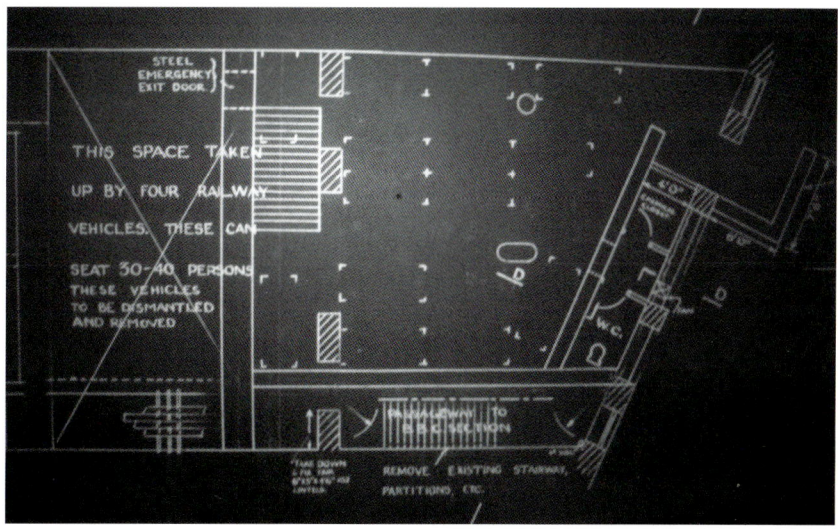

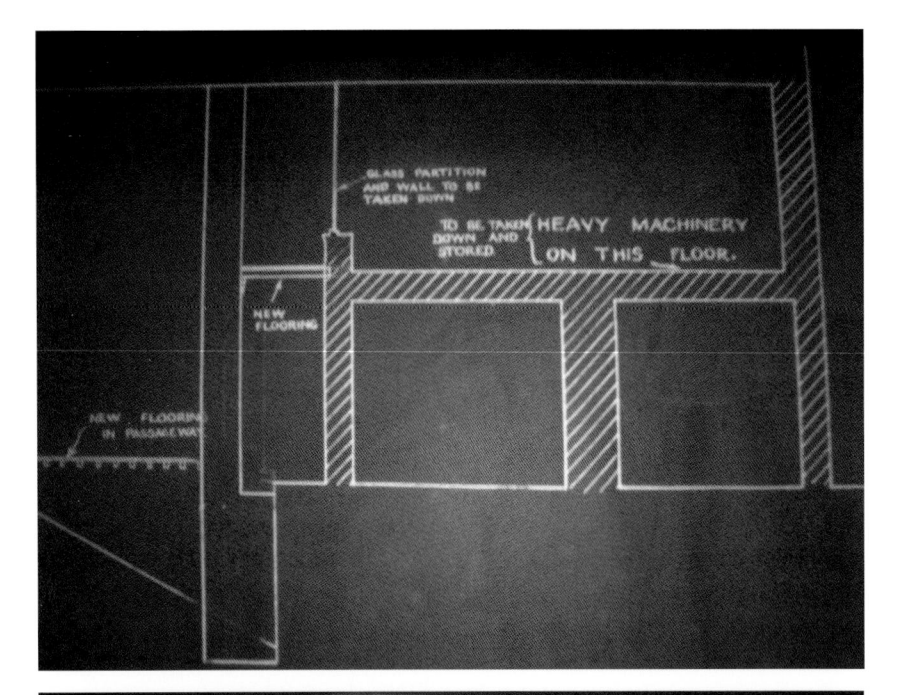

10.15. It must have been difficult smashing the Crossley engines and getting them out. The flywheel and crank would have been removed first. They must have been dropped down after the glass partition was taken away and before the front façade modified. Machinery was removed in April 1941. (BRO 42054.G.Drawer 2/08153)

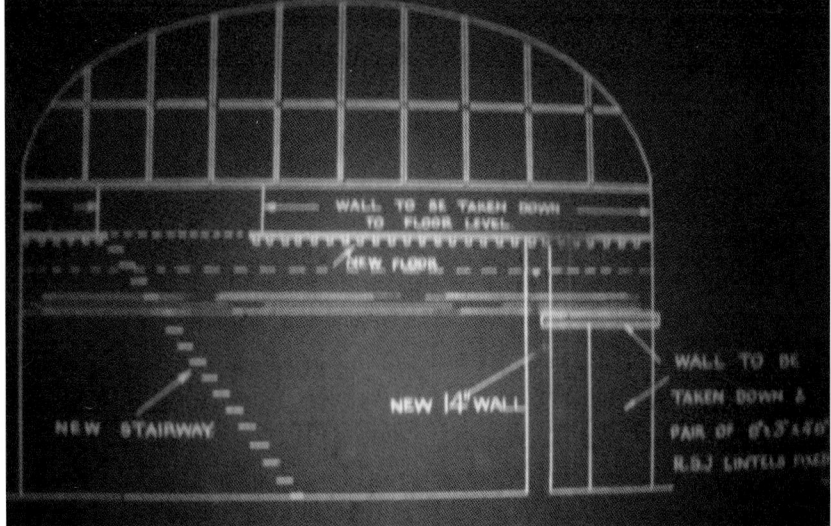

10.16. Interesting that the floor was made stronger.
(BRO 42054.G.Drawer 2/08153)

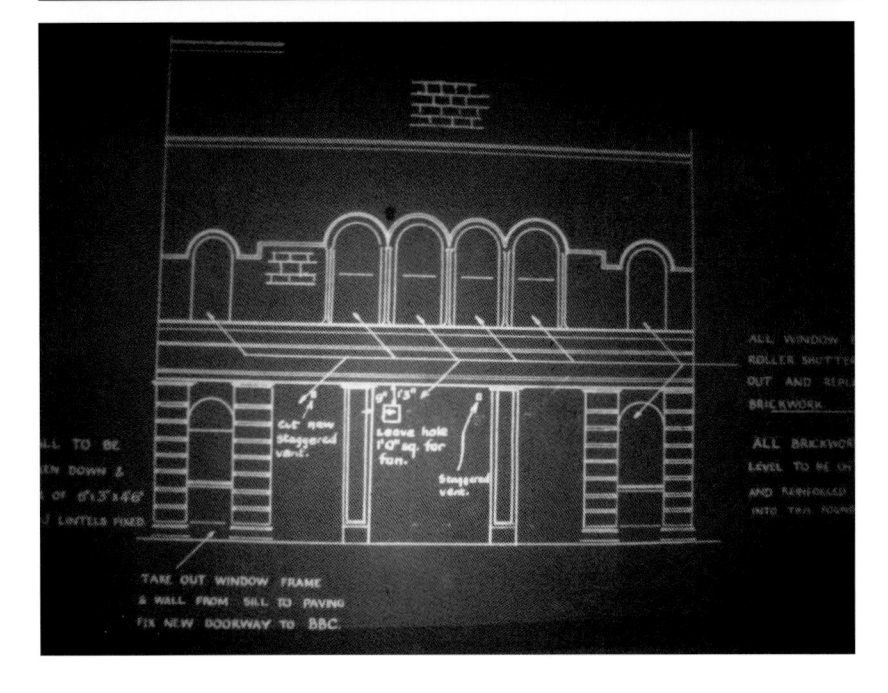

10.17. All window frames and roller shutters to be taken out and replaced by 14in brickwork for extra bomb protection. All brickwork built from ground level to be on foundation 9in concrete and reinforced with rods bonded into this foundation. (BRO 42054.G.Drawer 2/08153)

10.18. 4 May 1940: the tunnel being converted into shelters. Only six men shown working. The pre-cast stairs have yet to be laid on the right-hand side as the railway line is still showing. There is no electricity at the time of conversion. Hurricane lights, candles and torches were used during the war. (A domestic supply was laid down the stairs of the tunnel later.) There is still a huge amount of rubble under the ledges, particularly in Shelter 2, far more than expected. It is possible that some rubble had been put there before the conversion was started. (*Western Daily Press*)

AN AIR RAID SHELTER is being made out of the tunnel through which the old Clifton Rocks Railway used to run between Hotwells and Clifton, Bristol.

10.8 Clifton Rocks Railway conversion begins

On 28 August 1940, the Ministry of works wished to discuss the schedule of condition, alterations and further works with BTCC.

On 3 September 1940, conversion work for the Barrage Balloon Squadron and the ARP was completed. There was now a first-aid post, three shelters, stairs to the shelters, offices and workshop for the Squadron. It is amazing that such complicated works could have been achieved in so short a period of time – between May and September – but in June 1940, bombs had already dropped on the beach in Portishead; the big bombing raids were in November; and a bomb dropped above the tunnel in December – so it was all just in time.

On 7 September 1940, the Office of Works drew attention to the poor condition of the skylight at the top of the Railway. Apparently the Shelter for 140 people was provided at a cost of £1,700. The Ministry of Works' surveyor agreed to replace the skylight on the top station by a simpler structure.

On 12 September 1940, there was confirmation by BTCC to Imperial Airways that BTCC wanted to rent out the Railway at £100 a year – any alterations subject to approval of the BTCC surveyor.

10.19. Having removed the cars, there was no method to get up and down the tunnel so stairs were built on top of the railway lines. Cables can sometimes be seen under the stairs.

10.20. Date on top station roof showing when works completed. It is amazing that such complicated works could have been achieved in only 4 months – so short a period of time – and that the work was of such good quality.

10.21. Shelter 1 for Barrage Balloon staff working in the Hotel. The concrete ledges are hollow, with lino on the top to offer some sort of protection from the cold. Rubble underneath is visible. (Neil McCoubery, Bristol Photographic Society)

10.9 The BBC want the use of a tunnel
At the beginning of November 1940, BBC variety and music were using a site in central Bristol. That was fine until the Nazis conquered France, and Bristol came within easy range of enemy bombers, after which Gerald Daly (first mentioned in section 10.5) looked for a safer place.

The old railway tunnel [Portway tunnel 2] next to that where Bristol archives were kept [Portway tunnel 1] seemed a likely place for the orchestra from the shelter point of view, but how would an orchestra sound in the narrow confines of a railway tunnel? We decided to try and Sir Adrian assembled his orchestra of about sixty players then in the old tunnel. A record was made and to our amazement the musical quality was far better than expected. So we decided to put the matter to my bosses in London. The Director General, then F Ogilvie [Sir Frederick Wolff Ogilvie was Director-General of the BBC from 19 July 1938 (aged 45) to 26 January 1942], came down a week or two later and I took him through the length of the tunnel holding up a Aladdin paraffin lamp. Alas however there had been an unpleasant raid on Bristol [24 November 1940, the start of Bristol blitz] and particularly Avonmouth (about six miles away) and the tunnel was crowded with refugees, so we could hardly move. The city authorities offered to get rid of them, but we could not take it and gave up the idea.

However, the BBC lease on Tunnel 2 of the Port and Pier Railway was drawn up.

On 6 December 1940, incendiary bombs dropped on Tuffleigh House garden. There were people sheltering in the Railway tunnel who heard a thud and noted some mortar dropping (see Mike Farr in chapter 12.4). Children went looking for shrapnel in the morning. Gerald Daly writes:

The Imperial Airways people had, I suppose, a bit of a conscience about the BBC and sometimes invited the Director of the BBC and myself to dine with them. On one occasion while we were having dinner some enemy aircraft began bombing the Suspension Bridge, which was a favourite target for bombers. The raid became pretty hot and the Imperial Airways officials suggested that we repaired to their shelters. This we did and found that their shelter was the upper portion of the Clifton Rocks Railway. I was at once interested in the rest of the tunnel, especially when they confirmed that there were about sixty feet of rock between them and the surface.

On 28 December 1940, M.A. Small (a member of the shelter committee) wrote to Queen Mary informing her that the BBC wanted to use the Portway tunnel – which had 2,000 aircraft and munitions workers and their families sheltering in there. He considered their needs were greater than the BBC's. They had been told to vacate it on 1 January. The Lord Mayor then received a letter from the Queen's lady-in-waiting who considered it would be a catastrophe if the shelter was destroyed with 1,500 occupants. She agreed with the Emergency Committee that they already had adequate accommodation nearer home. The Emergency Committee considered that it should be converted to a shelter for 200 people with the BBC using the rest (apparently they had already started work on part of the tunnel). There was no made floor, so it was wet and muddy, and water was dripping from the roof. The BBC needed it for a bombproof reserve broadcasting station.

10.10 Prolonged correspondence between the BBC, Ministry of Works, BTCC and town clerk to get a lease and avoid restoring the Railway

Correspondence was prolonged (3 months to do initial negotiations), because the BBC did not want to restore the railway to original condition after they had finished with it. It took three months to do the structural work. The following information comes from the BBC Written Archives Centre, Caversham Park, Reading, and from the Bristol Records Office. Archives Centre, Caversham Park, Reading.

On 9 January 1941, the BBC decided to use Clifton Rocks Railway instead of the Portway Tunnel since the city could not get the people out. Services would be a lot easier. The apparatus designed for the Portway tunnel would be fine.[7]

On 14 January 1941, preliminary observations were made: it was not possible to provide two-deck accommodation owing to the lack of head room and the slope; it would have ventilation difficulties; it would only supply one third of the accommodation of the Portway tunnel; it would be obliged to keep the emergency staircase because of the large shelters above.

These would be the service stairs to BBC accommodation. It was wildly impractical if one wanted anything like the accommodation provided by the Portway tunnel, but then it may be possible to use Clifton Rocks Railway and one third of the Portway Tunnel space. The Corporation may need some of the Clifton Rocks Railway space allocated to the BBC. They delayed preparation of the drawings as the result 'may please no one'.

On 17 January 1941, it was not possible to brick off Clifton Rocks Railway tunnel until the legal position resolved. *No reason for referring Portway scheme as a fiasco.* (Caversham).

On 21 January 1941, the *Western Daily Press* announced that a portion of Clifton Rocks Railway would be made available as shelter accommodation for a limited number of people, as well as in the Portway tunnel for 400 people (priority for women, children and the aged). Admission would be by ticket.

On 21 January 1941, a decision was made to abandon using Portway tunnel since the BBC could not eject the people using it as an air-raid shelter. They decided to use Clifton Rocks Railway for engineering functions and install a control room, recording room, transmitter room and a studio using a diesel generator to provide an emergency power supply.

Gerald Daly writes:

> Although there was some opposition, the idea seemed to have general approval. The need for accommodation no longer mattered as far as the Symphony Orchestra was concerned for the bombs had got too much for them in Bristol and they fled to Bedford. The Variety department had also left Bristol – first of all we had transferred them to various halls and hotels in Weston-super-Mare, twenty miles away, and they had in turn fled to Bangor in North Wales (Variety and Music moved to Bangor in February 1941). It was said that Hitler made a point of following the BBC's Variety department wherever they went because of the funny nasty things the comics of that department said about him on the air. Tommy Handley, Arthur Askey, Kenneth Horne and others were the main culprits.
> So although there were still some London departments staying in Bristol such as Schools, Children's Hour and Admin, the departments requiring large halls etc. being no longer there, the new BBC station would not require large halls. This made the decision for the Tunnel station easier.

On 29 January 1941, BOAC write to the BBC about their plans, passed by both their civil engineer and the ARP officer. The construction of mutual walls and an entrance would be settled with the city engineer and approved by the Ministry of Works and Buildings. Gerald Daly writes:

> Before a decision was reached to go ahead, there were various meetings in Bristol between the London officials concerned and ourselves. I well remember the final meeting when the decision was come to go ahead with the project because it lasted all night.
>
> These meetings were held in the Regional Director's office – Mr Jely de Lotbinière – and about three am the latter said that as a decision was now reached, and the discussion would now be largely technical, he would go to bed. We all had our beds in our offices during the war. We were amused because when Lobby, as we called him, got into his bed, owing to his height, about 6.5 ft, his feet – bare – stuck out at the bottom. That impressed the meeting on my memory.
>
> The next step was to get the owners of the Tunnel together and to get their permission. There was a snag here because the actual ownership of the now disused Tunnel was complicated. We had hitherto dealt with the local city authorities, particularly the City Engineer, a Mr Bennet (see section 10.1).

On 3 February 1941, the Ministry of Works wrote to the BBC. They confirmed the scheme must be examined by their surveyors. They stated that the premises had to be restored to their original condition to the satisfaction of a BTCC architect. They were prepared to grant sub-tenancy for the portion of the tunnel and a proportionate rental to be fixed by them and similar conditions under which tenancy held by BTCC. BBC would have to pay any reasonable expenses. They could then authorise BBC to go ahead. Clifton Rocks Railway machinery and carriages at the bottom of the tunnel would have to be removed. Bristol Corporation was wanting to use the bottom station. *There must be a joint undertaking from the BBC and the Corporation that the machinery and carriages will be replaced at the end of the tenancy if required by the lessors.*

Gerald Daly writes:

> The firm which had originally built the Tunnel railway were the Bristol Tramways Company (the Chairman was Sir George White who had also started the Bristol Airplane Company, now he BAC with Concorde association). But they had leased the property from the Bristol Merchant Venturers, an old city concern going back to the Middle Ages and responsible for slave tradery, pirates and privateers and voyages of discovery to America. Cabot was, I think, financed by them. Also concerned with the Tunnel premises were the Downs Committee of the Bristol Corporation, who controlled any activity at all to do with Bristol Downs – a very upstage crowd indeed.
>
> With the help of the Bristol local authority, we invited everyone concerned whom we could find to a meeting in the actual tunnel to get their approval of the BBC scheme. I well remember this meeting. Representatives of the various people I have mentioned turned up together with others interested. The principal one of the latter was Sir Hugh Ellis, the Regional Commissioner of the West of England (he became virtual dictator of that area should there be an invasion and the main London government out of touch). [He had been chief of Civil Defence.]
>
> I remember we sat on some odd chairs collected for the occasion in the lower part of the Tunnel, with water dripping down on the heads and knees of these distinguished gentlemen. The place was partially lit by one paraffin Aladdin lamp, and the surrounding gloom was quite eerie. The three old trams at the bottom of the incline could be dimly seen in the shadow. The discussion went on through the afternoon from 2 pm to four sort of thing – everybody seemed to want to have their say. At last Sir Hugh Ellis said 'I'm fed up with sitting here with the drips coming down on my bald head – has anyone any valid reason for not letting the BBC have their broadcasting station here in the Tunnel? If not I agree to the idea and the meeting is at an end.'

On 6 February 1941, the BBC confirmed that they accepted the terms of the Ministry of Works. On 13 February a BBC internal memo requested information about the lease. They could not start work until the Ministry had formally approved the drawings, and so would write to the Ministry of Works to ask consent for the proposals to proceed and to submit a scheme for ventilation in due course. And on 18 February BTCC wrote to the town clerk sending the deeds and other correspondence with the Officer of Works, Merchant Venturers, E.T. Parker and BTCC.

Then on 18 February 1941, there was another BBC memo: they wanted the town clerk to enquire of the Merchant Venturers and have them waive the 'restoration clause'. Another memo suggests the city had cold feet over the clause and hoped that, in the next three days, the city would remove the trucks and machinery, build a party wall and get busy.

10.22. The four cars at the bottom, the rails being lifted and cut at the beginning of BBC work. This photo appeared in the *Bristol Evening World* 20 March 1946. The bottom station partition is in the background. Publicity about the conversion was avoided until the war was over. (*Bristol Evening Post*)

10.23. 'At the other end of each chamber was a trap door down into the chamber below for an alternative way out.' Norman Morse in the trapdoor (see chapter 13). (Gerald Daly)

10.11 BBC conversion starts
Gerald Daly writes:

> At out meetings with our London staff we had already designed roughly the way in which the Tunnel Fortress, as it was somewhat romantically called, should be laid out. The idea would be to have a series of rooms or chambers one above the other conforming to the gradient of the Tunnel ascent.
> These would be at the top the transmitter room housing the local Bristol transmitter for the town and environs, a communications transmitter to keep in touch with the rest of the BBC stations up and down the British Isles in case the telephone links were disrupted, and a spare transmitter. Underneath this would be a recording room with various types of recording equipment including film sound recording equipment and space to store records. Under this would be the main control room with control room equipment and landline terminations to other BBC stations and numerous transmitters including overseas services. We were transmitting programmes in about forty different languages all over the world.
> Under this would be the canteen, power room and stores

to give us our own power if the mains electricity failed and we were cut right off.

The entrance to these chambers would be by means of the staircase, which from the beginning of the Tunnel had been there.

A good deal of excavation had to be carried out on the solid rock and a special tunnel engineer was employed to do this particular work.

Three alternatives for how the control room would operate were discussed:
• There were sufficient local ends between tunnel and Post Office telephone exchange with suitable switching arrangements, to allow the Post Office to divert SB (manual switchboard) circuits into the tunnel in an emergency (this restricted output from programme producing units in Bristol if the cable was damaged in Whiteladies Road).
• Route all SB circuits into the tunnel and then into Whiteladies Road so they could be intercepted (this was too inflexible).
• Use the control room in the tunnel as the main SB switching centre (this was preferred since it used fewer Post Office circuits, and the control room would always be manned by sufficient staff if Whiteladies Road was out of action. Whiteladies Road control room would then regarded as the reserve SB switching centre.

Frank Gillard[8] wrote about the canteen and 'ozoneator':

Another of the smaller rooms was fitted up as a canteen. Sufficient food to last three months was stored in the tunnel. Huge tanks contained emergency water supplies. Others held enough fuel to run the engines for many weeks on end. A special ventilation plant was put in, with intake and extractor fans and an ozoneator. The tunnel was made immune even from gas attack.

The removal of cars began on 28 February. The City Corporation was considering buying out BTCC for a nominal sum but only if the Merchant Venturers and the Hotel were prepared to dispose of 'restoration rights'. If the city could acquire the leasehold it would be the most satisfactory solution for all and so the BBC declared a willingness to share the cost of release (£500) if the Balloon Barrage Command and Overseas Airways could agree to come in.

On 13 March 1941, a letter from Ministry of Works to BTCC regarding BBC, agreed to indemnify claims from occupying the tunnel. They could see no objection to present proposals. On 14 March, there was a long BBC internal memo about the current position: the Hotel pays ground rent of £50 per year; the Ministry of Works holds a sub-lease of £100 per year and have let to BOAC, Balloon Barrage Command, BBC and Bristol Corporation. The covenants under which BTCC must 'maintain' a railway have been passed onto the Ministry of Works who are seeking an indemnity.

The Corporation are frightened of the covenants, since they feel the Spa Hotel might squeeze them and extract a substantial amount of money in lieu of restoration – for no-one in their senses would ever suppose that the railway will ever run again.

The Corporation asked them to release the covenants. The town clerk, Ministry of Works and BBC met to agree how to negotiate with the Hotel – either to pay them £500 to dispose of the covenant or remit the £50 ground to give up their interest in the tunnel, (as would BTCC), leaving the corporation as holders of the lease from the Merchant Venturers. Either way it should be pointed out that if BTCC:

chose to assign their interest to a 'man of straw' [a person not intended to have a genuine beneficial interest in a property] the Hotel might lose their £50 ground rent as well as the benefit of the covenant and this might encourage them not to open their mouths too wide.

The BBC told the town clerk that the Corporation would almost certainly take their share in the negotiations – maximum liability for all concerned should not be more than £1600. They would want in return for their share permanent rights in their part of the tunnel at a peppercorn rent. They considered the best solution was for the City to buy out the Hotel ground rent – with luck for £1,250, of which the BBC's share would be £250. For that they would quit any responsibility to have to restore, save the present rental of say £20 per year, and have permanent right over premises upon which they were quite likely to spend £10,000. (Four days later an internal memo put the costs at £5,150 rather than £10,000.)

On 24 March 1941, 39 Post Office Circuits were taken up including:
• 14 established between the tunnel and Whiteladies Road (to extend studio items, two for trunk line, six PBX (private branch exchange) to tunnel for interstation calls and calls between offices in Whiteladies Road and the tunnel control room),
• five between tunnel and other studio premises (two

to Clifton Parish Hall which was at the end of the Fosseway, two to St Johns) and Redland Police,
• Eleven between tunnel and central exchange (three Washford, three Clevedon, one Weston control, two Weston music).

On 25 March 1941, we discovered, the Hotel would not release their rights for less than £1,600. On 1 April the purchase of the ground rent for £1,500 for 30 years was agreed in principle by the town clerk. It included the extinction of covenants. On 16 April, the bill for dismantling, breaking up and removing of cars was £56 11 2d. The struggle was not over. By 28 April the offered lease had shrunk to 21 years. Since the BBC reckoned they were the only lessees likely to stay after the war, they were not keen to sign – concerned that the Corporation might seal up the tunnel after the war and not maintain it.

On 23 May 1941, plans held by the electricity board showed that a 400-volt electric cable was laid down on the north side of the tunnel between 23 May and 5 June 1941. The junction was just outside the main gate of the top station. A spur in a 4in pipe went to the generator in the BBC section. The other junction was just outside the gate of the bottom station. Most of this cable was removed in October 2016 because the wall supports had rotted, so there were big sags in the heavy cable. In 2008 there were spectacular sparks from this cable at the top station and the street lights in Hotwell Road went out.

10.12 Still wrangling about the lease
In July 1941, the recording room was constructed to house the pair of Philips-Miller machines in use at Whiteladies Road, designated recording channel FB/1 (Film, Bristol /1), and these were moved to the tunnel, together with their two equipment racks (known as 'bays').

On 23 July 1941, the deed of surrender by BTCC was completed and on 29 July the town clerk issued a three-page statement of terms and history: that BBC would not be responsible for reinstatement; that the Ministry of Works had breached the covenant of keeping the tunnel fit for use as a railway; that the Hotel demanded any use not to cause a nuisance to its business; the Merchants demanded the entrances to be kept in repair (with the Ministry of Works to use the top entrance, and the BBC the bottom). In turn the Corporation would grant a lease for 21 years at a nominal rent, and contribute £400. The BBC asked for an additional clause: that 'the vendors, their successor etc shall not at any time unreasonably object to the presence of a slung aerial and transmitter hut erected

10.24. Four core round stranded copper conductors each of 0.1 sq. in. Gutta percha insulation. Jute serving Lead sheath. Bitumenised jute yarn bedding. Steel tape armour. Bitumenised jute yarn outer serving. Low voltage (230/400 volt), 4 core, 3 phase. Used for alternating current distribution. (Author)

on Spa Hotel property'. A BBC memo deemed the terms fair and, if another war was to come in the next 21 years, the money would not have been spent in vain. They had slung the aerial from the roof of the Hotel down to the entrance of the tunnel (see the location in chapter 13), and built a small wooden shed to hold various apparatus. On 2 August 1941, the city engineer requested an electric cable be laid of 14 kW to supply heating and lighting to the two chambers in the bottom station and thus create an air-raid shelter.

On 14 August 1941, the town cerk had stated that the Hotel objected to the hut and aerial but his solicitor advised that the Hotel should take it up with the Government department concerned with requisitioning the Hotel. It was nothing to do with the conveyance of the tunnel. They would be glad to get the conveyance disposed of without further delay. A BBC internal memo agreed it would be wise to waive the proviso about the hut and aerial. The town clerk wanted half the cost of repairing the tunnel entrance- this was considered reasonable considering they were only paying nominal rent.

10.13 Transmitting starts
On 13 September 1941, the RCA 1 E Group H transmitter was transferred to the tunnel's transmitter room from no. 91 Pembroke Road; Pembroke Road closed. Transmitting started the following day (clearly they had a good wireman as it was not just a matter of putting a mains lead to the equipment, and linking up all the Post Office cables in the control room, recti-

10.25. A kettle to put on a stove was found in the section – no electric kettles. Perhaps the transmitting engineers did not warrant better facilities.

fiers and transformers would also have to be installed).

Gerald Daly:

> The whole project took six months to finish and I moved the main engineering staff down to the tunnel in 1941. Henceforth for the rest of the war this was the nerve centre of the BBC in the West of England. Through the Tunnel control room for the next four years passed all the programmes of the BBC to home and overseas. We went back to the prewar control room in Whiteladies Road at the end of the war. We held onto the Tunnel for nearly another two years – we only paid a peppercorn rent to the Bristol Corporation for renting it. I think it was about £5 a year. Then we removed all our equipment and the City took it over – the chambers of course remain, and it could be useful perhaps in a nuclear war. [In fact the CRR remained a BBC Deferred Facility until 1960 — the author was probably reluctant to mention this at the time of writing.] We of course made it gas proof, and with all that rock above even a nuclear explosion would have little effect I think.

10.26. The derelict cave is on the left-hand side holding the 1877 grotto, suitably close to the BBC entrance to the tunnel.

On 28 January 1942, Gerald Daly asked the Ministry of Works (since they had requisitioned the Hotel), to confirm there was no objection to a modified Group H transmitter aerial and mast, lead-sheet earthing on the flat roof, and barbed-wire security fencing on the flat roof of the ballroom, provided that no damage was sustained to structure.

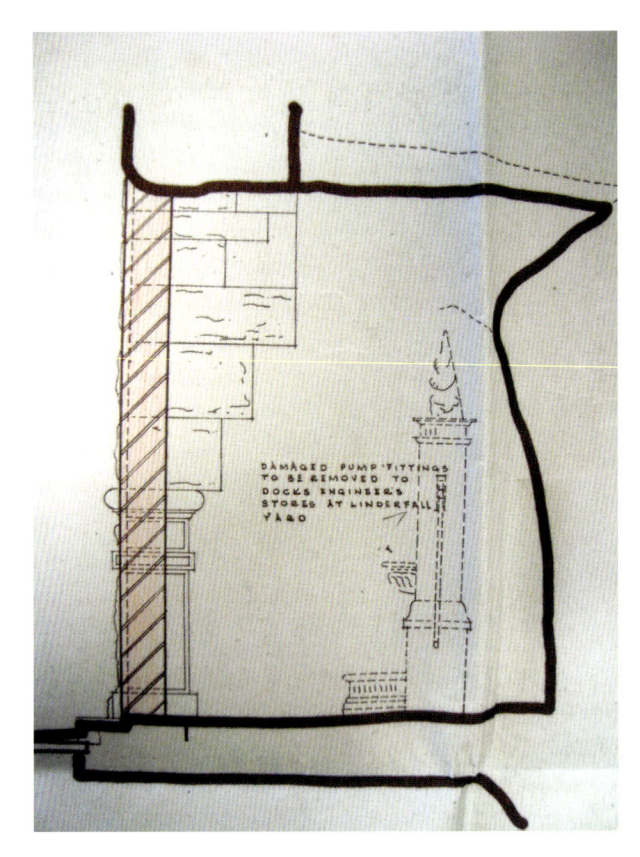

10.27. A plan shows the fountain. Also see plate 1.8 showing a photograph of the fountain – the pump fittings have vanished. (BBC Archives, Caversham R35/207)

10.14 Still more delay with the lease

On 16 October 1941, BTCC reminded the town clerk that the lease still had no end date. Two months later on 12 December 1941, the town clerk wrote to the BBC to confirm the delay in completing the lease was due to the fact that the treasury solicitor had not yet approved the lease from the Office of Works. The Office of Works did not wish to pay their share until the lease was approved. One month later, on 12 January 1942, BTCC confirmed with their treasury the town clerk's response. BTCC was to carry on with accounts payable and receivable. On 2 March 1942, a letter from the town clerk to the treasury solicitor referred to it as a 'complicated and troublesome case'. On 5 June 1942, the town clerk finally settled the form of documents for lease with Hotel and BTCC to complete on 1 July 1942. Purchase money was to be paid to the Hotel and the rent payment of £50 would cease. BTCC would pay rent up to 1 July 1942, which could be paid to the Hotel solicitor or the Corporation solicitor. The amount would be reimbursed by the

Corporation. BTCC preferred to pay the Hotel. The rent was due to BTCC from director of lands and accommodation. There would be three quarters of the amount due at the rate of £100 per annum, which worked out to be £76 12s 11d. BTCC asked the director if the payment was held in abeyance pending the transfer of BTCC's interest to the Corporation. On 4 July 1942, the date of deed of surrender, the BBC and Corporation woould each pay 1 shilling yearly rent on the bottom part of the Railway. Imperial Airways would continue paying £100 per year. Considerable delay had been caused in connection with the sub-lease.

10.15 More space needed

On 5 November 1941, a BBC memo declared that in view of the worsening war situation, main control should work full-time from the tunnel, rather than during blackout periods and air-raids (or invasion). Thus the canteen was not big enough (they would install a bigger one in six weeks – the time it takes to obtain the electrical equipment). They also wanted to house more administrative staff: the engineer-in-charge, the London lines engineer, recorded programmes engineers, a chief clerk, programme director, talks assistant, two producers, two announcers, liaison offices etc plus general office and duplicating staff. They also required a store room, two rest rooms, separate male and female sleeping quarters and lavatory accommodation. They found a large 12-room house for sale at £750 (no. 380 Hotwell Road) and fitted electric lighting, a gas cooker, and oil heaters. It was in use from March 1942 to 1945.

On 30 December 1941, the BBC confirmed that all essential work was complete. The transmitter, the control room (transferred from Broadcasting House), and emergency studio were all functioning.

10.16 Damp wall

On 13 August 1942, the 27in brick wall separating BBC section from the Overseas Airway shelter 3 was found to be very damp [it still is, since the wall goes up to the tunnel roof and this is the lowest shelter so it just runs down the wall], with moisture was finding its way into transmitter room. The BBC suggested adding a zinc and felt damp-proof course [still visible] to stop percolation level with the transmitter room ceiling, to avoid possible damage to possible apparatus. The proposal was placed before the owners, Bristol Corporation. The request was also sent to the town clerk from the BBC solicitor four days later. Nine days later on 21 August 1942, the town clerk approved and wished to inspect when done. The BBC was told to contact BOAC before put to hand. This happened the next day. On 9 September, the BBC engineer met BOAC's civil engineer to discuss work the partition wall, assuring them there would not be an accumulation of water on the BOAC side [this is the one location where one always gets wet feet.]. The district surveyor of the Ministry of Works was also consulted since the tunnel was owned by the Ministry. BBC letters were written by their legal department rather than by engineers. BBC contractors did the work and presumably the BBC paid for it.

10.17 Drips in BBC section contaminated by sewage?

On 6 November 1942, a bacterial laboratory report was requested on a quart bottle of water from the Tunnel to see if there were any signs of contamination. Some was slightly polluted but there was a considerable amount of time before it reached the tunnel. The BBC civil engineer sent the report of water to the Corporation and wondered if BOAC was interested too, since there was a lot of water. There was then confirmation that water leaking was not the result of a broken sewer so no further action was taken. This is very pertinent since I can reassure visitors doing tunnel trips about the quality of the water.

10.18 BBC tenancy of the cave

In November 1941 there appears first mention of the Hotwell spring alcove (the 'cave') to be used as an oil store. Originally the cave, owned by the Port of Bristol, housed the pumping apparatus to the Hotwell spring, and would make a safer store than the cellar of no. 31 Whiteladies Road, where the petrol and oil were currently stored. An agreement was reached. The entrance was bricked up and a door provided; the old iron pump shaft and iron basin were removed to Underfall Yard (the nearby Bristol dock maintenance workshop). The pump fittings were never seen again. The cave was used until September 1946 when, no longer needed for a store, it was returned to the Port of Bristol.

10.19 Closing down the station by the BBC

This again, was not as easy as expected. So many letters, so many departments, so much advice, so many inconsistent prices, and still worries about reinstatement.

The celebration of the end of the World War II in Europe, VE Day was on 8 May 1945. Articles appeared in the *Illustrated London News* on 6 April 1946, and the BBC *Year Book* 1946[7] about the role of Clifton

Rocks Railway as the BBC's secret wartime fortress.

On 28 July 1945, the BBC implemented its peacetime broadcasting plan (see chapter 13). The Group H transmitter network which was in use for the Home Service closed down.

On 20 October 1945, there was an internal memo from BBC lines department, Brock House, to Gerald Daly and others: This was confirmation of a proposal to keep the tunnel as museum piece. There was no point in retaining all telephones and lines, so they suggested keeping one or two private wires to Whiteladies Road control room, and one exchange line terminating directly on an instrument. It enabled the existing PBX installation and a number of private wires to be recovered and save rental costs. Miss Otley (announcer) still wished the use of direct line at 410 Hotwell Road.

On 7 December 1945, a memo from the BBC director of premises and stores at 46 Portland Place declared that the West region had no interest in retaining the tunnel because of heavy costs of £2520 per annum. The suggestion of retaining it as a museum would reduce costs of reinstatement. The telephone rental cost would be reduced. SB and intercom was on a five-year contract to December 1946. Director General approval was needed for financing to maintain it as museum piece. A summary of costs at various times is to be found in table 1 in section 10.21.

On 20 February 1946, a memo from BBC head of building department queried the termination of lease and the building work required to remove the engineering equipment. There was no information about reinstatement. What happens to the ventilation plant? On the following day, a memo from the BBC director of premises and stores said that querying obligations regarding reinstatement was referred to as a dark subject. It needed coordinating but no one has said it should be given up. It was difficult to find out more than lease terms from the landlord. A week later, a memo from assistant director of the BBC legal department confirmed the terms of the lease are for 21 years from 3 July 1942 at 1s 0d pa. There was no provision for an earlier break. They were asked to contribute one half of the cost of keeping the lower entrance in repair. There was no liability for reinstatement at end of the lease unless there was early termination. The Corporation might make it a condition to make a contribution.

On 8 March 1946, a memo from the BBC head of equipment agreed there was no great difficulty in dismantling and removing equipment. They required an estimate to remove part of the transmitter room wall, the blast wall at the bottom of the stairs, to lower equipment to the bottom to take to Hampton House, remove standby diesel plant. Ventilation plant and trunking were a problem. There was a separate room with two motor driven fans, air filter and trunking to all offices. These offices coould not be used without a working ventilation system. They should check with Bristol Corporation if they wished to use the rooms. If use was not required, the BBC should remove only the motors, fans and air filter, and leave the trunking.

On 8 March 1946, there was a quote from the Squires company to BBC Evesham for removal of transmitting gear etc., to cut away the blast wall; rig up blocks, pulleys and ropes to get equipment down the stairs; cut away partition walls to remove equipment and then make good. To also remove transmitter units, 14 panels from control room and the Lister Diesel Engine EA6K 1126 and convey to Evesham would cost £104. It does not include removing the storage tank under floor (for the generator diesel), or the removal of pianos (note the plural, easier to get them out than in owing to the stairs) from the recording room. The blast walls were still there, as was the storage tank.

On 20 March 1946, a memo from BBC head of equipment confirmed the quote. They required a decision on the ventilation system and the storage tank. There was no point in removing low frequency equipment in advance as it was necessary to leave heating and ventilation until the transmitters were removed. Allow £150 to cover transport etc. Interestingly, an article in *Bristol Evening World*[8] states that the whole conversion had cost £20,000. Two days later, a memo from BBC head of building agreed the £150 contingency. They were worried that the quote was only for removing technical equipment and not for reinstatement, which would be required by Bristol Corporation. They needed to inquire about the fate of ventilation system and storage tank.

On 26 March 1946, a memo from BBC director of premises and stores reminded them that rooms in the tunnel were not habitable without ventilating plant and trunking. This must be taken into account in a claim for reinstatement. He suggested it was better to continue the lease for the next 17 years to absolve liability for rates or maintenance. The nominal rent was only 1 shilling per annum, so this was a more economical solution and it was ideal security accommodation for emergency use. The tunnel was permanently manned and handled all SB traffic between November 1941 and August 1943. Since then it had been on a care and maintenance basis. So, the advice was to continue the lease, remove the equipment and leave the ventilating plant which was of no use. The cost of £150 for building work and £30 to remove the equipment to store would allocated to other schemes.

On 30 March 1946, the BBC head of engineering confirmed legal negotiations with Bristol Corporation were proceeding. He needed approval to transfer two bays of redundant, obsolete AC test equipment (book value £400) to install in the main Bristol premises. If returned to store it would be broken up for scrap. This was agreed.

On 1 April 1946, the BBC director of legal department wrote to the Corporation to see if they would accept a surrender of lease. They were prepared to leave a considerable amount of ventilation and trunking at an agreed price. On 3 May 1946, a month later, the Corporation replied. They agreed to surrender of lease provided the tunnel was reinstated to the satisfaction of the city engineer and with ventilation and trunking removed. Oh dear.

On 12 June 1946, the BBC engineering division referred the estimate revenue savings of £1,825 (as on 7 December 1945 in table 1 in section 10.21), and confirmed the stance of 26 March, that it was more economical to continue with the lease for the next 17 years to avoid liability for reinstatement, but remove equipment except for the ventilation plant. They still needed £150 for building work.

On 21 June 1946, a BBC income tax adviser suggested that to avoid the question of future liability, the BBC should remove their equipment as soon as possible, since it was unlikely that any claim would arise when removed. If they intended to keep it as a 'museum piece', care would have to be taken to avoid giving income tax authorities any handle on claiming the premises were occupied.

On 25 June 1946, the director of BBC legal department considered a tenant was usually able to reclaim income tax from a landlord by deducting from rent, but since the rent was only nominal, there was no right of recovery. Tax was not payable if unoccupied, so if they ceased to occupy the tunnel, schedule A tax probably would not payable. On 27 June 1946, the Corporation wrote to the BBC legal department: they did not wish to put the BBC to unnecessary expense of removing plant and fittings in reinstatement, provided the BBC would leave behind the ventilation plant and ancillary fittings in working order. The tunnel could then be used for a limited period by Corporation who would then accept responsibility for final removal and disposal.

On 2 July 1946, the BBC director of premises and stores advised that one month's notice should be given to terminate.

On 16 July, the BBC head of building advised that the BBC should hand over ventilating plant in toto to the Corporation *on condition they waive all claims for reinstatement* (cost of the plant was £1600. Only fans, motors and control panels were of use to the BBC). Electrical equipment would need rewiring as it was only suitable for the Bristol supply system. Taking this into account, and the cost of removal, the value of plant was nil. The next day the BBC head of engineering agreed that after negotiations, they would transfer two bays of obsolete AC test equipment to the main Bristol premises. He referred to the estimate given already.

On 19 July 1946, BBC legal department said they would not dream of negotiating with the Corporation. They were glad to know the ventilation plant would remain on site, as by handing it over to Corporation, they escaped all liability for dilapidation and the lease could be surrendered. On 30 July, the director of premises and stores approved the termination of the lease, and handed over ventilation plant in lieu of reinstatement.

On 7 August 1946, in a memo to the finance secretary at 318, The Langham, London the calculation of savings of £2,252 was reduced to £2,044.

10.20 Change of mind – the BBC need to continue to use the tunnel

Just when they thought it had all been sorted out after eight months of discussion.

On 28 August 1946, the BBC senior superintendent engineer issued an urgent memo: there is a crisis over 514 metres frequency. Do not remove or dismantle any Group H transmitters. He gave the instruction to hold up the completion of legal negotiations with the Corporation. The ex-Group H transmitter in the tunnel was required to broadcast the Third Programme from 29 September 1946. More details are in chapter 13.

On 19 October 1946, the Corporation responded to a letter of 8 October from the BBC legal department: the tunnel wa of little use unless they could have possession of the whole. The Ministry of Works had already finished their tenancy (of the top half of the tunnel) – for how long would the BBC require their part? Only then could the Corporation make arrangements for the future use of the property.

On 23 October, an internal memo showed that the BBC legal department confirmed to the Corporation

Table 10.2. BBC expenditure.

BBC Revenue Expenditure	7 December 1945	28 March 1950	16 March 1955 Deferred	16 March 1955 Existing
telephone	£36 6s 6d,	£5 17s 9d		
water	£10 16s 0d			
rates and tax	£105	£207	£55	£241
rent	1s	1s	£10	
technical assistants grade II	3 men £156 each £695 3s 6d	2 men £350 each		
Engineering Division:				
power, light, heat	£375	£276	£60	£248
household maintenance		£20	£5	£20
building maintenance	£30	£36	Inc masts £20	Inc masts £40
plant maintenance	£100	£60		
SB and intercom	£1,320	£35	Valves £10	
total	£1,825	£1,058 17s 9d	£160	£549

they wished to postpone the surrender of the lease, and wondered about the response to the Corporation's letter. 'In the present delicate state of affairs it is difficult to answer the town clerk's question, but glad if given a line of reply.' There is a handwritten note that there will be a permanent C-programme transmitter in Bristol, and they intend to get a new site but expect delays of at least 12 months. A week later, the BBC legal department wrote to the Corporation confirming international difficulties due to the use of wavelength 514m for the Third Programme, so will be using this wavelength at reduced power. The result is that more transmitting stations are necessary, and they propose to find a suitable site and build a new station. Until ready they will continue to use tunnel. It would not be ready until the end of 1947. 'We are sorry to inconvenience you in this way.'

On 1 January 1947, the *Western Daily Press* reported that the Lord Mayor, Lady Mayoress and Sheriff visited the BBC installation in Clifton Rocks Railway.

On 28 March 1950, a memo from BBC head of engineering services stateed that under the Copenhagen Wave Plan, and with the Bristol Third Programme transmitter closed, a decision was required as to the surrender of premises and associated annual rates, including staff (Technical Assistants Grade II) allocated to Bristol under the Third Programme for monitoring purposes. Operative rates were £1058 17s 9d (they had halved since telephone usage had decreased, rates had increased, and they would probably be employing two men rather than three). See the summary of costs at various times in table 1 section 10.21.

The Third Programme transmitter was closed due to a new main transmitter coming into operation at Daventry, but the BBC still wanted to use the tunnel as a temporary relay station (see chapter 13), so more negotiations and memos continued until 1960, as can be seen in section 10.21.

10.21 Earmarked by the Home Office
While the BBC negotiations were being carried out, the Home Office earmarked the whole of Clifton Rocks railway on 16 June 1949[9] to put on the Civil Defence register of Underground Accommodation for air-raid shelters as well as the Portway tunnel and other tunnels in the Devon (Ilfracombe and Plymouth) and Gloucestershire (Bitton) area. The Civil Defence Corps (CDC) was a civilian volunteer organisation established in Great Britain in 1949 to mobilise and take local control of the affected area in the aftermath of a major national emergency, principally envisaged as being a Cold War nuclear attack. Each corps authority established its own division of the corps. The Railway tunnel was described as being damp, on an inclined surface and used as an air-raid shelter during the war. There was 6,930 feet floor area underground according to Edmond Mason for Lands, Bristol.

On 11 April 1951, CDC was awaiting the development officer, who filled out a questionnaire on 25 February 1952, and prepared a thorough report on 26 August 1952. The questionnaire covered class of property, access, and entrances (it commented that there was a badly cracked wall at the bottom entrance to be partially rebuilt), workings (the tunnel all usable; no visible distortion of brick lining; no debris; moderate dripping of water which drips into canvas chutes and then into the Corporation drain at the foot of the cliff. The BOAC electric light was stripped out (on the south side), but the BBC portion still functioned. It was generally humid. BBC ventilation was provided by a plenum system (a system of mechanical ventilation in which fresh air is forced into the spaces to be ventilated from a chamber). The average thickness of the ground above was 35ft. Dry rot was prevalent, and most of the timber was in a state of collapse except in the BBC section. It mentioned the BBC rooms were still occupied, and contained the wartime transmitters being maintained, but not in use. Chemical closets were removed. It mentions that the canvas over the shelters was put in to collect all the moisture . The shelter previously provided seating accommodation for about 550 persons, but if the Air Ministry and BBC portions become available, could be extended to 800. Only 60ft to 70ft was unoccupied. It was unlikely that 500lb or 1,000lb medium capacity or general purpose bombs would cause heavy casualties thought there may be spalling (breakage into smaller pieces) with a direct hit. With HE (high explosive) blast protection at the tunnel ends, the degree of safety was satisfactory. It would be a grade A shelter if protection was provided at each end. A substantial blast wall 3ft thick should be provided 20-30ft inside. It was owned by Bristol Corporation.

10.22 Back to the BBC negotiations
On 12 April 1954, a memo from the BBC solicitor considered that the Corporation's lease offer from 20 February 1946 would still hold good. They would now probably only agree to immediate surrender if premises were reinstated. (Caversham)

On 24 June 1954, a secret memo from the BBC director of administration considered there was no peace or wartime use for tunnel and that the Group H transmitter should be moved to a less vulnerable area. However, on 13 December, Gerald Daly wrote to the City Estates surveyor confirming a wish to relinquish BBC tenancy and to lease the first-aid room at the top of tunnel. Two days later, a response from Civil Defence headquarters asked when they wished for the handover of top station to adapt for broadcasting equipment. The Civil Defence approved of relinquishing the lower portion of the tunnel, provided agreement could be reached to leaving existing lighting, heating and ventilation in place and removing BBC equipment. They also wanted to know how much the BBC would ask for leaving the plant. The City Valuer would then be consulted on rental in the top portion. The Civil Defence also wanted to discuss the possibility of taking over the installed telephones.

Two weeks' later, on 29 December , the Corporation provisionally agreed a 10-year lease at £10 per annum being preferable, with an option to break after five or seven years. This was assumed not to start until 25 March 1955. A partition was needed to be built to form a passage past the first-aid room.

On 15 February 1955, a memo from the senior superintendent engineer of sound broadcasting confirmed the Corporation had agreed to lease the first aid room to house the Group H transmitter (for deferred facilties) from 25 March 1955 and to cease the present tenancy on 24 June 1955. They needed to calculate revenue savings, so to proceed and negotiate. It needed priority action to maximise savings. The next day, a memo from the BBC solicitor stated that financial approval needed to be obtained first, before approaching the Corporation. The Corporation would require lighting, heating and ventilation plant left. The solicitor was not clear why the matter was now urgent, since it had taken 8 months to respond. Pressure of other more important business means it could not be given special priority.

On 23 February 1955, the senior superintendent engineer of sound broadcasting stated in a memo that the delay was caused by protracted negotiations between Bristol staff, Bristol Corporation and the Civil Defence. There was an existing aerial mast on the first-aid room 40ft high and stayed at three points and an A.T. hut 9ft 6in long, 8ft wide and 9ft high also adjacent to first aid room (see chapter 13, plates 13.24 to 28). This was arranged on a goodwill rental basis.

On 16 March 1955, a memo arrived from the BBC assistant head of engineering about revenue expenditure. Deferred facilities would cost £160, surrendered rates on home sound direct £549. A deferred facility meeting on 8 March had given authority to complete legal negotiations. The next day, the BBC solicitor confirmed to the Corporation that they would leave lighting, heating, ventilation on bottom station on the understanding there would be no claim for reinstatement, and would like to take tenancy of first-aid post from 25 March 1955, and to maintain the aerial mast and hut on top of first aid room.

On 22 June 1955, City Estates stated that Civil Defence had agreed the lease of the first-aid room. The partition to form a passageway was needed. A nominal price of lighting, heating, ventilation on bottom station of £5 5s would be charged. Two days later the BBC solicitor in a memo wanted comments so he could proceed with the lease. On 12 July 1955, the solicitor confirms the BBC can take possession without waiting for the completion of a new lease.

The surrender of the bottom station lease was from 24 June 1955 – with Bristol Corporation granting a lease at the top of the tunnel in substitution.

On 31 October 1955 there was a memo from the BBC solicitor regarding the terms of the new lease: the landlords were the Lord Mayor, aldermen and burgesses; the entrance to the top station was via Sion Hill. There should be a term of 10 years from 25 March 1955, determinable by the Corporation at the end of the 5th or 7th year and with three months' notice. The Corporation would be responsible for keeping the premises well and substantially repaired and to construct, at its own expense, a partition to form a passageway. It is interesting to note that work to stabilise the bottom station by the Corporation began in 1957 – see chapter 14.

On 9 June 1959, there was a memo from senior superintendent engineer of sound broadcasting about surrender of leases for the mast and Group H site. On 10 July 1959, there was a letter to City Estates for the formal surrender of lease to be on 24 March 1960.

The BBC finally moved out. Ever since the tunnel has only been used for housing a telephone cable (not the original one), and the live 1941 electric cable.

Having looked for signs of the partition that the Civil Defence wanted – this is rather a mystery. There is evidence of ventilation on the top of the first-aid room – which has quite a small surface area – but this would have probably been necessary during the war as there were no windows. Lighting, heating, ventilation was left behind in the bottom station. There is no sign of reinstatement anywhere, either by the Air Ministry, the Corporation, or the BBC.

eleven

Barrage Balloons
Operation and Memories

This chapter gives the background of the use of barrage balloons,[1] and how Clifton Rocks Railway was involved. It has been written with help from David Wintle,[2] John Penny[3,4] and John Christopher.[5] This chapter has been included because the administrative headquarters of the Air Transport Auxiliary were in the Grand Spa Hotel, their staff air-raid shelter was in Clifton Rocks Railway, and there were barrage balloon offices and a workshop in Clifton Rocks Railway.

11.1 Introduction of the use of barrage balloons during the war

The first application of balloons used to provide a barrier or barrage of deadly cables hoisted high in the sky was not exclusive to WWII.[6] The Germans had started experimenting in 1917. The earliest known allied barrage was raised at Venice by the Italians, and approximately seventy balloons were flown from rafts floated around the city at heights of up to 10,000ft. Their primary function was to keep enemy aircraft flying high in order to reduce the accuracy of their bombing. Whilst a dive-bomber could enter the barrage, it could not pull out of the dive without hitting a cable. The strong, steel cables were capable of causing sufficient damage to disable or bring down an attacking aircraft. The original concept of an apron of cables had not worked well during experiments in 1918 because the cables were too heavy and had a tendency to sag. The idea of a ring of balloons encircling a city was also abandoned around 1937 because modern aircraft could have flown over the protecting outer screen and then come in low. It was decided that balloons randomly dotted around at a height of 10,000ft (any higher meant that they would be difficult to handle and would need longer and heavier cables) would keep bombers high enough to reduce their accuracy, and to force them into zones where anti-aircraft guns and fighters could locate them.

Balloons were 68ft long with a diameter of 27ft. They were filled with hydrogen and had a capacity of 19,000 cu.ft. A trailer load of 40 cylinders was needed to inflate just one. Their shape made them stable. They were kept head to wind by three stabilizers, which were also filled with air fed in by an air scoop in the lower fin. The panels were made of two sheets of fine linen glued together with a thin film of rubber solution and another film over the outside, which was then given a silver coating. This was then cut into shaped panels and machine-stitched together to form the balloon with a one-inch strip glued over the stitching to prevent any gas loss. Several pieces of balloon fabric have been found in Clifton Rocks Railway in the designated barrage balloon section (see the plans 10.7 and 10.8). The balloons flown around Bristol were designed for a maximum flying altitude of 5,000ft. Winch speed meant that balloons required 11 minutes to reach that height. At times, the winch could develop quite a high charge of static, and in order to avoid an electric shock, the winch operator and the rest of the balloon crew always wore rubber footwear. When having to enter or leave the winch case, the operator would jump on or off the winch so that his body did not provide a contact between the winch and the ground. Rubber footwear was also necessary so as to avoid igniting any leaking hydrogen gas. Often a circle of ash or bricks was laid down to provide a flat, free-draining area. Balloons were difficult to handle in windy weather, and lightning was another hazard.

Balloon Command was part of Royal Air Force Command, which was responsible for controlling all the United Kingdom-based barrage balloon units during World War II. It was formed on 1 November 1938 at RAF Stanmore Park in Middlesex.

The Air Transport Auxiliary[7] (ATA) was a British civilian organisation set up during World War II and headquartered at White Waltham Airfield, that ferried new, repaired and damaged military aircraft between factories, assembly plants, transatlantic delivery points, maintenance units (MUs), scrapyards, and active service squadrons and airfields, but not to naval aircraft

11.1. Public demonstration of a barrage balloon flying in 1939. Barrage balloons were supposed to make it difficult for enemy raiders who would fear crashing into the balloons, or into the steel cables that attached them to winches on the ground for raising and lowering them. (*Bristol Times* 12 January 2006)

carriers. It also flew service personnel on urgent duty from one place to another and performed some air ambulance work. Notably, many of its pilots were women, and from 1943 they received equal pay to their male co-workers, a first for the British government. The administration of the ATA fell to Gerard d'Erlanger, a director of British Airways (see section 11.3). On 24 November 1939, the British Overseas Airways Corporation (BOAC) was created by Act of Parliament to become the British state airline, formed by the merger of Imperial Airways and British Airways. The companies had been operating together since war was declared on 3 September 1939, when their operations were evacuated from the London area to Bristol. On 1 April 1940, BOAC started operations as a single company. During the war the airline operated as directed by the Secretary of State for Air, initially as the transport service for the RAF, with no requirement to act commercially.

11.2 Balloon Command comes to Bristol

A defensive survey of the Bristol area had already been carried out in July 1938, and it had been decided that the presence of Filton aerodrome prevented the use of balloons to cover the whole area. Two small independent barrage layouts were proposed to protect the harbour installations at Bristol City Docks and Avonmouth Docks. Later, a separate barrage was authorised for Filton to protect the airfield and Bristol Aeroplane Company's works.

On 12 October 1938, an appeal was made in Bristol for 21 officers (aged 32-50) and 750 men (aged 38-50) for the proposed three new balloon barrage squadrons located in Bristol and district. Each would have 24 balloons, seven officers and 250 men and would form part of the local Auxiliary Air Force unit at Filton.

On 1 November 1938, Balloon Command was formed under Fighter Command, and established provincial barrages for Bristol, Cardiff and twelve other cities. Provincial Balloon Groups were established shortly afterwards. On 16 January 1939, the area extending from Bristol to Gloucester, Swindon, Salisbury, Yeovil, Weston-super-Mare and Exeter was the responsibility of no.11 Balloon Centre, which was established at the HQ Territorial Association and Air Force Association at Clifton. Flight Lieutenant Dempster was appointed adjutant.

On 9 August 1939, the new camp at Pucklechurch became the permanent home for No.11 Balloon Centre, and HQ for its three squadrons.

11.2. Public demonstrations of a barrage balloon flying took place at Sea Walls, Durdham Down in order to stimulate recruitment in Bristol. They were inflated by men of the Royal Air Force. A huge group of spectators gathered despite the rain. (*Western Daily Press* 29 March 1939)

11.3 Clifton Grand Spa Hotel becomes the main registry for Air Transport Auxiliary

In September 1939, the whole of the Clifton Grand Spa Hotel was requisitioned by the British Air Transport Auxiliary (ATA). Nos. 1-3 Caledonia Place were also requisitioned (and the author's garage[8] which also belonged to the Hotel at the time, was used to store their vehicles). The chairman and deputy staff as well as the engineering department were on the first floor. Gerard d'Erlanger (of the French-English banking family and known as 'Pops' by his staff) was the founder and head. ATA was responsible for transport of dispatches, mail, news etc. D'Erlanger[9] and Gerald Daly (see chapters 10 and 13) from the BBC, would sometimes dine together. After the war, d'Erlanger became managing director and later chairman of BEA (British European Airways), a member of the Air Transport Advisory Council, and deputy chairman and then chairman of BOAC. Viscount Knollys became the BOAC chairman operating from the Hotel from March 1943.[10] The memories of people who worked for BOAC in the Hotel can be found in section 11.12. Accounts were on the second floor, the post room in the basement; the legal department was based in Tuffleigh House. No-one slept at the Hotel. Air crew were vaccinated (for smallpox, tetanus and typhoid, and possibly yellow fever) in the medical department at Clifton Park. The reserved occupation offices were in the Pump Room,[11] where foreigners – chiefly Polish men – were to be processed for employment. The reserved occupation offices also arranged for aircraft delivery/return by the Air Transport Auxiliaries. Personnel were not allowed to carry a camera.

For information about Air Raid Precautions (ARP) and Clifton Rocks Railway, see chapter 12.

11.4 Squadrons embodied

War was now imminent and on 25 August 1939, 927

11.3. Imperial Airways badge. A silver commercial pilot's badge, centred with blue enamelled Imperial Airways shield on wings within an elaborate silver laurel wreath. It has a Birmingham silver hallmark for 1938.

11.4. The reverse is stamped: N.PIPER, B.O.A.C. CLIFTON BRISTOL OAAQ 186/3.

Squadron was embodied, followed by 929 on 29 August, and 928 the following day, after which the squadrons left Pucklechurch to occupy their wartime site. The squadrons formed skeleton barrages around the Bristol City, Portishead and Avonmouth Docks. Balloon squadrons normally comprised three flights of eight balloons, but this often varied.

927 Squadron was the first to move, commencing deployment at Avonmouth on 27 August, and by 5 September had balloons flying at 10 Sites, while three other sites were occupied but not yet operational. The Squadron's headquarters moved to The Chalet, Henbury, on 11 September, while A Flight's HQ was set up at no. 6 Kingsweston Lane, Shirehampton; B Flight at no. 205 Avonmouth Road, Avonmouth; and C Flight at Cowley Farm, Kingsweston.

On 7 September 1939, the locally-recruited 928 (County of Gloucester) A Flight took over four sites from A Flight 927 Squadron and began deploying its balloons at Clifton, as well as at sites across the River Avon at Pill, Portbury and Easton in Gordano, and moved its headquarters from Portbury to nearby no. 3 Caledonia Place, Clifton.

On 18 September 1939, 929 Squadron commenced deployment in the Bristol area with B Flight occupying Site 40 (Robinson's Athletic Ground, Bedminster) and Site 30, (Victoria Park, Bedminster); while C Flight was deployed at Site 15 (Athletic Ground, Whitehall) and Site 16 (St George's Park). The new Squadron HQ was established at 57 Victoria Street, Bristol, and although A Flight remained at Pucklechurch as a Reserve, and for further training, B Flight's HQ was established at St John's Lane, Bedminster, and C Flight at Whitehall athletic ground, and then to the car park at Eastville greyhound track. Finally, with all the squadrons now deployed, No.11 Balloon Centre issued its first practice 'Attack Alarm' on 27 September.

October 1939 saw a number of changes take place with regard to the local balloon barrages, starting on the 1st, when 928 Squadrons A Flight HQ was set up at Sheephouse Farm, Easton in Gordano, while on the 10th, very high winds were reported in the vicinity of Avonmouth and 2 Balloon Barrage, complete with motor lorry attached, blew into the docks. On the 19th, no.11 Balloon Centre was ordered to prepare to move 929 Squadron, complete with equipment, to Queensferry, to protect the Forth Bridge. A Flight 928 Squadron was re-designated D Flight 927 Squadron, bringing 927 Squadron up to four flights but leaving 928 Squadron with two flights only. 23 October saw the departure of 929 Squadron for Scotland, and by the 30th the HQ of C Flight, 928 Squadron is known to have set up in Felixstowe Cottage, Litfield Road, Clifton.

The next change took place on 22 November 1939, when 928 Squadron hauled down their balloons pending orders to leave the Bristol area, departing for RAF Felixstowe in Suffolk two days later. This left only 927 Squadron to man the Bristol and Avonmouth barrages.

A replacement, 951 Squadron, was, however, formed at Pucklechurch on 15 December, and on 30 December they took over as their Squadron HQ the offices formerly occupied by 929 Squadron at that station.

On 20 January 1940, 927 Squadron had 32 balloons deployed at Avonmouth (10 inflated, 22 deflated on site) and 50 balloons at Bristol (33 inflated, 17 deflated on site). It was not until 30 March, that 951 Squadron was able to take control of the Bristol Barrage when their Squadron HQ was moved to no. 3 Caledonia Place, Clifton, the old HQ of 928 Squadron.

On 30 March 1940, 951 Squadron from Cardiff took over the Bristol City Docks barrage and established its headquarters at 928 Squadron's old premises at Caledonia Place. By August 1940, they had 40 balloons flying around Bristol. Balloons in Clifton were flown at the lookout point at the bottom of Sion Hill (the tender was parked in Caledonia Place), at the Observatory (where anti-aircraft guns were sited along with dummies made out of telegraph poles), at the top of Bridge Valley Road and at the water tower on the Downs. Unfortunately, no maps now exist showing balloon locations, consequently, it is impossible to say where individual balloons were flying at any given time, bearing in mind that their positions changed, protecting the aeroplane works at Filton, and the Bristol, Avonmouth and Portishead Docks between September 1939 and July 1944.

Balloon barrages at Filton and Avonmouth by early June 1940 were a most sensible precaution, for it was on the night of 19 June 1940 that the *Luftwaffe* carried out its first attack on the West Country. The attack force of at least seven Heinkel HE 111s of III/KG 27 based at Merville in Northern France were briefed to attack the Bristol Aeroplane Company at Filton, in addition to harbour installations at Avonmouth and Southampton. The attack was not a success, the nearest bombs falling at Portishead at around 2:15am, but this was just the first of a long series of raids that were to last until May 1944.

One of the biggest risks for the balloon barrages throughout the war was the danger of a lightning strike during a thunder storm, and probably the most serious incident of this type took place in the Bristol area on the night of 26 July 1940, when between 00.30 hrs and 00.47am, a total of 28 balloons were struck by lightning and brought down in flames. Those destroyed were 19 of 927 Squadron at Avonmouth; four of 935 Squadron at Filton; and five of 951 Squadron at Bristol.

> Squadron Equipment Location in August 1940:
> 912 24 balloons Brockworth
> 927 32 balloons Avonmouth
> 935 24 balloons Filton
> 951 40 balloons Bristol
> 957 24 balloons Yeovil

By the summer of 1940, aircraft factories were singled out for special protection from barrage balloons, and on 29 May, no.32 Group ordered 935 (County of Glamorgan) Squadron to move from Cardiff to Filton with 16 balloons, winches and crews.

On 31 December 1940, a new policy for the use of the small, very low altitude Admiralty Mk VI Kite Balloons in land barrages was introduced. These balloons had been designed to protect shipping and harbours. The plan called for the substitution of Mk IV balloons for LZ (low zone) balloons in barrages protecting certain aircraft factories, and this included Filton. The idea was to ultimately withdraw 60% of LZ balloons and substitute with Mk IVs – two Mk IVs to one LZ.

11.5 Balloon locations

Wartime records give details of incidents and casualties of the barrages, but are recorded by squadron and site number which give little or no idea of the locations. The memories section (from page 174) has recorded where people remembered balloons. David Wintle has identified 29 sites he could remember. The ones nearest to the Railway were the one by the Observatory, top of Bridge Valley Road by the Promenade and Proctor's fountain, and one on the dockside behind the Hotwell Road.

11.6 Wartime telegraph posts

11.5. There is a still a line of four 1939 telegraph posts leading from the Observatory (which has been used as a lookout since at least the Iron Age) to the Suspension Bridge booths, which would have continued down to the Avon Gorge Hotel. This enabled the balloon operators to communicate with Hotel headquarters. The Observatory was also requisitioned. (Author)

11.7 Permission granted to use Clifton Rocks Railway

On 25 March 1940, Imperial Airways gained permission to construct an office suite and use part of the upper section of Clifton Rocks Railway for storage and air-raid shelters, having started investigations on 7 September 1939. The plans can be seen in chapter 10, plate 10.7.

On 1 April 1940, Imperial Airways amalgamated with British Airways to become British Overseas Airways Company. Later plans for the Railway refer to the Air Ministry and BOAC. Passes were used to gain access to the shelter refer to BOAC – one can be seen in chapter 12, plates 12.5 and 12.6.

On 3 September 1940 (we found a date on the roof of the top station of Clifton Rocks Railway – see plate 10.20), the top section and chambers in the tunnel were completed using 14in thick brick walls, and put to use. The first chamber was isolated because it was only 12ft below ground, and this is where the top station railway artefacts were dumped. The second was for the barrage balloon section where there were offices and a workshop. One 31ft shelter for Imperial Airways staff and two 91ft shelters were built further down, and the last 91ft was used later by the BBC. Two staircases were built either side of the tunnel. Ministry of Works for Imperial Airways rented the top station for use as a first-aid post, completed in September 1940. They paid £100 per annum to the Tramways Company.

Staff in the Hotel did drill practice in the top refuge area; they sheltered in the tunnel if a daylight air raid was imminent.

11.6. The section in the tunnel has two floors (the height of the tunnel is 18'6", width 28') so that the tunnel roof can be accessed. There are strips of wood on the roof which could have been used to suspend the material. The plan for this section can be seen in plate 10.8. The two small sections by the tunnel wall are water closets, the adjacent rooms have an unknown use. The barrage balloon material was found on the floor in the little corridor above the central steps.

11.7. Repairing a barrage balloon. Balloons would have been too big to inflate in the tunnel, but they were repaired there. Panels were overlapped and taped. The solvents used in balloon repair (see plate 11.13) were mainly benzene based and carcinogenic. This is why there was a ventilation system in the tunnel as well to remove the fumes. All fabric workers had the 'benefit' of a free pint of milk a day because of the ingestion of the fumes from the solvents. All the electrical fittings in this part of the tunnel are spark-resistant brass, and cables are covered in lead (see plates 11.10 and 11.11). (*Bristol Times* 12 January 2006)

11.8. The ventilation hole for the fan (there is one on each side of the tunnel).

11.9. Peter Garwood in section 11.10 describes how a deflated balloon could be moved from one spot to another and this photograph shows women workers doing this admirably. It must still have been a monumental struggle carrying balloons needing repair down about 100 steps in a 4ft-wide passageway. (Stephen Rowson collection).

On 24 November 1940, the first large-scale raid on Bristol took place when 135 enemy bombers attacked the city between 6:30pm and 11:00pm. No. 951 Squadron recorded where flares were dropped, where balloons lost height or balloons were adrift (one with 2,600 feet of cable, another with 4,500 feet of cable), ripped or hit by splinters.

On 1 December 1940, a bomb landed on Tuffleigh House garden adjacent to the Railway, when shelters were in use. Only a dull thud could be heard, and the children who had been sheltering in the tunnel went looking for shrapnel in the morning.

11.8 More Balloon Command activity and the end of the war

On 14 January 1942, 951 Squadron was absorbed by 927 Squadron, which then covered Avonmouth. No.3 Caledonia Place could either have been vacated at around that time, or perhaps taken over as a Flight, rather than Squadron, HQ. On 14 April 1943, 927 and 935 covering Filton, combined to form 927/935 Squadron, with an HQ at Sneyd Park.

No successful large-scale raids had been carried out against the area since the heavy attacks on Bath and Weston-super-Mare in the summer of 1942, and the last of all, attempted against the harbour installations at Bristol by a force of 91 bombers on the night of 14 May 1944, was a complete failure. Only one aircraft succeeded in even finding Bristol, dropping its three sticks of high explosive bombs at Abbots Leigh, Kings Weston, and Headley Park at around 02:00am. In Bristol the 'all-clear' that night sounded at 03:07am; signalling the departure of the last Luftwaffe bomber to threaten the area.

927/935 squadron which had 32 officers (including ten WAAF's) and 972 other ranks (including 678 WAAF's) on its strength on 30 June 1944, and was made up of at least six flights, continued to operate the local barrages until 11:59pm on 12 July 1944 when all balloons in the Bristol area were declared non-operational prior to their move to combat V1 flying bombs (doodlebugs) on the East Coast, where an extensive barrage of 2,000 balloons was planned.

On 8 May 1945, the war in Europe ended.

On 18 September 1948, the Grand Spa Hotel reopened. The Pump Room was opened for dancing on 2 October. It is not known how much compensation the Hotel would have received for the 10 years it was closed.

11.9 Barrage Balloon artefacts

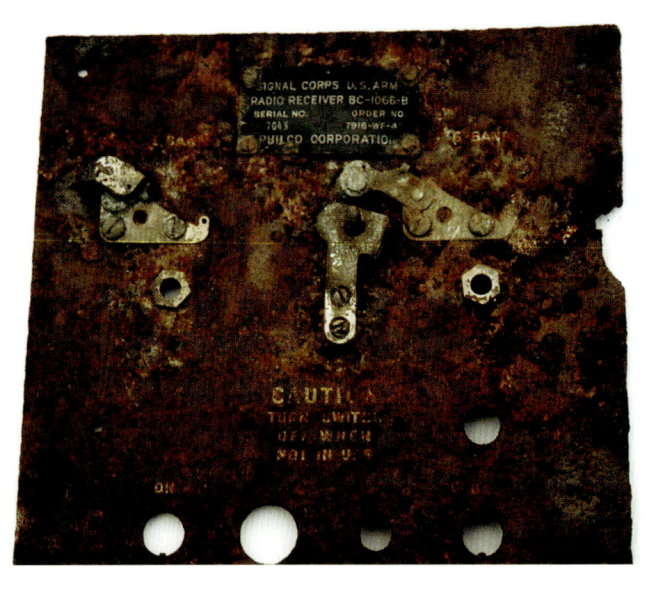

11.11. Spark-proof brass cable fittings. Cable covered in lead.

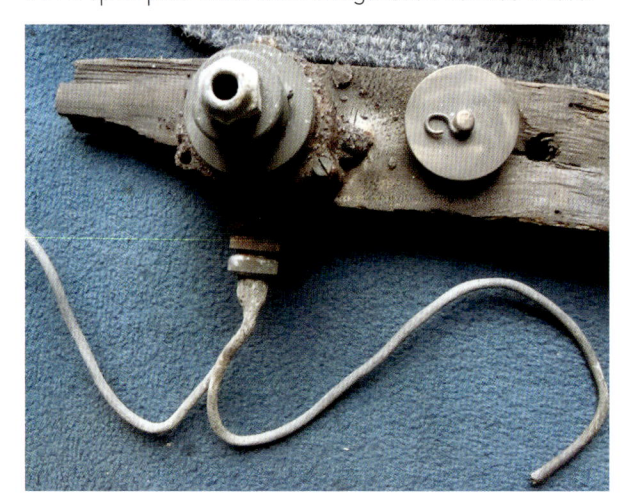

11.10. 22 Signal Corps US Army radio receiver panel BC-1066-B Philco Corporation (found in the rubble on the top station rails). This is a two-band 1940 receiver for testing aircraft IFF (Identification Friend or Foe) sets. A ground transmitter sends a transmission to the aircraft, which in turn switches a transmitter in the aircraft to 'send'. The ground station receives the aircraft transmission which enables the operator to identify the aircraft. The receiver was battery powered. 7.4in. x 7.1in.

11.12. Brass connector. The brass bung with chain attached is employed when the socket is not in use (the socket unscrews). Highly flammable hydrogen was used in balloons.

11.13. An aluminium and brass turn switch, 2.7in x 2.8in. Found in the balloon office area.

11.14. Balloon material (rubberised to the touch, silvery grey). Found upstairs in the barrage balloon area.

11.10 Memories – barrage balloons

Barrage balloons were distinctive and often had names. I have not found anyone who worked in the Barrage Balloon section. This section is followed by memories of people who worked for BOAC in the Hotel.

David Wintle

As a schoolboy, he used to watch barrage balloons in Clifton riding the Bristol sky.

'In April 1941, my father was discharged medically unfit for further service, but through the good offices of the *Auxiliary Fire Service* (AFS), he was successful in obtaining the post of Caretaker of a large house on the Promenade at Clifton that had been requisitioned as offices for the City Treasurer's Department, whose offices in the City had been badly damaged during the Blitz on Bristol. My family moved from Hillfields to Clifton, where much to my delight there were balloon sites at each end of the Promenade. Balloon watching became almost routine as I had to pass the site at the top of the Promenade four times a day going back and forth to school. Consequently, I was then able to watch many of the activities that went on. One night in May, the balloon was blown into trees on the side of the site and the following morning on the way to school I spent quite some time taking in the scene. Coming home at the end of the morning, I could see the balloon had been removed but two remnants had been left behind, as probably they were not worth salvaging and these I also still have as a reminder of those fascinating times. This particular balloon was unusual as, instead of being the normal silver, the body of the balloon containing the hydrogen had been over-sprayed a golden colour. Just what the purpose was I can only think was either that it had been used as some sort of experiment in camouflage or perhaps the fabric had become porous and had been over coated to seal it, but why this particular colour?

The site at the bottom of the Promenade was at the top of Bridge Valley Road in a lovely clearing surrounded by trees, and once more on this site I watched a balloon being inflated, again on a Saturday morning when I did not have to go to school. This time though, I was able to watch the whole process and see it change from being a big limp sort of bubble into a shapely balloon as it rose into the wind and the fins fill with air. There were two other sites on the Downs which I used to visit from time to time, one of these being on a very exposed site right on the edge of the Avon Gorge at a spot known as 'the Sea Walls'.

11.15. Books were published to make balloons seem less threatening to children.

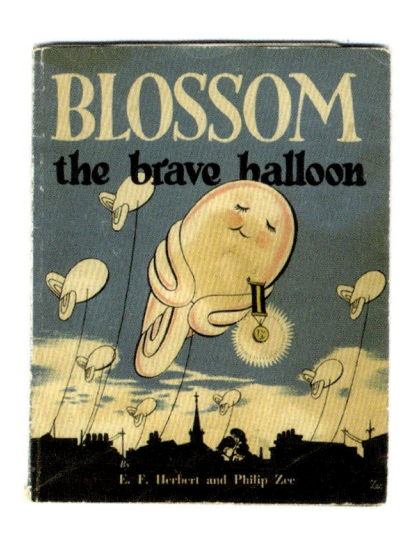

There was absolutely no shelter from the wind from whatever direction it blew from but I am not aware that any balloons were lost from there. The next site round the barrage was at the top of Blackboy Hill with the junction of Westbury Road behind a copse of trees giving some shelter from Southerly winds. Both these sites had Fordson Sussex winches, unlike the other two sites which had trailer-type winches. I will not bore you with the other sites I gradually became aware of, except to mention that in February of 1942 I was recommended for the newly opened Technical School at Bedminster Bridge, and three afternoons a week I had to go to the workshops at what was previously Temple Technical School, which was situated just off a major road called Temple Way. Here another balloon had been sited on a small vacant plot of land only just large enough to take a balloon. When it was bedded down you could almost touch it but when it was hauled up and down on a windy day it sometimes scraped the side of the adjoining furniture warehouse due to being in such a confined space.

I noted many improvements both to the sites, and the balloons during my many visits to the different sites. One of the most notable was when, on moving to Clifton, balloons were no longer flown straight off the back of the winch, but the cable instead of going vertically up to the balloon, was paid out horizontally and round a pulley wheel concreted securely into the ground and then up to the balloon with the winch itself moved well away from the actual bed. The next so say improvement was that ash was laid all over the bed, covering the grass, no doubt to avoid working in mud in very wet weather and also to avoid vehicles getting bogged down. Another surprise that I had when visiting the two sites at each end of the Promenade was to find the trailer type winches had been lifted off the actual trailer and mounted on two strips of solid concrete about 18 to 24 inches high with a reduction in the height of the winch from the ground, no doubt making it considerably easier to jump onto the winch. The purpose of this I never found out, and can only assume that perhaps the trailers were put to another use. [If anyone can enlighten me I would be very grateful and also if the winches were lifted by crane or by means of some form of jacks.] The next obvious improvement was that the handling lines of the balloon, instead of being left hanging free when it was flown, were coiled up and gathered together just behind where the flying wires were attached to the cable and placed in a canvas holder similar to the bags that held ballast but now replaced by concrete breeze blocks. A tail line was also fitted from the three attachment points between the fins with a loop in the line which went over a wooden peg attached to another line joined to a pulley, which could run along a mooring cable that went round the whole bed and held a foot or so above the ground on short wooden posts, the balloon being left free to face into the prevailing wind. When the balloon was at 'stand by' it was held by several short cables called a 'cradle' and the main cable allowed to go slack. When the balloon had to be flown, the main cable just had to be wound in a little so that the balloon could be released from the cradle, the tail line released from over the peg and the balloon sent up almost in a matter of seconds. Another major improvement made was that bricks were laid in a circle around and outwards from the main anchorage to form a paved bed which provided a first class surface to work on which was also free draining.

Lightning was the worst enemy the balloons faced, even more so than the Germans. It only took a few flashes of lightning to decimate a barrage, and there was the cost of replacement. To overcome this, a lightning conductor was held in a wooden frame mounted on the top of the nose with a wire running round the balloon covered by a strip of fabric that I presume was connected to the cable via the flying wires. It was a sad sight to see balloons go up in a mass of flames and be gone in a matter of seconds, leaving a trail of black smoke marking the descent of the remnants and the cable towards the ground. When it happened at night however, it was quite spectacular.

The one area where improvements did not seem to occur to me were in the domestic arrangements. When I compare the conditions in which I lived while in the RAF, to those on the balloon sites, it makes me wonder just how they managed. All the sites that I remember in Bristol just had the one standard wooden hut with no special washing, cooking or toilet facilities, which I can only assume must have been carried out within the confines of the hut. [If anyone can enlighten me further on this angle I would be most pleased to hear from them.] I guess that water and electricity must have been laid on.

Before I end, I have one other happy memory which I will relate. The Balloon Squadron, No.927 which was the main Bristol squadron, had a Flight or Squadron office in Caledonia Place in Clifton. I happened to be passing by it one evening when a Fordson Sussex winch happened to be parked unattended on the opposite side of the road. It was most unusual to see a winch parked in the street rather than on a site, so I took the opportunity to take a good close up look at it. There was no wire mesh door on the side of the cage. Sometimes these were left on and tied up when on site so as to avoid the trouble of holding them up while at the same time climbing into the cage but quite a lot of winches seemed to have had them removed. Fortunately for me this one was missing, so having spent a few minutes seeing just what was inside the cage I wondered whether to take a chance and climb in and sit in the winch operator's seat.

As I could see no one about in the Flight Office, I thought I might be able to get away with it. After all, I was never likely to get such an opportunity again and probably the worse that could happen was someone shouting out 'Get down out of there' or words to that effect. Having decided that it was worth taking the risk, I climbed up in and sat down. No one saw me as nothing happened and I spent a happy ten minutes or so pretending I was a winch operator. Now when I see photographs of a Fordson Sussex, I have the satisfaction of knowing that I have actually sat in one.'

Paul Adams
Born February 1939, lived at 24 Royal York Gardens. Paul remembers the barrage balloons were sited at the children's playground by the Observatory.

Stephen Alexander
Born 1924, lived at no. 29 Victoria Square. He counted 42 air balloons in the sky from the roof of his house on 4 August 1940, 54 on 29 August; saw balloons go up on 12 August and land on 4 September 1940.

P.F.H. Clarke
A Clifton College pupil who was a Local Defence Volunteer. 'On 24 June 1940, on the first night of bombing two German bombers circled at a colossal

height of about 20,000ft, and came down into the clouds conveniently just above the balloons (8,000ft). They dropped 20 high-explosive bombs causing three deaths in Brislington and putting Temple Meads out of action for a few hours (eight bombs fell on the lines or on surrounding platforms), six bombs in Old Market and killed two people by the eye hospital. The raid lasted 70 minutes; not a single anti-aircraft gun fired and not a single fighter went up.'

Mona Duguid
Born 1915. 'On 20 July 1940, there were balloons everywhere. Often, during a bad thunderstorm, they would be struck down by lightning, burst into flame and slowly drift down, usually burnt out by the time they reached the ground.'

Mike Farr
Born 1931, lived at Sion Hill. Recalls balloons below the observatory, at Keepers Green in Leigh Woods and Fountain Hill (Fairyland). He remembers Big Bertha – the name of the gun on the Observatory.

Tony Fowles
Born 1931, he lived at the top flat in no. 7 Canynge Square. 'There was a barrage balloon by the junction of Suspension Bridge Road and Observatory Road. There were several people with it, and being a small boy he was always very curious about it. Sometimes it was up, others down.'

Peter Garwood
'Be careful because the solvents used in barrage balloon repair were mainly benzene-based, which is carcinogenic. There must have had some ventilation system in the tunnel [there was] to remove the fumes. All fabric workers had a free pint of milk a day because of the ingestion of the fumes from the solvents.

Deflated barrage balloons could be manhandled by a "sausage" of people (see plate 11.9), each holding either side of the rolled-up balloon. This was probably no more trickier than moving a modern tarpaulin from one spot to another, but requiring a bit of concentrated organising and practice. I was intrigued by the "anti-spark" precautions. I am not convinced there were sewing machines in there.'

Donald Lear
Remembers barrage balloons on the Downs and Brandon Hill.

Brian Long
The following is a fictional story based on fact from Brian Long's personal experiences when, as a small boy, he ran errands for the all-girl crew on the Marsh (the stretch of grass that ran parallel with Winterstoke Road, Ashton) with their barrage balloon. The barrage balloon site was there for some years, but became an all-girl crew only in 1943. The site was disbanded, and the personnel re-posted in 1945 simply because it was realised that barrage balloons were totally useless against the dreaded rockets falling on the Home Counties, though fortunately these awesome weapons never reached this part of the country.

With a tremendous roar, the autumn wind of 1940 once again threw itself like a nuclear blast at the huge, bulbous thing suspended over the Marsh. Instantly a loud crack was heard, making it evident that the winch cable had snapped causing the silver-coloured giant to rise slowly but silently from its mooring site.
Frantic yells from the RAF girls echoed around the camp

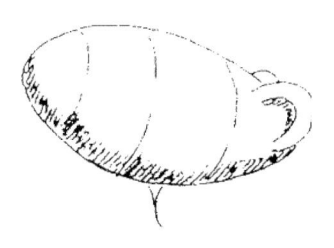

11.15.

as they came rushing from their billet. They gathered to stand with worried faces as they stared up at the elephantine monster that was now, thankfully floating away from the city centre, with its severed cable trailing like some angry, twisting snake.
'Oh, well, it's goodbye to Charlie now,' said one, giving it a sympathetic wave. 'Will our Spitfires shoot it down when it reaches open country?' asked another. 'Maybe,' replied Sergeant Gibbs, a tall ginger-haired girl, who added, 'as Charlie is heading towards the Mendips he should be reasonably harmless. He could explode, though, if he soars to a very high altitude — you know, the pressure and all that.' The girls nodded in answer, only to receive immediate orders from her to clean up the site and replace the broken winch cable in readiness for the enquiry–cum inspection that would soon follow from the higher ups, as Sergeant Gibbs put it.

11.16.

Within the hour the job was done, and the girls were soon back in their billet, waiting for the arrival of a new balloon. They did feel somewhat disappointed at having lost Charlie, since they were proud of the work they had been doing protecting Bedminster and Ashton from low flying enemy aircraft. The girls began wondering about the balloon sites at Purdown, Durdham Down, Dundry and all the others. Had they lost any, they wondered.

As the day wore on the girls began to get bored, and moped about because they couldn't do their proper job, until Sergeant Gibbs told them that the delivery of a new balloon would be within twenty-four hours. The cheers on hearing this echoed throughout the billet and their spirits were soon back to normal.

'Shall we call it Clark, after Clark Gable, when it comes?' squealed one, dreamy eyed, as she fondly fingered a photograph of the famous film star which was gummed to the door of her locker. Sergeant Gibbs shook her head, saying, 'I think that in the light of what's happened it would be best to name our next balloon Winston. After all, their names and numbers are put on the register for identification, so I think choosing the Christian name of Mr Churchill, our Prime Minister, is appropriate and patriotic, don't you?' The rest agreed, but they could hardly wait for their new baby to arrive, deflated of course, on the back of a transporter lorry.

Later that evening, Dreamy-eyes called out, in a muffled voice from under her bed clothes, 'Roll on Operation Winston, your mummies are waiting to make you into a big, strong boy and put you to work, so that you can protect us all from those nasty bombs.' Someone sniggered and muttered, Silly sausage.

Snores soon followed, with Sergeant Gibbs finally turning out the lights to mark the end of a dutiful, but unfortunate, day for those brave, hard-working balloon girls on the Marsh.

Iris Mitchell
Remembers balloons by the Suspension Bridge in a children's play area.

Marina Rich
'The barrage balloon tethered by the Suspension Bridge was very high. It was shiny and silver with big ears. It was fabulous.'

Tony Riddell
He remembered a barrage balloon by the Observatory and one at Clanage Road, Bower Ashton.

Barbara Salter
'Barrage balloons were in a dog shelter across the River Avon and one got loose and landed in mud.'

Stan Snook
Born 1914. He can remember the barrage balloons. They were moved around since they were mobile. He can remember a thunderstorm when many were lost when they had been too slow at getting them down.

11.11 BOAC memories
Jean Gillingham
Born in 1922, Jean was an employee of BOAC. She grew up in no. 1 Victoria Terrace. During the war she said that BOAC had offices (reserved occupation offices?) 'beneath the dining room' (the Pump Room) and that they were not really allowed to know what went on in the tunnel. For some reason she was involved with finding jobs for 'foreigners' – chiefly Polish men. They turned up at the office to be processed. One of their responsibilities was to make arrangements for aircraft delivery/ return by the Air Transport Auxiliaries. They managed to get hold of soap stamped BBC from the tunnel. It was rationed so it was a treat. She can remember beds, tables and chairs there.

Local people did use the tunnel as an air-raid shelter but Jean's family preferred to go home. There was a lot of bombing. Jean could remember that Sir Winston Churchill was smuggled 'by the back way' to the tunnel. He also visited Eisenhower at Clifton College where American troops were billeted. [American troops were billeted from October 1942 when General Omar Bradley moved in and the College became the US First Army HQ.]

After the war, Amy Johnson came on private visits to the hotel. There was a special code signal to announce her imminent arrival and they had to run a hot bath for her. She could remember Derek McCulloch, head of the BBC (a BBC Radio presenter and producer. He became known as Uncle Mac in *Children's Favourites* and *Children's Hour*, and the voice of Larry the Lamb in *Toytown*. He was head of children's broadcasting for the BBC from 1933 until 1951) and events at St Andrews Hall (on Merchants Road). The BBC had offices in Druid Stoke Avenue and later in Stoke Bishop.

Patricia Jean Morris
(from letters supplied by her son Ewan MacLeod)

Patricia was born in 1919 and died in 2014. Her wartime letters disclose a very particular and rapidly changing world. She arrived in Bristol for the very first time on 31 March 1939. She notes the anniversary date in a letter written exactly four years later, and she recalls spending her last night on the Isle of Wight at Thorncroft with her grandmother before she left, because her own parents were away in France. She was just 19-years-old when she first came to Bristol. She had recently qualified on a secretarial course in London, and her first job in Bristol was at the Burden Institute in Bristol, working as secretary to Professor Golla, which was where she met Ewan's father (Leslie – a scientist working at the Burden Neurological Institute at Frenchay. He was also a gunner with the Homeguard AA Battery 215 throughout the war). She left the Institute not long afterwards to begin work with BOAC, the newly created merger of Imperial Airways and British Airways. She worked for BOAC in Bristol from 1940 to 1 July 1946.

Mother left the Burden Institute to work for BOAC under the terms of an Essential Works Order' (EWO). The work was deemed to be of national importance, and she was prohibited from leaving BOAC unless she transferred to other work of higher priority, or was going to have a baby. The EWO meant that her potential wartime enlistment into the armed services was deferred indefinitely.

Mother recalls more than once that her first visit to the cinema with Leslie to watch a film together was on 21 April 1939 when they watched *The Young In Heart* (1938) starring Janet Gaynor and Douglas Fairbanks jnr. at the Odeon in Bristol. They frequently watched films at the Embassy and the Whiteladies Picture House as well. Mother once told me that when she first came to Bristol, she used to watch foreign language films in French and German at the Plaza/Academy in Stokes Croft, one of the very few cinemas in Bristol to show subtitled feature films at that time.

Shortly after BOAC was created in 1940, they moved their HQ to Bristol, setting up in the Grand Spa Hotel overlooking the Clifton Gorge and Suspension Bridge, and then later out at Whitchurch field in south Bristol which became an important staging post in WWII civilian travel. It was often referred to in wartime news bulletins as 'an airfield at a town in the west country'. BOAC subsequently moved to Hurn airport for the duration of the war in 1944, before moving to Northolt and Heathrow in 1946. Much of the daily background in her letters relates to her working life as a clerical officer within BOAC. Mother wrote letters most often when she was sent away to other BOAC stations, notably Leuchars in St Andrews Scotland, and Hurn airport near Bournemouth. She was also seconded up to Airways House in Victoria London on more than one occasion. In both Bristol, and later at Hurn she was often working as PA to Capt. J.C. Harrington who was the head of operations at BOAC, #2 Line during the war, and for a number of years afterwards. The task of managing the BOAC fleet was complex in the extreme. The merger of companies had created an air fleet of highly diverse planes including the long-haul Short S.23 flying boats of Imperial Airways and the land based Douglas DC2 and DC3s of British Airways. BOAC became an umbrella operator for Qantas which was flying to Australia and New Zealand, and they had also inherited at least six airliners and staff from the Dutch KLM fleet that managed to escape the German Invasion of Holland in 1940. BOAC had also gained the services of a number of refugee pilots from Poland, Denmark and Norway and were also clandestinely operating high speed blockade runner flights from Leuchars to Stockholm in specially converted De Havilland Mosquitos bringing back valuable industrial ball bearings, and VIP refugees such as Danish nuclear physicist Nils Bohr, who had escaped from the Gestapo into neutral Sweden in 1943.

One important and dangerous BOAC route was from Whitchurch to Lisbon in Portugal, a service that the emigre KLM fleet of planes had contracted to run. They had to fly over the Bay of Biscay and were often attacked by German fighters even though they were in international airspace. One tragic loss mentioned in the mother's letters was that of flight 777-A on 1 June 1943 when a KLM DC-3 was shot down over the sea en route to Whitchurch from Lisbon with the loss of

17 lives including that of film actor Leslie Howard. There was much speculation both then and since that the German Abwehr may have mistakenly believed that Winston Churchill was onboard returning from the Cairo conference.

The letters invoke a wartime Britain of continuous and escalating shortages. Both food and clothing were quite strictly rationed. In one letter mother refers to enjoying a succulent orange that had been given to her. This was a special treat indeed, oranges and bananas had vanished completely for most people by 1943. In one letter she was concerned about mislaying a coat, and in another letter she was highly agitated because her case with all her spare clothing had been mislaid during a transport to Hurn. Her upset becomes more vivid when you understand that clothing rations had been reduced to just 18 points per annum by 1944, and that a single item such as a winter overcoat would consume all of this allowance. In another letter mother recounts donating her clothing coupons to her mother because it was her parents 25th wedding anniversary on 6 June 1943. Mother quite frequently refers to purchasing cigarettes for father, Craven A as a rule. Tobacco was not officially rationed, but cigarettes were increasingly hard to come by. Many supplies had disappeared under the counter and were kept for regular customers, or found their way onto the black market at five times the normal price. Mother who did not smoke made a point of finding supplies of Craven A for father throughout these war years (he later switched to Players Senior Service). Even paper was in short supply. Mother frequently had to struggle to find blank paper on which to write her letters using mismatched remainders of office stationery, hotel writing cards and paper, and even airmail letter pads from BOAC stores. Mother quite frequently mentions hunting around to find copies of *Punch* and other periodicals that father enjoyed. Once again, wartime shortages of paper and newsprint meant that many periodicals were produced in sharply limited print runs, and many books dropped out of publication altogether.

The country was at war and civilian populations were under attack with frequent air-raid alerts. One letter written by mother on 24 November 1940 mentions the first heavy blitz attack on Bristol. She was very alarmed because Leslie was missing from work just after the raid. One letter refers to hearing the AA guns at Purdown firing at night when mother was living in Chesterfield Road. St Andrews, and another letter mentions hearing the different sounds of the Portbury and Avonmouth batteries when she moved up nearer the Clifton Downs. In one letter mother was back at her parents home Windyridge on the Isle of Wight convalescing from a period of ill health. She mentions that she had inadvertently broken the radio set so her father 'Guggy' could no longer listen to his much loved BBC radio news bulletins, which was unfortunate timing to say the least because the letter is dated 6 June 1944 which was the date of the D-Day landings in Normandy.

'Over the next couple of days mother wrote of seeing vast numbers of aircraft flying over the island en route to France and of hearing distant heavy gunfire in the channel. She also saw a crippled V1 flying bomb that only narrowly cleared the house and came down just over the hill on Mount Pleasant near Carisbrooke.

One becomes conscious of the vast dislocation of everyone's lives that was caused by WWII and of just how long it took to reverse the mass induction into military service and return so many people to ordinary civilian life, even long after it had all ended. Aunty Tom for example was still in the WRENs well into November 1945, and Philip was not released from naval service until 1946 even though there was nothing much for him to do after Xmas 1944. He was apparently billeted on a training ship HMS *Dauntless* moored in Rosyth which by that stage was so overcrowded with RNVR personnel that he was having to sleep in a hammock on a communal mess deck according to one letter. Mother's own clerical work at BOAC did not become any easier with the end of hostilities. She recounts one experience of a meeting where she was supposed to be taking shorthand notes. The meeting was a showdown between representatives of Rolls Royce and BOAC's own chief engineer. Apparently the head-blocks of Rolls Royce T24/2 engines used on Qantas flights were failing after only 120 hours of service which meant the engines had to be rebuilt or replaced after every single flight. The discussion degenerated into such a violent argument between the men that mother was hard put to produce any printable memorandum of the meeting.'

The letters provide a fascinating snapshot of the entertainments and diversions on offer in the wartime years. There was no television service. The experimental TV broadcasts by the BBC from Alexandra Palace in the mid-1930s had been shut down in September 1939 just before WWII broke out amid fears that the high power VHF transmitters would provide perfect guidance beams for inbound German bombers attacking London. The BBC Television service did not resume broadcasting until 7 June 1946. For the duration of the war listeners relied solely on the BBC

Home and Imperial radio services for entertainment and news. Mother was particularly fond of classical music and radio play broadcasts. Cinemas were initially shut down across the country after war broke out over fears about air attacks, until the government realised the powerful role that the cinema industry could play both as a morale builder and a potent way of spreading public safety information and government-controlled propaganda. Cinemas were reopened for the duration of the war in December 1940 and were very popular. Cinema going became a shared passion for mother and Leslie throughout the war. She often mentions lengthy queues to get into screenings both in Bristol and in Newport, IOW and she quite often mentions particular films she had enjoyed at some length. She was for example powerfully impressed by the film adaptations of Daphne Du Maurier's novels *Rebecca* and *Frenchman's Creek*. In one letter from 1945, mother spends three pages trying to explain the plot intricacies of a Bette Davis film called *The Great Lie* that had caught her attention. She was much less impressed with Marlene Dietrich's film *The Gardens of Allah*.

The letters contain many little snippets of fascinating information and conundrums of various sorts. In several letters mother refers to a regular problem with her wristwatch becoming magnetised and running fast which vexed her enormously because she was a highly punctual person who could not abide not knowing what the correct time was. The letters disclose that Leslie successfully cured the problem by demagnetising the wristwatch with a degaussing tool he had in his laboratory at the Burden Institute.

'What intrigued me was why mother's wristwatch was becoming magnetised in the first place ? The explanation appears to be that it was an unforeseen complication arising from the fact that she was a very frequent flyer who made many trips in aeroplanes in the course of her work with BOAC. The balance springs of watches have fine coils that can stick to each other if they become magnetised, effectively shortening the length and period of the spring, which makes them run faster. The problem is exacerbated because these springs are often made of a special alloy that is intended to self-compensate for thermal variations by employing internal self-magnetisation to compensate for thermal lengthening of the hairspring. Climbing rapidly to higher altitudes in an aeroplane through the earth's magnetic field upsets the magnetic self-compensation of the balance spring.'

Doreen Whitfield
Born 1926. Speaking in August 2005.
'I have tried to remember and research as much information as possible for you, recalling many happy memories of my Bristol days, before my transfer to London and the London Airport offices of British Airways – formally BOAC and Imperial Airways – in 1946, when I was 20 years old.

I was a member of the BOAC staff from 1943 until 1953 and was initially employed in the BOAC offices, which took over the whole of the Grand Spa Hotel (now the Avon Gorge Hotel), and also in the Legal Department, based in Tuffleigh, a house at the rear of the Hotel then occupied by Mrs Patrick and family. Some of my work involved looking through correspondence regarding Air Transport Command activities covered by some BOAC aircrew – for this I had to sign the Official Secrets Act.

As far as I can ascertain there was never a swimming pool in the basement of the Hotel. Before the 1939-45 war the basement area was the pump room and later became a ballroom with stage for an orchestra. During my time at the Grand Spa there was an entrance from beside and behind the reception desk of the Hotel, into the top of the Clifton Rocks Railway, where staff could enter the tunnel if a daylight air raid was imminent. As far as I can remember it was only used for a staff practice, but it may have been used by staff prior to my appointment.

The shelter was extremely cold; the steps were very steep (impossible for disabled personnel) and must have been very uncomfortable for the local population, who could shelter there during night raids.

The bosses were on the first floor, I was based on the third floor. I only knew about the barrage balloons in Redland. We were not allowed to stray too far. The Pump Room was full of desks.

The BBC built studios into the tunnel, which could be used by the corporation as an 'air-raid proof' studio and transmitter facilities. We had nothing to do with the BBC.

As a former Bristolian, I have found remembering and researching extremely interesting, and it has brought back happy memories of spending our lunch breaks shopping in Clifton, walking up to the Observatory and occasional visits to our favourite pub, the Portcullis, then owned by a sweet, plump elderly lady. I regret that I have no photographs of Bristol or Clifton during the war – cameras were not allowed. We were also banned from using the terrace adjacent to my office, because of wartime restrictions. Nor do I have any artefacts of the period – just memories.

British Airways and Imperial Airways amalgamated effectively on 1 April 1940 with the HQ at Airways Terminal, Victoria, London. I am not sure of the date of transfer of staff from Airways Terminal but it was probably soon after the merger of the two airlines. I had been trained as a nanny but did not like it.

I was working with the Forestry Commission (on low pay Civil Service salary) when I was told of an office vacancy within BOAC. A lady was already working at the Grand Spa was also doing Voluntary Service with the Air Raid Rescue Service stationed near the top of Blackboy Hill. My father was also a volunteer there throughout the war as well as having a fulltime position in the financial department of WD & HO Wills. She mentioned to my father that there was a position with BOAC I might be interested in with a higher salary plus a London weighting addition (a consideration for the staff who were transferred from London). I applied for the job, had an interview with Miss Thompson at BOAC Registries. Her husband also worked at Wills and knew my father slightly. I became a member of the Registry staff in 1943 and transferred to London with the majority of staff in March 1946.

My boss was Miss Thompson (formally in charge at Registrars with the BBC) and her deputy was Miss Harris, both returned to Airways Terminal Head Office in 1946. They were in charge of all the registries at the Mall in Trafalgar House, also at West Mall, and with the medical department at Clifton Park, where air crew were vaccinated.

The BOAC chairman at the time I was at the Grand Spa Hotel was Viscount Knollys who was introduced to all the Registry staff when he was appointed.

The only time I sheltered in the tunnel was during an evacuation practice at 11:00am. I took a cup of tea with me. This was in the first shelter [Shelter 1]. There may have been real emergency sheltering before I was employed at the Spa – nobody from BOAC actually worked in the tunnel to my knowledge. The BBC were already there. I left BOAC in 1953.'

The next chapter deals with peoples' experiences using Clifton Rocks Railway and the Portway tunnel as a shelter. There are more wonderful memories. These are the next group of people to use the tunnel, which was ready for them at the beginning of September 1940.

twelve

Air-raid Shelter
Operation and Memories

This chapter deals with operation of shelters by the Air Raid Precaution (ARP) team, details about the Clifton Rocks Railway and the Portway shelter facilities, the amazing collection of artefacts found in the Railway under stairs and shelters, and memories of the Clifton Rocks Railway shelter, the Portway shelter, and of people in the near vicinity which give a good insight into what it was actually like. The Portway shelter has been included to describe the totally different experience for people desperate to hide in an underground railway tunnel nearby.

12.1 Air-raid Precaution operation

ARP Wardens were responsible for local reconnaissance and reporting, leadership, organisation, guidance and control of the general public. Wardens would also advise survivors on the locations of rest and food centres, and other welfare facilities.[1] More details can be found in section 10.2. There was also a regional centre split into divisions such as the Clifton Division run by the Civil Defence (in 1941 Civil Defence replaced the ARP), which dealt with air-raid messages, and essential services such as water, electricity, gas, telephones, roads and sewers. Information concerning an air-raid incident was given by police and wardens to the appropriate divisional report centre, which then passed the details to its own action depot.[2] ARP had charge of the equipment; the Emergency Committee determined policy and control of operations. There were two warden post signs found in the railway (plate 12.38). The wardens mentioned in memories lived locally and were either too old or not eligible to go to war, or worked for BOAC:

Fred Adams: (fish merchant – too old) York Gardens
Fred Beacham: West Mall
Miss Clinch: Hotwells
Wilfrid Hanks: (BOAC)
Mr Thomas: (BOAC) 28 Caledonia Place
Mr Crane

12.1 Cloth badges were sewn on Wardens' overalls. They also wore red ARP on black arm bands, or later a yellow Civil Defence armband. (CRR Trust)

12.2. Wardens also wore silver ARP badges. (Author collection)

In the early part of the war the service had no recognizable uniform. Members would generally wear civilian clothes, typically boiler suits. As uniforms became available from February 1941, the service was issued with dark-blue battledress and berets. Wardens not issued with a uniform would wear with a blue armband with 'Civil Defence' in yellow. Insignia included a circular breast badge worn on the left pocket incorporating the letters 'CD' topped by a king's crown in yellow on dark-blue or black. A smaller badge with yellow circle around the CD and crown was used for the beret. Shoulder flashes would identify the individual's type of service. Additionally there were instructor badges, and first-aid badges worn on the lower sleeves with a red chevron for each year of service. Rank was indicated by yellow bars (2½ inches x ¼ or ¾ inches) or chevrons:

Controller – two narrow over one broad
Chief Warden – one narrow over one broad
Deputy Chief Warden – one broad
Divisional Warden – three narrow
Head or Post Warden – three chevrons (sometimes beneath a star)
Senior Warden – one or two chevrons.

12.2 Clifton Rocks Railway Shelter
ARP here was run by the Air Ministry. Passes (see plates 12.3 and 12.4) were obtained from the Air Ministry at the Grand Spa Hotel, and wardens were able to get space on the ledges for their families. Passes were free. Two wardens were stationed at the Railway top station all night to inspect shelter passes and direct people.

Shelter 1 was for Air Ministry staff, with nine ledges for 72 people sitting. It was used during the day. Shelter 2, for Clifton residents, had 22 ledges for 176 people sitting, as did shelter 3 for Hotwells residents. One could only stay in Shelter 2 and 3 at night from 6.00pm – people ate before they came and left in the morning. When a raid was on, the shelter lights were dipped three times. When the 'all-clear' sounded, tunnel lights were dipped once. During the day, the wardens cleared the toilets and litter.

The passes that were allocated seem to have gone to houses without basements. Residents of Alma Vale Terrace, Sion Lane, York Place, Glendale, St Vincent's Parade, Freeland Place and Princess Victoria Street were certainly given tickets for specific areas. One person said that if one did not have a ticket then one could sleep on the stairs; Donald Lear told the harrowing story that his mother had been turned away since she did not have a ticket, even though she was nine months' pregnant. She gave birth on Joy Hill. The lady up the road from me in Princess Victoria Street did not want a pass – she said it was too claustrophobic and she would rather die in her home.

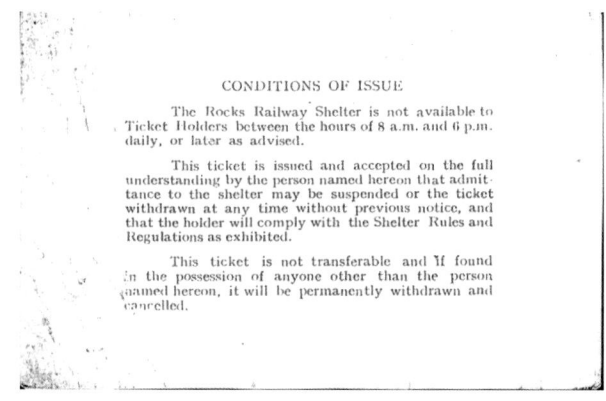

12.3 Front of a shelter pass. This identified the location of the ledge. One came back to the same one. (Peter Davey collection)

12.4 The reverse side of the shelter pass, specifying when it can not be used.

Home locations of shelter users
Bellevue Crescent 1, Rodney Place 1
Caledonia Mews 1, Sion Lane 1
Caledonia Place 1, Southmead 1
Canynge Square 1, St Vincent's Parade 3
Freeland Place 1, West Mall 1
Glendale 1, York Gardens 2
Portland Place 1
Princess Victoria Street 1
Polygon 1
Love Street, Hotwells 1
Bedminster 1

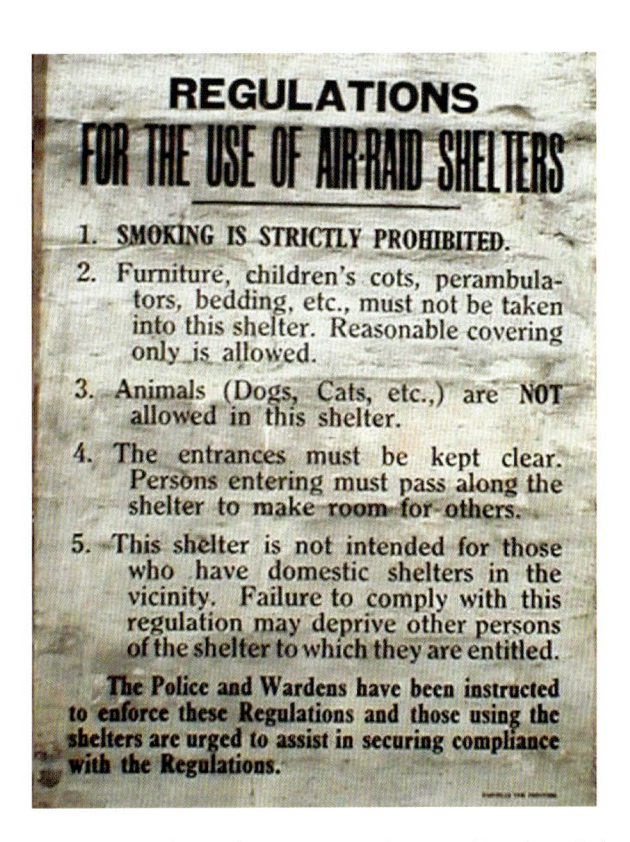

12.5. Air Raid Regulations. Note that smoking is strictly prohibited, though the rule did not seem to be enforced. (CRRT)

We found cigarette packets, and we were told that one person had brought a dog in, so clearly regulations were not always obeyed.

12.3 Shelter Artefacts

An article published in 2007[3] identified itmes we found under the ledges. It is a fascinating glimpse into everyday life under wartime conditions with evidence of many locally manufactured items. It was surprising how many labels survived in the the damp conditions. Just what does one bring when one is going to spend all night in the tunnel?

category	Number of items
beverage	79
cigarette	7
domestic	80
Shelter sign etc	11
shelter, electrical	6
toy	10

Table 12.1. Categories of items found

We wanted to know what life was like in the tunnels, so we searched for items under the sleeping ledges and the stairs to see what the residents left behind during their wartime stay. With refuge for 352 people, and 72 people in the smaller space, it was a big area to search. Also, access was difficult because of the large amount of rubble under the ledges, and the narrowness of the openings between the ledges. It is hard to search under the ledges – one can reach under to a certain extent but one needs to be quite slender to actually roll oneself under and it can be difficult to get back out. The gap under the ledge is about 60cm and the gradient just over 1:2. We have devised several techniques using litter pickers, garden hoes and extendable forks in inaccessible places. We found the bottom refuge area to be the wettest, especially in winter, and must have been unpleasant to sleep in. The canvas shelters over the shelter walls must have been very necessary, with water draining from the canvas into wooden troughs to drainpipes at the bottom of each section. Only the stairs were lit by electricity; the refuges were lit by hurricane lamps and small items would be easily lost. Bottles tended to roll down the slope under the ledges, sometimes stopped by rubble. China breaks and disperses. During the day litter was cleared away by the wardens, but, since people were allowed to leave their belongings behind, only rubbish in designated places would have been removed.

The refuge ledges have yielded many bottles and china (mostly broken), as well as more poignant items such as civilian gas masks (still with charcoal granules inside), a pair of children's glasses, shoes, a lead soldier, balls and a doll's head. It is a fascinating insight into social life in the 1940s.

12.6. Toilet seat found in the toilet area of shelter 3. (Author)

Post-war, many people have entered the area and broken things accidentally or deliberately, removed things and added things (such as disposable nappies - these are probably a relic of 2001 when *An Inspector Calls* played for three performanaces in Shelter 2).

12.7. One technique of looking under the ledges. Not practical if people were trying to sleep. (Author)

12.3.1 Beverages

More bottles were found for soft drinks than for any other beverage, as can be seen in table 12.2. Nearly the same number of beer as milk bottles were found. Soft drinks may have been the most popular beverage, but beer was undoubtedly popular. More smaller bottles than larger bottles were found (the top of a flagon of Georges beer was exceptional, perhaps because a bigger deposit was paid on larger bottles encouraging people to return them, or that it was easier to carry small items). Most drinks were made by local firms with many bottles still showing the remains of labels. Mixing hot drinks was also evident but less practical. We did, however, find an Acme thermos flask. The Acme vacuum flask company was established in 1931.[4] When war broke out, virtually all of the company's capacity was given over to military requirements. Table 12.3 indicates the number of bottles for each manufacturer. Bristol United was the most common beer bottle found (plate 12.9). At least one bottle still has beer in it. Only one bottle was from Georges Brewery.

The Spackman and Gosling Curfew sherry bottle (plate 12.10) is fascinating because its label presumably encourages drinking in curfew time. Curfew wines (port and sherry) were distributed locally from the Quadrant at the top of Princess Victoria Street in Clifton. Astral (short for Australia) was an Empire Curfew wine.

Brooke and Prudencio[5] (who made lemonade and ginger beer between 1860 and 1960) had many styles of bottles (plates 12.12 to 12.14). This was undoubtedly the most popular brand drunk in the tunnel and is the longest running mineral water firm. Their trademark was the Neptune statue (now on St Augustine's Parade). Also in evidence are lemonade bottles from MAPS in Fishponds, and milk bottles from many local dairies. Liquid Coffee Essence was the only alternative to ground coffee (instant coffee powder only introduced in 1950).

Milk bottles were mostly from Clifton dairies; one wonders what one from Abbots Leigh was doing there.

Type	1/4 pt	1/2 pt	1 pt	2 pt	Number of bottles
Soft drink		23	8		31
Milk		12	5		17
Beer		14	2	1	17
Coffee essence		4			4
Ginger ale	1	3			4
Spirits	3 miniature	1			4
Sherry		3			3
Ginger beer			2		2
Cider		1			1
Horlicks		1			1

Table 12.2. Beverage bottles found so far.

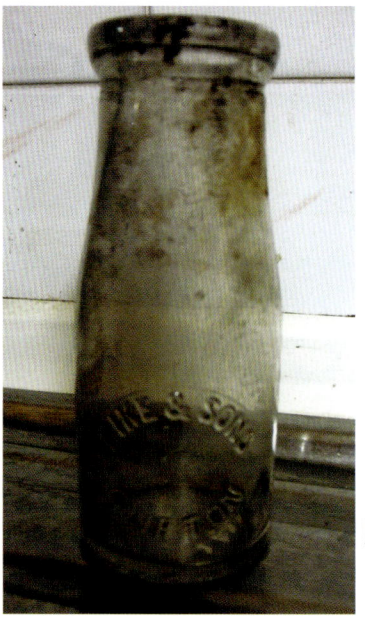

12.8 Pike of Clifton 1/3 pint milk bottle. (CRR Trust)

12.9 Bristol United Beer. (CRR Trust)

Table 12.3. Numbers of bottles found by manufacturer.

Dairies	Number of bottles
H.W. Case, Cotham, Kingsdown, Westbury Park, Bishopston	7
Hornbys Dairy, Whatley Rd, Clifton (many branches)	3
Pike, 10 Sunderland Place, St Pauls Rd, Clifton	2
Bristol Cooperative Society, 28 Regent St, Clifton	1
Bristol Dairies	1
Glen Farm, Abbots Leigh	1
Safety First Milk Association (school milk bottle)	1
Wigmore, 163 Mina Rd, Baptist Mills	1
Beer	
Bristol United, Lewins Mead	13
Georges, Bath St, Victoria St	2
Soft drink	
Brooke and Prudencio, 124-132 Newfoundland Rd, St Pauls	15
MAPS (squash and cordial), 13-15 Victoria Pk, Fishponds	5
Idris, White Hart Lane, London	4
King, York St, St Pauls	3
Corona, Wales	1
Keystone (aerated), 13 Victoria Pk, Fishponds	1
Tizer, 2 Mile Hill Road	1
Ginger ale	
Schweppes, 50 Queen Charlotte St	4
Ginger beer	
Brooke and Prudencio, 124-132 Newfoundland Rd, St Pauls	2
Sherry	
Astral Sherry, Spackman and Gosling, 42-44 Welsh Back, Bristol	1
Wyld, 2 Bristol Bridge	1
Spirits	
Martini	1

12.10. Astral Curfew Sherry. (CRR Trust)

12.11. Maps of Fishponds. (CRRTrust)

12.12. Brooke & Prudencio bottle with Neptune engraved on the glass. (CRR Trust)

12.13. Brooke & Prudencio lemonade. (CRR Trust)

12.14. Brooke & Prudencio ginger beer (stoneware bottle made by Price & Sons). (CRR Trust)

12.15. Bon coffee and chicory essence by Brooke Bond. (CRR Trust)

12.16 Coffee and chicory essence by SCWS (Scottish Co-operative Wholesale Society, Shieldhall), (CRR Trust)

12.17. Horlicks bedtime milky drink. (CRR Trust)

Guinness Bottle[6]

During one of our recent weekly group trips down the tunnel, we were going slowly back up the 300 stairs and noticed a bottle underneath. We managed to fish it out and were intrigued to find it was a Bristol United Brewery bottle with part of a Guinness label. The Guinness Collectors' Club dated the bottle as pre-1945. The label fell out of use in 1953. They kindly sent me the picture of the label shown in plate 12.18. Bottlers used mostly recycled bottles during the war due to shortages.[7] Bristol United beer was made between 1889 and 1961, and was merged with Georges Brewery in 1956. They had a bottling plant at Avonmouth where the Guinness was unloaded and probably bottled.

12.18. Guiness label stating where bottled. (Guinness Collectors' Club)

12.19 Bristol United bottle with Guinness label. (CRR Trust)

12.3.2 Cigarettes

We found a number of cigarette cartons. It was surprising to find them considering how damp it was under the ledges. The regulations also specifically stated 'Smoking is strictly prohibited'. (see plate 12.5). Woodbine cigarettes were the cheapest at 2d per carton of five in 1939, and appealed mostly to men.[8]

12.3.3 Domestic artefacts found

Everyday items such as medicines, flit sprays, Jeyes disinfectant, a Corporation of Bristol pencil, inkwell, fountain pen and ink bottle were found – see table 12.5. Only odd shoes were found. Even though they were only sleeping there during the night, people cared enough about their appearance to bring perfume and face powder.

There was an urn at the top of the steps dispensing cocoa. It was run by the WRVS and was free. Residents provided the cups. Many more cups than saucers or plates were found, and most were broken. One must surmise that saucers were considered unnecessary baggage, or that they did not break so easily (having broken a cup there was no point in taking it home). Some of the cups we found were very pretty (plates 12.21, 12.22). The two sugar bowls were probably from the hotel as they were the same style as four cups in Shelter 1 where we know the BOAC staff did their drill practice during the day. There was no insignia on them, so they could have been hotel cups. The Bristol Vitrite 'Venice' cup (plate 12.24) was made by Pountneys Bristol Pottery and Co at Fishponds (which continued production to 1969). *Vitrite* (glass-

Table 12.4. Cigarette cartons found.

Manufacturer	Inscription	Number of packets
Camel	Camel	2
Carreras	Craven and Virginia	1
Wills	Star 10	4
Wills	Woodbine	2
Wills	Embassy	1

like) was the mark used for hotel china. In view of the large number of cups and milk bottles found, maybe a ready supply of hot water was available for cold shelterers, and not just from Thermos flasks.

Table 12.5. Summary of domestic items found.

cup	26	pencil	4
plate	5	fountain pen	1
Sugar bowl	2	ink bottle	2
tumbler	1	gas mask	1
saucer	2	gas mask bag	1
glass bowl	1	glasses	1
brooch	1	hand-brush	2
1/2d coin	3	jam jar	1
6d coin	1	small, horsehair mattress frame	1
Half-crown coin	1	page of *Western Daily Press* 8 March 1944	1
comb	1	paste bottle	1
disinfectant bottles (Harpic, Jeyes)	2	scissors	1
eye dropper	1	Terrys mixed drops sweet bottle (Plate 12.24)	1
medicine bottles (milk of magnesia, glycerole, gammon cough, Parkinson's lemon balsam.	9	baby's sandal	1
vaseline jar	1	shoe	3
phial	2	saucepan lid	
face powder lid (powder cream Potter & Moore England Rachel)	1		
small perfume bottle (gold medal eau de cologne)	1		

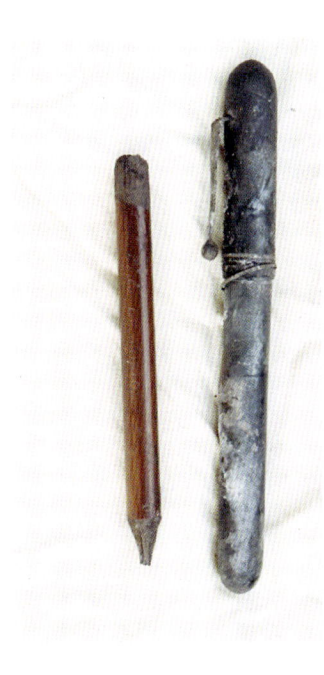

12.20. Fountain pen and pencil. We also found a bottle of Stephen's ink. (CRR Trust)

12.21. Tuscan china cup, decorated with a poppy design. (CRR Trust)

12.22. Porcelain cup. There was a blue crown on the base. (CRR Trust)

12.23. Made by Count Thun factory in Czechoslovakia 1918 through the 1930s. (CRR Trust)

12.24. Pountney cup from Hotel found in Shelter 1. (CRR Trust)

12.25. Terry's fruit drops bottle (CRR Trust)

12.26. Acme thermos flask. (CRR Trust)

12.27. Rachel face powder cream by Potter & Moore England. (CRR Trust)

12.28. Gold medal eau de cologne. (CRR Trust)

12.29. Miscellaneous shoes, bones and Tiny Tim poodle food. (CRR Trust)

12.3.4 Toys found

The dolls were broken but the lead figures must have been difficult to find under the ledges due to their colour and being small. It is interesting that the figures were WWI – valuable resources were not used to make WWII soldiers until after the war. The balls would have bounced and been equally difficult to find.

Table 12.6. Toys found under the shelter ledges.

lead aeroplane, a Bristol Blenheim?	1
Sorbo ball	2
doll	2
seated lead figure	1
lead soldier	1
toy gun	1

12.3.5 Signs

We found many signs under the ledges and stairs, which are now slowly being reassembled. Most were in Shelter 1 (the BOAC shelter) under the first ledge. We were particularly pleased to find 'Shelter 1' in the top refuge area, and 'Shelter 3 (Plate 12.35) in the lowest one. The signs for 'Women's toilets' (plate 12.36) and 'Men's toilets' have arrows pointing different ways (in the BBC section the doors were signed for 'Ladies' and 'Gentlemen' so clearly a different class of people), and we also found signs for 'Management' and 'Emergency'. We had thought the complete sign was for 'Emergency Management', but later finds proved the sign was for 'Emergency Exit' (plate 12.37). This seems strange since one could only go up the stairs to get out (users of the ledges entered from the top, and the BBC from the bottom – there was a heavy steel blast door at the top of the BBC section which was normally shut). Possibly, as people without a pass were originally allowed to sleep on the steps, only one set of stairs was used. We then found part of a sign saying 'only', so this must have been related to the 'Management' sign. We wondered why there had to be a place just for the management when there were also at least three 'Warden's Post' signs (plate 12.38).

12.30 Model of a Bristol Blenheim and a doll. (CRR Trust)

12.31 WWI lead figures. (CRR Trust)

12.32. A domino, broken doll, toy pistol and a sixpence. (CRR Trust)

12.33. Child's spectacles with one side clear and other side opaque glass (not allowing light to pass through) for squint Amblyopia (*lazy eye*). (CRR Trust)

12.35. Shelter 3. (CRR Trust)

12.36. Women's toilets. (CRR Trust)

12.37. Emergency exit. (CRR Trust)

12.38. Warden's post. (CRR Trust)

12.3.6 Artefact Summary

The recovery and cataloguing of the items told us a great deal about life in the shelter. What was particularly surprising was the huge variety of items found and the story behind each one. Did the large glass bowl get broken before or after its contents were eaten? Why were so many cups found? Why were so many single shoes left behind? The items help one imagine people writing letters, children playing with their toys and doing puzzles by hurricane lamp. We know that many people stayed every night and felt safe surrounded by everyday items.

12.4 Memories of people using Clifton Rocks Railway as a shelter

I have been very lucky to speak with many people who sheltered in the tunnel (though the numbers were fewer than those who travelled on the railway), who give invaluable insight into matters that we had not considered. *Secret Underground* was recorded in 2005 with Chris Serle as the presenter and shown on ITV1 West, and was an excellent source of information. It was lovely to listen to them reminiscing together on experiences from over 60 years ago. The lights kept going out due to the heavy demand on the power supply in the tunnel so we had to have someone on trip duty.

12.39. Chris Serle with Tony Riddell, Barbara Salter, Raymond Wade and Mike Farr. (Mike Edwards)

12.4.1 Secret Underground Bristol 2005

Present: Tony Riddell, Barbara Salter, Mike Farr and Raymond Wade, who were about ten-years-old when they last came to shelter in the tunnel over sixty years ago, and are still good friends today. These people were interviewed again later, so more information from each appears later. Transcript follows:

> Tony: You have got to remember I was 8 or 9 years of age. It was an adventure as far as we were concerned that bombs were dropping outside.
> It was not packed. Records seem to suggest it was. There always seemed to be plenty of space available. I used to share a step with Barbara.
> Ray and Mike: We were sheltering inside when a bomb fell. We heard a big heavy thump. We knew it was a bomb. All bits of plaster came down. I can remember it as clear as day. We went out the following morning to see where it struck. Sure enough, it had struck the garden outside of Tuffleigh House. I think the house was empty at the time. It was pretty overhead. I don't know the line of the tunnel in relationship to the garden. It was to the right of where we were sheltering. The shelter worked well else we would not be here now.

Ray said they could not come in without a pass. Originally they allowed people in ad hoc but as it filled up, people got used to coming, so they had passes if they lived in Clifton. His dad had a pass [his father's pass is plate 12.3]. What did it feel like? Ray said: We would be at school, come home, have a meal, mother packed up a bag of sandwiches, filled up a flask and we traipsed down here. We were allocated a space on a beam. We left bedding, came about 6-7:00pm. We stayed overnight and went out next morning, every night. Because of the steps arrangement we knew people above and below. We did not wander around much. It was not cold. There was no heating. Maybe the BBC generators helped keep us warm. We got heat from the hurricane lamps we had – we would put our hands round them.

The photos show a tent covering? We don't remember it. It would have been put up for damp and bits dropping down. It must have been put up later where the steps were, due to water dripping. It was very damp on either side. We sat on the steps at the bottom (Barbara lived in St Vincent's Parade). It was damp there. They would open the door and let so many in to the top. We had to queue on damp steps. Barbara's mother sometimes stayed on the damp steps all night long when the sirens had gone. Barbara stayed with Tony's family on the ledge.

Mike's father was fire-watching. He regularly used to come and tell us what was happening. He told his brother and him to look at this, and took us to Royal York Crescent.

All the docks were on fire. It was worth seeing, but not a nice sight, the whole docks were lit up. It was an air raid in conjunction with Bedminster. It was probably the same night the sugar warehouse went up near the docks by Cumberland Basin. There was a magnificent blue and orange light due to the sugar. The show was better than any firework display. One bad night particularly affecting Stokes Croft, Raymond's mum's mum lived there, and we walked to see granny. The Park Row ice rink was burnt out struck by incendiaries, the theatre was gone. We could only see pipes with water. We picked our way over to get to granny. She was OK. An experience you don't forget. There was lots of shelter around but Clifton Rocks Railway was one of the best.

Barbara had a bunk at the bottom after a while [she was in the bottom station after the BBC moved in; the access can be seen in plate 12.40]. Mum had a key (she was second key holder for the bottom if there was no warden, and opened up if the warden was not around).

We had blankets. When we first came in to use the shelter, Hotwells people came in at the bottom to get to the steps/ledges. This was before the BBC moved in. The big tanks of the cars were there. We had to walk through the carriages to get to the other side. Some of the cars were cut in half. Tony nearly fell into the tank when he got pushed off edge during the rush but his mother was hanging on to him and rescued him.

There was no entertainment, no Vera Lynn. Some ladies formed a band whose signature tune was 'Fays Away'. Someone had a sense of humour. It was not a hardship as portrayed. Since we were children we were OK. It was hard and worrying for the parents. The shelter was more organised than in the Portway. It was better conditions than under Bridge Valley Road. It was terrible there. People flocked there at night time. They lived like cavemen. Water since there was no concrete on the floor, as it was where the railway lines had been pulled up. Morale was bad to start with. War was bad to start with. People were scared – they would run from houses anywhere until it was more organised. People wanted to go underground as it seemed safer – there were no surface shelters, only under the stairs, or where it seemed safer. Caves were safe to go in, then later on, war got better organised, and Clifton Rocks Railway was set up for us.

We waited until 6:00pm or when sirens went, before we went to shelter and stayed all night. Some waited till 6:00am, some would wait for the all-clear to go home. If you were local (like Sion Place) you could go home and go back to bed like Mike. Raymond's father was a home guard, so often he was on guard all night, so he was happy that the children were safe so he could do his work. He worked all night, or got in early in the morning, and then had to go to work in the morning. He often went 24 to 48 hours with no sleep, and bad meal times. People did

12.40. After the BBC moved in, St Vincent's Parade residents entered via a modified separate entrance on the right-hand side. The porch on the bottom façade was for extra protection. This entrance would only have taken one into the bottom station. Plans can be seen in chapter 10.7, plate 10.16. The 2ft thick wall inside separating the BBC from the bottom station would not have been there before April 1941. This area would have been a bomb separation area originally. The cars were removed in February 1941 (see plate 10.22). (*Illustrated London News*)

not the times with very little sleep. Children were resilient and did not understand. We had no focus on danger. It was fun watching planes going over and the soldiers and guns. It was a game with dire consequences. Clifton Rocks Railway saved a lot of lives. Us children had to go to school next day. It was a quality place. It earned its keep. Cold [filming was in November]. Brings back many memories.

12.4.2 Individual memories of the shelter
Paul Adams

Born 4 February 1939 at 24 West Mall, a nursing home. There were six children in his family plus mother and father. He lived at 24 York Gardens. He lived in the whole house as a family house. It was not in as good condition as it is now. It cost £500.

His father was born in 1892, he had fought in Somme, so he was too old to go to war again. He had an office at home in the front room. They were not allowed in. He worked for Andrews of Bath, who supplied hospital stainless steel. He was in the special police based at Brandon Hill during the war, but mainly a warden. He died in 1951. He would swear at German bombers if he saw them fly by. Paul never really found out what his father did as a warden. He never spoke about WWI at all. He finished up as Deputy Provost Marshal in Cologne.

The youngest child was born in 1944, the eldest was called up in 1943 to the Far East. By the time he came back his sister had left to get married. His family was never all together due to age difference.

Paul grew up in the area. He remembers stables opposite 97 Princess Victoria Street, (Pat 'Peggy') with a horse and cart. There was only one car in York Gardens. He had a bicycle.

The housing was run down. Wellington Terrace was dreadful. He remembered gas lighting, and being lit using the ladder rest. There was a community spirit since there were families. It was rather fun, and safe to go out. He left house in the morning and came back when it got dark.

He was a school boy at La Retraite catholic school (Emmaus House, Clifton Hill) even though he was a protestant, since it was the closest. They had to pay a fee. He remembers the four or five nuns who came back from war, who had been in prisoner-of-war camp. There was a big celebration when they came back. His brother Donald went to Cathedral School as a singer (he became principal singer with D'Oyly Carte). He remembers going to the cathedral in 1948 to hear him. He sat by Vaughan Williams who was a very funny man. He had a watch which chimed.

He has a newspaper clipping from 3 January 1941 which describes what happened when his house was bombed (just the front part so it was patched up):

C.F. Adams' Home hit in Bristol Blitz. Rescue Squad Dig for Family – But they were Sheltering Elsewhere.

It can now be revealed that during the recent air raid on Bristol, the home of Mr C. Fred Adams was hit by a bomb. His large family had been in the habit of sheltering in a room under the stairs which caught the main force of the blast and was wrecked. But Providence had guided them that night to another shelter. Rescue men, who were not aware of this, began digging beneath the debris under the impression that Mrs Adams and the six children were there.

Mr Adams, who is hon. secretary of the National Federation of Inland Wholesale Fishmongers told 'F.T.G.' that it was the best thing he ever did when he insisted on his family going that night to cellars in another part of the district. It is certain, he said, that many of them would have been killed had they been using their usual shelter. In reply to the question 'How are you all feeling now?' he said 'We have been living like hermits in caves, but we are all quite well, thank you, and carrying on unbroken in spirit and very much alive to the fact that victory is on our side.'

By now it is probable that Mr Adams office is in working order again.

He remembers housing where WH Smith was, queues to get bread, but no animosity, no drugs, no drunks, no wine, just weak beer (Georges, and Bristol United). In 1946 the Fyffe boat brought in first bananas to Avonmouth – every child was given one. He did not know how to eat or peel it. No one was overweight. He still has his ration books.

He played at the quarry playground by the Observatory (it was deeper then). It was filled in in 1950 by 5ft to 10ft. There was a cave. He played football there. He was not allowed to go to the docks as his parents considered there were slums below Cornwallis Crescent (which was very run down) but did when he could. He used to watch the war ships going through the locks. He remembers three Seaman's Missions in Hotwells. He would also cross the river to get to the railway tow path. He had a family ticket to go to Zoo. He would go to a café (there were no restaurants) for a cup of tea in cracked cup or in enamel mugs. He never played on Christchurch Green much except for playing cricket by the war memorials (they put wickets on top and used a soft ball), swimming in Abbots Leigh pool, in the Victorian baths which were not heated (Clifton Lido) and Jacobs Wells baths.

He used the Railway to shelter. They had a pass. The top was very neglected for a long, long time, even when he got called up. He remembers going down the steep steps. He had sheltered under stairs at home and in the cellar of 12 Royal York Crescent (owned by the local milkman). There were two shelters on the grass above 29 Sion Hill which they did not use. He used the Railway shelter from January 1941 to the end of war (after his house was bombed). His sister was slow to comb her hair and he would tell her to hurry up. He used it every night.

He knew Reginald Wade whose pass we have (his father was disabled).

They had tea before they went (the main meal was at lunchtime every day). He often had bread and dripping (he is still healthy, despite this) or sandwiches before they went down.

The sanitary arrangements were dustbins behind hessian sacking. It was not very nice.

They used the little shelter or the bottom of the large ones. They went down about 4-5:00pm when the sirens went. They went back to same one on right hand side, so they used the right-hand side steps – he did not go into top station to access the other steps, maybe because it was a first-aid post. It was very dark, with flickery lights, some had candles, others had hurricane lights. The warden was posted at the top (his father was a warden which is why they got space on the ledges). When the bomb dropped at the top, they all had to go out via Hotwell Road.

The whole family was in there except father who was a warden. He would take cat naps. Paul would take a toy car, and fight for who took a favourite teddy. He remembers cocoa at the top. They all used to rush in. It was very wet down the walls and steps. People brought cats and dogs in which made a noise. There was a black dog near his ledge, even though regulations stated otherwise, and people coughing.

Sometime they staged plays with Hitler in, who they used to boo. There was much audience participation. Sometimes there was singsongs, but not often. These were at the small area at the bottom of the ledges.

He remembers the barrage balloon at the Observatory. Recruitment was at Gloucester Row in tents on the grass just after war began.

Bomb sites 5-6 York Gardens (he went hunting for shrapnel and would get told off), Wellington Terrace (people killed there), two houses in the middle of Royal York Crescent (14, 15), Regent Street, the bottom of Hensmans Hill below La Retraite (Emmaus House) where a friend's husband Mr Stevenson was killed and which also destroyed two houses in Cornwallis Crescent on the side of the lane- there are now modern houses (25-27). Cordeaux all gone (large shop at the top of Regent Street). All bombing activities were at night, not during the day. He remembers the searchlights. No lessons were interrupted at school.

When a bomb damaged houses, they were patched up with concrete if inhabitable. Wardens repaired them. He remembers water tanks on Christchurch Green past the lookout point, and in Victoria Square for putting out fires. He remembers a DKW (German off road jeep) going up and down Bridge Valley Road.

He can remember Americans at Clifton College playing baseball at the Close. They would give him gum chum, Butterkist and Life Savers (like Polo Mints). He met them when he went to the Zoo. They adopted one – gave him a second home to be friends of the family. He remembers the prisoner-of-war camp over at Abbotts Leigh – Germans? Italians? The prisoners had diamonds and circles on their clothing to identify them. Americans were based there before it became a prisoner-of-war camp.

He remembers a plane with a Polish pilot crashing in Long Ashton. There was a big crash and a bang. The pilot survived. They saw it happen out of the window. He can remember his mother fall where water poured over pavement when holding him, when they were going to the shelter. Sirens would go.

They had two street parties at end of the war – one in Royal York Crescent garden (where there were two tennis courts), and one in the Mall Gardens.

Valerie Barrett
Born 1939, and lived in Portland Cottage off Portland Place off West Mall.

She lived in Mall Cottage behind a large gate and up a lane. It had one room on the ground floor, two rooms above, an attic and a cellar. It was gas lit and had an outside loo. There was a wooden door to the sweetshop next door, but the girl who lived there never gave them any sweets. Her mother worked in a greengrocer shop on the corner. She washed in a sink in the garden. She had a very happy childhood.

One day she went to the NAAFI (Navy, Army and Air Force Institutes – an organisation created by the British government in 1921 to run recreational establishments needed by the British Armed Forces, and to sell goods to servicemen and their families) in Whiteladies Road opposite the BBC. Americans were billeted there and she was given an ice cream by all the soldiers so she told the girl next door. As well as ice cream, the soldiers also had lots of food and butter and sausages there too. They ate things from the shop. There was vegetable butter and sugar, but no meat. Her mother used to deliver whiskey to a big house in Leigh Woods and would be given pears, blackberries and flour.

She was nine-months-old when war broke out, and sheltered in the tunnel every night until war ended. Her sister was three years younger thhan her. Grandmother brought them up because her mother worked very hard. Her father went off with another woman just after her sister was born. The courts awarded them 5/- a week. The house cost 4/- to rent. She would go to pay the rent with money wrapped in paper. Her mother had to work hard to get them educated and pay for school uniform. She went to school in Princess Victoria Street, and later to Colston Girls – there was a bus stop opposite Gloucester Row which took them to Cotham Hill. The bus would return to Christ Church. Her godfather owned Keir Garage in Gloucester Row. Her grandfather was a painter on the Suspension Bridge. He also did the gold leaf in Goldney House. Her family originated from Redruth and had been miners.

She would run down Caledonia Place to the Railway.

One day her grandmother was carrying her sister, when some shrapnel just missed them. They would run after tea when the sirens went. She thought the sirens were at the Observatory [they were at Bridge House]. There was no time to get anything like a drink. The latest they went down was about 8:00pm because she was woken up to go. They had to leave the dog at home because animals were not allowed in the shelter. (Paul Adams commented that there was a dog near him so they did not always obey, like the 'no cigarettes'.) She was always relieved to get home and find the dog was still alive.

She remembers being given a black rag doll for Christmas in the Railway. She can remember going down lots of steps down the left-hand side and the bottom opened out. There was a man at the top of the steps who guided them where to go. He checked the passes.

Her mother, grandmother, sister and her were all on the same ledge so they could not lie down. They had no permit, but Mr Bishop who owned the grocers shop had organised that they could stay. They would wrap up warm, but it was very warm in the Railway. It was not too dark in there – in fact quite well lit but then since she was used to gas lights and candles at home then, it was probably better than she was used to. It was quite quiet – you could not hear too much except for dull thuds every now and then. They would sleep and chat and grandmother would tell stories, and sing nursery rhymes. They would keep themselves to themselves. She could hear a constant thudding and wondered if there was a machine room nearby – maybe there was some kind of air conditioning. The walls were damp and it smelt musty and damp. She does not remember the toilets, but then she was used to using an outside toilet anyway. There was a lot of people in the shelter. She was in there the first Christmas it was open.

Trevor Beacham
Born in 1935, and lived in West Mall. There were five in the family. His relatives who lived at York Gardens also used the shelters. He had a pass, and took blankets and pillows. His grandfather was Fred Beacham, a warden who checked passes at the top of the steps. He remembers balloons just below the Observatory.

John Bidgood's brother
John was born in 1925. During the war, John and his brother lived in Bedminster. John's brother was very frightened of the air raids that hit Bristol, and went missing. He had been injured in the air raid on BAC and Rolls Royce when the first night air raid was starting. He left the house with the flares coming down, it was over a month before we saw him again. His nerves were shattered. Apparently he had walked to the Railway and sheltered there every night.

Colin Britton
His mother and three children sheltered in the Railway. Colin was six. He lived in Freeland Place, and went to school in Princess Victoria Street. Their doctor got them a pass since his mother was pregnant. He took small toys, books, crayons. They made tea with a vacuum flask. They used hurricane lights. He was evacuated in 1942 to Devon with his brother and sisters. He had bunk beds; he had a shelter number. He used to sign in at 6:00pm and sign out at 6:00am.

Maureen Budd
Born in 1930. Maureen lived in the Polygon. There were five children in the family. They used the Portway tunnel three times before using Clifton Rocks Railway spasmodically twice a week between 1940 and 1941. She remembers a warning on the radio that they may be bombing Bristol. It was moonlight and a clear night. They left nothing behind. They took sandwiches, thermos, blanket and books. There were no refreshments available, and no sanitary arrangements. It was lit by electricity down the sides. Initially there was no organisation, though her older sisters remember bossy wardens later. She remembers two to three dozen people in there. She went with mother and three younger members of the family regularly. Father went fire watching, her older sisters were out with their boy friends. Books and chatting kept them calm and occupied. She remembers cigarette smoke. They put on a show once and dressed up as gypsies. They were in Shelter 2. They had a pass – they had to queue up in the hotel to get it. She remembers the 'all-clear' in the morning. They sat up all night. She remembers having a row with her sister when she ate her bacon sandwich because they were scarce.

Margaret Bygrave
Born 1932, she lived in Southmead She was eight, and went down to the bottom shelter. There were wood beds and they slept in their clothes, the toilet was a bucket behind hessian sheets . It was lit by candle and hurricane light. They took their own hot drinks. There was a person on the gate to check passes. Margaret was evacuated to Shepton Mallett.

Rosemary Clinch
Rosemary lived in Hotwells, and attended the Dowry

School and Holy Trinity Sunday School. 'You had to have tickets to get in and we kids didn't want to take our clothes off at night in case of a raid. My eldest sister who was a warden had the answer, "I've got your tickets" she'd say. We much preferred our own shelter in the back garden even if it was always filling up with water. If that happened we had to get under the kitchen table.'

M. Crosier
Born 1937, he lived in St Vincent's Parade and would be woken up to go into the shelter at the bottom station. He remembered hearing planes flying overhead. Key-holders were Mr and Mrs Pugh in the Colonnade.

Jenny Evans
I was born in 1942 and my brother in 1944. My mother lived in 8 York Gardens, Clifton and brought us to the Railway for safety. We were two babies in a pram who spent hours here. We had Mickey Mouse gas masks. A friend gave me some dripping. I didn't have time to take it home so brought it to the Railway. Mother always told me about the Railway.

Mike Farr
Born in 1931, Mike went on the Railway many times with his family before it closed when he was three years old. Mike has vivid memories of being in the tunnel when he was ten or eleven years old and especially of a bomb landing almost directly over the tunnel in December 1940. The blast was big enough to knock all of the small white stalactites off the ceiling and onto those sheltering in there. It shook everything. After the bomb incident, Mike went to find the crater the next morning and go hunting for shrapnel. The bomb had landed in the garden of Tuffleigh House, in Princes Lane. The crater was about 6ft deep and 6ft across. Mike doesn't remember the canvas sheets being there when this happened, but remembers them in subsequent air raids.

He lived in Sion Lane until 1955. The rent was 5 shillings per week. There was no electricity or bathroom. The landlord offered to put electricity in the house if they paid half, but his father had no money so they survived on gas lights, oil lights when the gas mantle failed. The AA kept bikes in the garage below and there was hay above. He went to the school in Princess Victoria Street (now the library). He used to go to Clifton Parish Hall where there was a skittle alley. He was chased with a horsewhip for playing in the Mall Gardens and gathering conkers. When the railings were taken away during the war he no longer had that problem. Most of Clifton was poor and dowdy. Wellington Terrace was rough and the Coronation Tap was tatty. Princes buildings and Sion Hill were posh. No. 6 Caledonia Place sold for £600 – big houses were not wanted. The Grand Spa was a top-notch hotel with a 'flunkey' (doorman) outside and revolving doors. He loved running through the doors but the flunkey would stick his foot out to stop the door revolving. Imperial Airways occupied the Grand Spa Hotel and nos. 1-3 Caledonia Place. The men wore uniform and he thought at first that they were from the RAF.

Mike entered the tunnel from the top end; the entrance to the tunnel was to the side since the railings were longer then. The steps were wide enough to sleep on and they used to take their Thermos flasks, torches and food. The tunnel was dim. The warden was on the top steps at the entrance. Initially those with no refuge tickets slept on the stairs. He felt safe since it was quiet. Bedding was left there, since they slept on same shelf. Children could sleep two per shelf. They could enter 6:00pm and leave at 6:00am or earlier if the 'all-clear' lights had flashed.

More shelters were in the railway tunnels in Bridge Valley Road where there was a shooting range recently. The mayoral treasures such as the mace were stored here. These tunnels were wet and smelt. The people who slept here lived like cavemen and stayed there all day and night. They used hurricane lamps in the tunnel. His father did not like them because they did not do any fire watching. Bunkers were built later. There were no air-raid shelters in Clifton to begin with. People would shelter in Princes Buildings and the cellars of Royal York Crescent. There were water tanks in West Mall and by the look out point in Sion Hill.

His father and brother used to go fire watching. He would watch from Royal York Crescent and see the docks bright orange. They were bombing the docks, the railway and the aeroplane company. Regent Street was badly hit and he remembers the fish and chip shop (at no. 19) being hit. Bombs sometimes dropped during the day. He used to love watching the aircraft dog fights when he was at school and his teachers urged him to shelter. They had lots of days off school when there was no coal or water. The sirens would start at 6:00pm and keep going until 6:00am the next morning, since there were 12-hour blitzes. He was one of seven children, but being the youngest his mother did not want him to be evacuated.

Mrs Ford
She lived in a terrace in Kingsdown – now demolished – a two-mile walk away. With the shelter at the top of St Michaels Hill closed she was allowed in the Railway shelter. She brought two children, Frank and Gladys.

B. Fouracres
He was about 10 years old at the time. His family lived in Bellevue Crescent and used the top big shelter where he sat rather than lay – it was 'rather exciting'. The warden would tell everyone when they could leave.

Tony Fowles
Born 1931, he lived at the top flat in no. 7 Canyngne Square, the rest of which was rented to BBC personnel. His father was a post-office worker who joined-up in 1940. There were three children; they were evacuated to Street, They often moved to different places such as Weston and Blackpool but always came back to 7 Canygne Square. He went to school in Princess Victoria Street and later to Pro-Cathedral, which was run by nuns (although they were not catholic) because the other school was full. They did not eat very well. Food was scarce. He once ate a whole lettuce on his own. When they could get a ticket they would go in at the top entrance, but would go in at the bottom too having run down the zig-zag. He remembers the slabs of concrete and the steps. The Railway was not lit well. He remembers one evening when an aeroplane machine-gunned them going down the zig-zag – his father protected him by throwing the bike over them. When the siren sounded (it was on the top of Bridge House) they would run across the Downs and down Sion Hill. There was a large cellar that went under the road at Canyngne Square to the opposite house but there was no light in it. There was a barrage balloon where the ice-cream van is now just by the junction of Suspension Bridge Road and Observatory Road. There were several people with it. Being a small boy, he was always very curious about it. Sometimes it was up, others down.

He would collect shrapnel – he does not remember what happened to it. He remembers when the corner of Regent Street got bombed, and Park Street.

His grandfather was stone deaf and could not hear the sirens or know where to go so often he would walk in the wrong direction and they would have to turn him round and make sure he was going the right way.

Donald Lear
He Lived in Love Street off Hotwell Road. There were 11 in the family. They would shelter anywhere they could. They were refused entry to the Railway shelter in 1940 during the blitz by the warden at the top. His Mother was nine months' pregnant and the baby was born that night on the pavement at Joy Hill, Hotwells.

Marina Rich (née Cohen)
Born 1938, she lived in the basement of 7 Rodney Place. The people in the flats upstairs would shelter in the basement together. When she was six, her uncle Wilfrid Hanks who worked for BOAC and was a warden, got tickets for the Railway shelter for her mother and the children. But her mother suffered claustrophobia and they went only once and never returned. Her brother (born 1943) was a babe-in-arms in a big gas mask suit with only his face peeping out. It was quite a way for them to walk.

She can remember walking down the steps and along and down and along until she got to their shelter place. She only remembers sitting rather than lying down with her brother in his gas suit. She took her gas mask with her. It was winter time so she was wrapped up warm. She remembers her mother being very tense.

She can remember the smells – musty and damp and of bodies. When she gets on a bus nowadays, certain body smells remind her of the Railway. There were not very many people in there. Everyone was sitting together. She did not know anyone and cannot recall any children though she is sure there must have been some there. She was in Shelter 2 on the left-hand side. She took her doll down as a comfort. It was quite light with lights on the wall. When in the shelter you could not see far – just see where you were.

The whistle of the bombs would 'put the breeze up' her mother. She preferred to be in her own surroundings with a glass of sherry to calm herself. She died at 62 and never really recovered from the war. Her father had fought in WWI and had been at the Somme and Ypres. He was very brave. She is strong like her father.

Two incendiary bombs fell on Rodney Place – one on the roof, the other on the garden. She remembers the air-raid siren (she does not know where it was but it was very loud), and the 'all-clear'. Days were not scary, unlike the nights. Her father put them out with sand and water from a neighbour. The grass never grew again.

She can remember a bomb hitting the corner of Princess Victoria Street (no. 3, Ross the greengrocer where the Boston Tea Party now stands), and flattening the site. She could also remember Windmill and Lewis (motor car agents at 10-15 Merchants Road) being bombed. She remembers the bombs dropping on Richmond Terrace and Granby Hill, and the garage at 54 Royal York Crescent as she remembers the worries about the petrol store.

She had a wonderful childhood. She and her friends from Victoria Square would go off to Albert's Pool for the day; she loved playing marbles and skipping and she had a big dolls house her father had made for her out of scrap wood. As children they built a den in the Gorge. Mike and Keith made a rope swing in Nightingale Valley and Nipper White used to walk up and down the chains on the bridge when the bridge-keeper was asleep.

She recalls Norah Fry (of the chocolate factory) who also lived in Rodney Place who wore very high heels and fur coat down to the ground. She and her brother went to school at the Primary school in Princess Victoria Street. Princess Victoria Street was tidy. She remembers Mr Ayres the grocer at the bottom left hand side of Princess Victoria Street.

Mr Cohen used to collect soda bottles from the back of the Spa Hotel and get 2 shillings for them as returns. Food was basic. They grew vegetables in the garden at the back of the house which stretched down to Waterloo Street where there was a baker's shop. Rats used to get in the grain which her father used to kill. Mother used to get a bit of beef which seemed to last for weeks. They also used to get fish.

She remembers the barrage balloon tethered by the Suspension Bridge very high. It was all shiny and silver with big ears. It was fabulous.

She lived in Rodney Place until she got married at 19 and then went to live in Royal York Crescent and then Tyndalls Park. She did not go dancing at the Hotel or the Glen (at the top of Blackboy Hill) since her father was very strict.

Tony Riddell
Born 1931, his family moved to a house in St Vincent's Parade, Hotwells, in 1940 since they thought it would be safer than living near the Whitchurch airfield. But as Tony pointed out, they were just as vulnerable since they were now close to the docks. Some of the houses were dilapidated, some quite good. His earliest impressions were very interesting (port/shipping etc). They moved to Cornwallis Crescent in 1941 and he went to the school in Princess Victoria Street. There was a good community spirit.

He only went up the zig-zag to get to school, never to get in the shelter, since to begin with he went in at the bottom gate, then when they moved house to Cornwallis Crescent, they went in at the top – which was a bit more difficult since they had further to go in the dark when they were in a hurry to get in the shelter. This was shortly after the BBC had taken over the bottom part of the railway and had the middle bit cut off. He sheltered in the Railway from 1940 to 1942, and before that in the old railway tunnel in Bridge Valley Road. There were wetter conditions under Bridge Valley Road. It was terrible there. People flocked there at night time. They lived like cavemen. Water, wet, damp and smelly. Morale was bad to start with. War was bad to start with. People were scared and would run from houses anywhere until it was more organised. People wanted to go underground as seemed safer. There were no surface shelters, only under the stairs or where it seemed safer. Later in war it got better organised, and Clifton Rocks Railway was set up for us. The shelter was more organised than in the Portway.

The Railway was not packed with people. Records seem to suggest it was. There always seemed to be plenty of space available. He used to share a step with Barbara who also lived in St Vincent's Parade. They all knew each other. The upper shelters were used for the 'posh lot' in Clifton. They played games to keep themselves occupied. In the beginning Tony's family used the shelter every night, later just when the sirens sounded. They had a pass, otherwise they would not be allowed in. They did not leave any items behind that he could remember. He can vaguely remember the canvas over the walls. They took blankets and a hurricane lamp. It was dark except for the hurricane lamps. It was noisy, damp and smelling of paraffin. They would sit on the steps or lie down on sheets and blankets. The gross area was divided into families and they were all neighbours, so they all knew each other. The ones he did not know were gatecrashers. Sanitary arrangements were primitive, if any. He remembers toilets being upstairs and downstairs.

When he first came in to use the shelter in 1940, Hotwells people came in at the bottom to get to the steps/ledges. The door was in the centre. There were two big doors leading into the reception hall. There was a deep archway. The big tanks of the cars were there. They had to walk through the carriages to get to the other side and to the concrete steps to get to the ledges. They were in the process of dismantling the cars [this must have been in March 1941], some were cut in half. They were placed in a line (see plate 10.22). Later the cars were dispersed and planks placed over the tanks. In a rush and push to shelter, Tony nearly fell into the tank but his mother rescued him. [The 2ft-thick wall would not have been there either –

this was added when the cars and machinery were removed in April.] The concrete steps were on the side at the bottom to get up to the ledges. Initially they used the lower shelter, later the upper shelter. He never used the shelter in the bottom station since by then he had moved house to Clifton, and went in at the top.

He used to play outside the railway and one day he had to shelter in the archway because a plane came by flying very low during the day and machine gunned the road. He saw the shells bounce along the Portway and he went and collected the cartridges which were still hot. He can remember when the bomb dropped by Tuffleigh House in December 1940. It was propped up on rocks in the garden. It exploded as soon as it came down.

Tony adds: 'You have got to remember I was eight or nine years of age. It was an adventure as far as we were concerned that bombs were dropping outside.' He was evacuated in 1942.

Barbara Salter
Born in Hotwells, Barbara lived in St Vincent's Parade for 52 years. She was an only child. Her father worked in the shipyard in Bristol Docks. He was as 'deaf as a post'. She remembers the first raid: they were shouting and bawling at him to shelter.

Before the Railway was available, they would hide under the stairs in a cupboard. Her family sheltered in the top station between 1940 to 1945. At the start, before the Railway was available, they would hide under the stairs in a cupboard. They never went to the tunnel under Bridge Valley Road. Her father did not come into the shelter with her, just mother and friends. They would be all ready to go every evening about 5:00pm to 6:00pm. They went up the zig-zag to meet with friends and children who lived in Clifton because they had passes. She remembers one night a friend had dropped her handbag on the zig-zag and all the contents fell out, and they had to try and find everything in the dark. They had to go back in the morning and see if they had found everything. They then waited until the sirens went, so they could go in. If the sirens did not go, they did not go in and went home. Wardens would open the door and let so many in to the top. You had to queue on damp steps to get in. The steps on the side were wet and not very wide. The wider steps [ledges] were drier. Inside, you could see the railway lines. There was a door half way up to gain access to the ledges. Barbara's mother sometimes stayed on damp steps all night long if she could not get into to the shelter area, while Barbara stayed with Tony Riddell's family on the middle ledges.

There was a lot of people in the shelter. They sat on the ledges. They kept their things dry by sitting on them. There was some lighting, but not brilliant. She took nothing besides blankets, and left nothing behind, including rubbish. She drank nothing – no one wandered around with bottles of water like they do nowadays. She ate nothing- she had eaten before they went. They were not into eating and drinking like people are nowadays. She didn't need to use the toilets as she didn't drink.

She kept herself occupied by sleeping or tried to sleep. She took no toys. She did not talk much with the other people, just her friends and people on ledges above and below. She could not hear any noise from outside. They went home when the 'all-clear' sounded. The warden came and told them when they could leave. Later in the war they ran along the Parade and went in through the gate at bottom (not the BBC gate) where Barbara had a bunk. She went back to same one on the ground floor. Her mother was the second key holder (the man at the end of St Vincent's Parade was the first key holder), so they could always get in and would open up the bottom if it was closed. The lighting was not brilliant. [The lighting would have been gas or hurricane light since there is no evidence of electricity downstairs.] There were not many children living in the Parade, so few cots were needed. They were aware the BBC was in the bottom but had no contact with them. Although they lived by the Suspension Bridge there were no big bombs there. She was frightened in a way but when you are young you don't always see danger. Rationing affected her as she had no sweets. She was not evacuated but her cousin was.

There were the barrage balloons where dog shelter was across the river. One got loose and landed in mud.

Norman Frank Thomas
Born in 1931 in Nailsea, Norman moved to Bedminster in 1935 then Royal York Crescent in 1942 and afterwards in the top floor flat of no. 28 Caledonia Place. He later moved to Almondsbury in 1944. Accommodation was rented. It was not unusual to keep moving house – wartime forced house moves. He had two sisters who were fourteen years older than him. He had experience of Anderson Shelters too.

'I have no photos or mementoes, just vivid, clear memories. I was not evacuated. I was lucky because I stayed with my family. At the beginning of the war, when we lived in Bedminster Down, we had an Anderson shelter in the back garden. This was dreadful because it was very cold, and a small confined space. Granny refused to go in it and stayed in the house. There were many near misses at Bedminster. It was in April 1941 when we experienced a huge night air raid. This went on for twelve hours from 6:00pm to 6am the next morning. No sooner had we had we got back in the house at midnight than the warning went again! There were bombs dropping everywhere. It was a nightmare. The raiders were preceded by pathfinder aircraft which illuminated the whole area by 1,000 flares on parachutes which slowly fell to the ground giving a pinkish glow everywhere. They lit the whole place up and then the bombs fell. One particular bomb fell nearby and I thought it had our name on it. It made neither a scream or a whistle just a massive rush of air and a huge explosion which took all the earth off the shelter roof. There was five of us in the shelter. The bomb landed somewhere at the back of us and I think there were casualties just up from us, but I never found out how many. When we eventually got back in to the house my grandmother, who had refused to go in the shelter, told us that all the doors in the house had been blown open by the blast, and the glass broken. I saw an incendiary bomb burning fiercely on the upstairs window ledge. Someone bravely knocked it to the ground. The sky was red with fires everywhere. The planes would fly so low you could see their shadows on the ground. There was smoke everywhere. I saw a German plane go over and I saw the pilot waving after dropping its load. I saw Bristol Blenheims, Brigands, Beaufighters at Whitchurch airfield. I did not have a spotter book but was very interested. I remember the Cumberland Basin fire at the sugar factory. I saw thousands of hosepipes. I have never seen so many before. The fire lasted several days. I went to see it during the day. I never did find any shrapnel even though I looked for it. I could hear it tinkling down at night. My father worked in Security in BOAC guarding Whitchurch Airport. He had been a printer before this but he applied for this job when he was unemployed. One night he saw torches flashing signals at the perimeter fence so he went to investigate. He was knocked unconscious and was found the next day in Dundry suffering from concussion. After the War he continued to work in Security for BAC. He did not get any medals.

'In 1942 we moved to Clifton and I went to Christ Church School in Princess Victoria Street I was the milk monitor. Once a week I went to Hotwells School for woodwork lessons. I hated arithmetic. It was hard to get pencils. In 1944 I went to Almondsbury Secondary school and left at the age of 14 in 1945 to become an office boy at BOAC at Filton. My education was poor due to four changes of school. At one school I was told to bring in board games because I couldn't study very well. I was very happy in Clifton despite it being wartime.

'Clifton was very tidy and the flat was up to date. There were few cars. My father used buses to get around and did not own a bicycle. He never did learn to drive. I had a bicycle. It was difficult to get food except on the black market. At the Clifton Caterer in the Mall we heard that Crunchies were in, but we had to queue for them. They were hidden under the counter. We could not get cheese and lived hand-to-mouth. I can remember Woolworths. We never lost electricity or gas. Winters during the war years were very severe and we were short of food. We could get fresh fruit but little else.

There was a big water tank in the Mall gardens in case there was a direct hit with incendiaries. I can remember the anti-aircraft guns at Purdown called Purdown Percy. Several balloons were in the area but I did not take much notice of them.

When returning from school in Clifton one day at about 3:30pm, there was a rattle of machine gun fire close to. Owing to the tall buildings, I could not see any planes and did not know what was happening. I was very frightened and ran down the road until I could find a doorway to shelter in. I was expecting bullets in the back. It turned out later that it was a British plane firing at a German plane.

I was fascinated by the revolving door at the Grand Spa Hotel. I used to go in and out about five times a day. I went to two BOAC wartime concerts at the Spa. I must have been a good boy since I was the only one there. One had Alfredo Campoli the famous Italian violinist playing, and the other had Doris Hare of *On the Buses* fame. I remember her singing 'All over the Place'. I don't think the concerts were for any special reason. Planes would fly people in. The concerts were in the ballroom. BOAC took over the Grand Spa Hotel and the disused Clifton Rocks Railway tunnel. I used the Railway refuge from the age of 11 (1942-44). It was an ideal shelter and we spent many nights down there. Sheltering in Rocks Railway was fantastic and we felt very safe especially compared to the Anderson Shelter. My father got us a pass to go in because he was a BOAC security officer, and we used it every night until we moved to Almondsbury in 1944. My father was a warden and fire marshal when we lived in Clifton. The warden at the Railway was called Mr Crane and he stayed at the top by the top gate. He looked at all the permits. You could not get in without a pass. I used to enter it by the Sion Hill gate, not from Princes Lane. My mother, sisters and I would all go down together about 4:00pm. The other wardens roamed around the area. I can also remember Mr and Mrs Davis of Albemarle Row in the shelter. When a raid was on the lights were dipped three times. When the all-clear was sounded the tunnel occupants were notified by the lights being dipped once.

12.41. Entrance to Portway Tunnel number 2. (Author)

The top station looks very much like it used to. I remember that there was an urn at the top of the steps which contained cocoa. I think it was run by the WRVS. I never used to have to pay. We provided the cups. We used to go down the left-hand steps because our ledge was on that side. Other people would go down the right-hand side. We never went down the corridor towards the ballroom. We just did what we were told. I feel reasonably comfortable sitting in the shelter. Memories come flooding back. We took everything with us every day, so did not leave anything behind. We took quite a few blankets and pillows to keep warm. I did not have a mattress but slept well. It was warmer than outside. I would bring a few toys and comics- something to read. I read Champion – which had war stories. I didn't mind reading about the war. I was right in it, being a civilian. I didn't take my marbles down, they would have got lost. I do not remember losing anything. I can remember the rubble under the ledges. We would eat before we came so did not bring food with us. I used to run up the stairs to get cocoa for my mother and sisters. Being a youngster I was quite fit. We were in the bottom shelter. I was about half-way down on the left-hand side. I walked down the central steps to get to my ledge. I do not remember anyone sleeping on the stairs. There was red lino on the ledges. I slept well, curled up on the ledge by myself. My mother and sister slept on different ledges. There were people coming and going all night. Some sat, and some lay, and scattered around when they wanted to sleep. All were reasonably well behaved, and cheerful. We rallied round each other.

I can remember a big Keep Out notice on the door to the BBC section on the right hand side. Not even Please! The door was black with white lettering. I never saw anyone going through the BBC door. I knew the BBC were in there.

It was damp rather than cold, with water running down the walls. It was wetter than now. It did not drip from above, just the sides, trickling through the rock. The tunnel had a white ceiling. The area went right across. It was very well lit with lots of electric lights. There was strip lighting across the stairs, and on the roof of the tunnel. There was the same intensity of lights on the steps as in the shelter. We did not take torches or candles. The lights would dip three times when there was a raid, and once for all-clear as we could not hear the sirens down here. We did not leave the tunnel in the morning till all-clear. We would hear the distant crump of bombs outside when we were in the shelter. It did not smell in the tunnel. I would take my rubbish home with me. There was always a big relief when the all-clear signal went. I cannot remember where the sirens were in Clifton, but I heard them. No incendiaries dropped on the top when I was there. I got to know the names of the people in the shelter. Some were aloof but we were all in the same boat. Mr and Mrs Austin from Canyngne Square were Canadians and brought a guitar and organised impromptu concerts. They had sombrero hats. We had a concert at Christmas 1942 in the tunnel. I sang 'Home on the Range' with Mrs Austin and can still sing it now and remember the words. There were seven verses but I can only remember one now. I stood up to sing, just like now. We had a captive audience. There was about 150-200 people. I would love to sing it to you.

12.42. No. 4 Victoria Square, Clifton, cut clean away.[10]

Oh, give me a home, where the buffalos roam
And the deer and the antelope play
Where nothing is heard a discouraging word
And the skies are not cloudy all day.
Home, home up on the range
Where the deer and the antelope play
Where nothing is heard a discouraging word
And the skies are not cloudy all day.

I can remember the Portias family.

There was no doubt that the Rocks Railway tunnel was a godsend to us and was impregnable to bombs. It was a happy episode for us in a grim part of the war. I felt safe. I was at Almondsbury near the end of the war (1944). There was less threat so it was less dangerous. There were more planes at Filton so we were winning. We did not have a street party at Almondsbury at the end of the war but we did at Filton.

At Almondsbury School, our headmaster Mr Marsh was very keen on letting us know what was happening on the Russian front and he put it all on the blackboard. We also had a school wireless on quite a lot. We did projects and we had a map showing where the Allies and Germans were. We were well informed. I used to listen to BBC Home Service and Uncle Mac, and Romany (kids in the countryside – a wildlife programme).

When out blackberry picking one day in September I saw 200 aircraft towing gliders flying directly over my head very low at about 11,000 feet. They were flying in echelon and they all had the Allied markings of black and white stripes under the wings. Amazing sight. They were going to Arnheim.

Next to our house at Almondsbury there was a search light unit and I was rather fascinated by it and hung around. However I was warned not to look into the lens or it would blind me. The cook at the searchlight site, a man called Alex Girgan used to give me tins of food to give to my mother. He was a very generous Scotsman. We opened our home to many service men and women at Almondsbury, also some Americans. One night we entertained 17 service personnel. We had them all sign the visitors' book – I still have this book today.

Raymond Wade
Born 1932, he lived in Princess Victoria Street. He had a tunnel pass which he gave to the Trust. He used the shelter from 1941-3 every day, in Shelter 2. There were many people in there. They took games and books to keep themselves occupied. It was smelly.

Hilary Thyer (née Wood)
Born 1936. She lived in Freeland Place. Her grandfather had the jewellery shop at bottom of Granby hill. My father was an air raid warden. He would tell people to shut their lights off and close the curtains.

At first they would hide under the stairs at home, but it got too dangerous so they came to the Railway. They used bottom shelter (on the right hand side). It was not well lit. She was able to lie down. Her mother, father, she and her sister all sheltered there. At first they came in from the bottom, but then they came in from the top. They were told they had been promoted! Some of the neighbours from Windsor Road hid in caves under the houses. They would come in after tea or when the sirens went. They went most nights. She would be told to get up, so would put on her little pixie hat to help keep warm. They would take torches since it was pitch black outside. She can remember planes in the Clanage, Bower Ashton (now the cricket field), and the sky when it was red.

12.5 Portway tunnel[9]
In December 1940, the mass observation report in the *Clifton Chronicle* stated:

There is no sign but two printed paper notices with an arrow pointing in. Just inside there is a blast wall of sandbags. As one enters, the stench is overpowering, a mixture of sandbags, urine, disinfectant, sweat and bedding. For about the first 50 yards one has to walk through water two inches deep, over bumps in the ground. On each side, beds are standing in water. Sometimes blankets or an old mac is placed on the mud. This is spring water and comes from a spring where people fill their kettles. It is a simple business to divert this water out but Bristol Corporation has refused them permission. Consequently all night long water is pouring into the shelter like a miniature waterfall. Farther along it is drier, and there is great congestion. About half a dozen families have tent arrangements of sackcloth thereby ensuring a certain stuffy privacy. A little over half way along there is another brick wall. Beyond this the walls are whitewashed and bunks four across have been built. The poorest and dirtiest people of them are using this end. The children are four to a bunk. Lighting is by candles and oil lamps. Some distance from the other end is a brick wall with sackcloth; on the other side are the closets labelled M & W. The closets are never empty for more than 30 seconds at a time; they have to serve 1,000 people. There is a stinking gag of chlorine. Beyond is the open air.

On 21 January 1941 the *Western Daily Press* announced that the disused Port and Pier Railway tunnel (tunnel number 2 – the entrance to be found just north of the traffic lights at Bridge Valley Road and shown in plate 12.41) had been provided officially to give shelter accommodation for 200 people, but it had become very overcrowded by people imagining it was bomb-proof. Sometimes as many as 3,000 crowded in at night into the unlit tunnel behind the shelter which was very wet and thought not to be safe due to faults in the rock and brick arch. The conditions had become a grave danger to heath, and entailed the risk of a large number of casualties in the event of a direct hit. It was under 40ft of rock and 525ft long.

The adjoining Tunnel 1, which was 216ft long, was already being used to store civic treasures and archives. An effort by the authorities to restrict numbers to 200 from 1 January 1941, using allocation cards failed and Mr St John Reade, a Clifton College master, Labour councillor and shelter marshall urged that the proposal to let part of it to the BBC should be abandoned. He was promptly sacked but continued to turn up for duty. Rev Mervyn Stockwood, vicar of All Saints in Clifton (who later became the Bishop of Southwark) protested at the way he had been treated. 800 women and children sat it out in the tunnel day and night until on 10 January 1940, the Chief Constable stated that he was not prepared to eject the people from the tunnel until the marshals had tried and failed. The tunnel users even tried to involve Queen Mary at Badminton but she passed the letter onto the Lord Mayor. The tunnel was then cleared and workmen cleansed the tunnel and built a new shelter at the end of the tunnel for another 200 people. Priority for the 400 tickets was given to women and children and the aged, most of whom lived in the neighbourhood.

12.6 Portway tunnel memories
Colin Boyce
Born 1928, he lived in 6 Oldfield Place, Hotwells. He sheltered in the Port and Pier Tunnel 2. His family (of five) was one of the first to use it. His grandfather had been the station master. His father worked at BAC. Initially they sheltered in relatives' cellars in Dowry Parade. His father suffered severely in WWI and remained seriously nervous for the rest of his life – the bombing on 31 October set him back again so they had to search for a safe shelter. They had no pass. They were there with no lights; there was rubble and rats. They took bedding and a pushchair. They slept in an old bed frame they took with them after the first time. They had a paraffin light and a torch. They went nightly during blitz period. There were no parties. They kept calm by sleeping. Colin was evacuated to Barnstaple.

Maureen Budd
Born 1930, she lived in the Polygon. She used the Portway tunnel three times in 1940 before using Clifton Rocks Railway. The Portway caves were 'ghastly'.

Ralph Smith
He was six when he went to the shelter. He lived in Bedminster. He was youngest of thirteen children, but at the time of the war there was just his brother, sister and him at home. During the war the family was told to go to the Portway shelter They presumably had nowhere to shelter nearer to home. They walked (via the Brunel swivel bridge and Ashton swing bridge) to the tunnel (quite a distance).

He had a cot in the tunnel. He was in there for three days. The walls were wet. They were allowed out for fresh air when the 'all-clear' sounded, and then they were allowed back in. Someone was playing the accordion and they tried to make it homely by having a sing-song including army songs and children's songs.

The wardens sealed the entrance up to stop the crowd trying to get in. There was lots of crying in the tunnel and his mother picked up another baby which was crying, which made him jealous. He could hear the planes overhead and the ack-ack (anti-aircraft) guns (which were on the downs and the Observatory). They did not know what was happening outside. The planes dropped a basket of bombs in the river trying to hit the Suspension Bridge. It was in December 1941.

The tunnel was lit by paraffin lamps. There was a toilet at the top end with a wooden seat. Just buckets, and newspapers tied with string for toilet paper. One would swill out with a bucket of water. He could remember the warden coming in.

They were then told to go back home. Bedminster was bombed to smithereens and he saw blood coming down with the rain. They had lost their home.

Their mother went to live with her sister and the three children were evacuated in February 1942. They were told they were going to Weston, which they were looking forward to. They went to Temple Meads station and were given Mickey Mouse gas masks. They had to get on the train and say goodbye to their mother. They were excited when they went past Parson Street station. They passed Weston and ended up in a hall in Bideford. He was crying. They were put on a double decker and put in a square. People then chose which children they wanted, and he was left behind as no one chose him. He was too small to work. His brother and sister went to separate locations. He ended up in a two-room cottage with his sister for two years and worked on a farm. He got on so well that when his parents picked him up, the family wanted to adopt him and he did not want to go. He remembered stirrup pumps, and two big air-raid shelters on the Downs. One was by Blackboy Hill, the other by the Zoo. He remembers barrage balloons at the look out point, Observatory and at the gulley.

He remembers the lights out regime where you had to switch the lights off and use a torch, and listening to the radio. He remembers walking along in Bedminster with no shoes and holes in his pants. When the siren went he hoped to hide away in a safe place. He remembers they had put sticking tape on their windows at home. The General Hospital was used for wounded soldiers.

Mrs Gregory
Used the Portway Tunnel. People were very orderly and patient. It was wet and cold. People sang. Babies cried. Her six week old brother got whooping cough. A lady doctor came to visit.

Keith House
He was about 10-years-old at the time. They lived in Bellevue Crescent. Keith remembers the damp walls in Bridge Valley Road tunnel. He remembers being covered in sores and subsequently finding out after the war that it was due to some bacteria in the tunnel.

12.7 General wartime memories
These memories are mostly from people who lived near Clifton Rocks Railway. They remind us what people were going through to add some context, and experiences of people sheltering in their basements. After all, air raids affected everyone, since one did not know the destination of the planes, merely that they were overhead.

Stephen Alexander
Born in 1924, he lived at no. 29 Victoria Square. He was 16 in 1940. He wrote a detailed account at the time.

'I moved to sleep on the ground floor because of air raids in July 1940 and to the basement in August where I had a camp bed against the stairs. Some people were ringing on the bell and asking for shelter as the shrapnel from our own guns was falling so heavily.

On 2 December 1940, while we were away at school, my mother had been writing letters when the siren started. She got up and crossed the window to call the students from Kings College, Cambridge who were billeted with her. Just then a large bomb landed in the Fosseway and all the glass came in behind her, she was unhurt. A warden then called and asked everyone to leave the house as they were worried about the gas main. My mother called the students and the two girls who helped her in the house, and they set out for St Vincent's, Clifton Park (the site of the Roman Catholic Cathedral) where her sister-in-law lived. However, they stopped on the way to leave the students with a friend in Lansdown Place. My mother said afterwards that she would not have done this had she known that the students were put to sleep on the top floor. My mother continued to St Vincent's but found that the front of the house was ablaze from incendiary bombs. She continued to 79 Pembroke Road where Mr Shepherd the gynaecologist lived. He took them in and they sheltered in the basement. His wife and family had been evacuated out of Bristol.

Dr and Mrs Price lived in Lansdown Place and that night they found that due to the blast from the Fosseway bomb they were unable to open their front door. This bomb must have been at the Queens Road end of the Fosseway, and due to the road being set down, the damage to the houses were less than might have been expected, the churchyard on the other side was also a help to minimise the damage, although all the slates of the houses, including no. 29 were blown off. Mrs. Price looked out of her window and saw the dentist and his wife from no. 28 walking by, she called out to them and they said they were going to the warden's post as they had also been turned out of their house for fear of the gas main.

On the night of 24/25 November 1940 when the Museum etc. was bombed a bomb landed near Richmond Park/Royal Park. A doctor's house there received a bomb in the garden and the goldfish were blown into the roof, they were still wriggling when found among the rubble.

The following day my mother, being unable to contact some old friends in Knowle by telephone, walked right across the city to find them alive and well. She then walked back again – such concern.'

Stephen wrote a marvellous diary July to August 1940 which describes how the day was normal, playing cricket, buying clothes, having a picnic, taking the dog for a walk, going to see a film, helped make jam for the Womens Institute and then describes how the very frequent air raids affected him:

'26 July; 27 July; 29 July; 30 July; 31 July; 1 August; 2 August: one as goes to bed, and another late at night, brother went to LDV (Home Guard) all night and slept all next day; 9 August, helped uncle sort out field dressings, bullets and paper for the LDV troops; 11 August, 15 of our bombers lost and 57 German [actual figures 30 British 35 German]; 12 August, new green noses fitted to gas masks. Bombs fell near Flax Bourton and on Bristol – 31 Germans and 9 British lost [actually 27 German and 20 British]; 13 August, hear whistle and air-raid siren during cricket match at Flax Bourton, so go in until 'all-clear'. Arrive back in Bristol during an air raid; 15 August, Bristol has bad night of air raids; 17 August, woken up with a tremendous crash at 2:30am and another at 3:00am so go down in the shelter until all-clear sounds. Big crater in Mud Lane (Flax Bourton). Paul cycled out from Bristol for weekends, so looked for more bomb holes, read, played tennis and the cow nearly ate a tennis ball. On the way to church at Wrington men were lifting an unexploded bomb out of a 20ft deep hole; 22 August, back in Bristol, an air raid at midnight, mummy awakes us at !:00pm as bombs falling so we sleep in the shelter; 'all-clear' at 4:00am. Paul went to LDV and had a parachutist scare, so back home later than usual; 24 August, air raid from 9:10:00pm to 5:00am so up most of night; 25 August, air raid from 9:00pm to 4:00am. Slept in the kitchen most of night and only woke once; 26 August, cleaned out the small larder to make an air-raid shelter, distemper and block up holes and read. Air raid at 10:00pm, a few planes over. 'All-clear' at 4:00am. Cat jumped on bed so shut in scullery. 27 August, air raid 10:00pm to 4:00pm; 28 August, very short air raid at 8:00am. Finish painting room. Ann and Paul went to a film, I go by bus to Flax Bourton to play cricket. Air raid 9:15pm to 4:30am; 29 August, air raid 9:30pm to 3:30pm; 30 August, air raid; 31 August, read and went for walk on the Observatory. 1 September, sort out books for the troops, go to the Zoo, a rather bad air raid, very noisy, and hear whistling bomb; 2 September, air raid quite bad; 3 September, air raid in the afternoon and another soon after 9:00pm; 4 September, read and bike, air raid early and at dinner time. Mr Anderson comes to put a prop in the old larder [to support the ceiling], read, go to tea at St Vincent's and see the baby. Fighting overhead in the afternoon with no air raid warning. Read in evening, very hot day. Air raid 9.20pm to 4.50 am; 5 September, read etc. all day. Air raid; 6 September, shop with Mummy and Anne, buy stuff for siren suit, and Mummy's dress. Siren goes when we are in Jolly's shop [Blackboy Hill], we go to their shelter, it lasts only ten minutes. Air raid; 7 September, take magazines round on bike and other odd jobs. Bombs drop without warning on a school, headmaster and some boys injured, also a whistling bomb [it was the Cathedral School but the report was incorrect as boys were not injured]. In the afternoon we were just setting off for a walk over the Suspension Bridge when Daddy arrives so do not go. Church bells [sound in the evening], Paul is called out for LDV but it was a false alarm. [This was the night of the invasion scare when all Home Guards and troops were called out, uncle at Cleeve refused to ring church bells until he had proper orders and he turned out to be right.]; 8 September, we were going to the early service but we were so late last night that we did not go. Go to the Zoo later and then to Flax Bourton and pick blackberries; 10 September, buy shoes then go on the Downs with Mummy, Daddy and Anne, saw balloons landing; 11 September, in the morning Anne hears her school certificate results [v. good]. Go to Bickley, pick apples, look at [new] calf, play tennis come home by 4.00pm Listen to Churchill's speech, he warns that an invasion is being prepared. [My father must have returned to London that day]; 12 September, go to *Yeomen of the Guard* at the Princes Theatre [soon to be bombed] very good. Get back for a late tea; 13 September, shop, fetch our new shoes and buy a book for Mummy's birthday. Go to Flax Bourton and pick blackberries, come home for tea. Paul goes at 7:00pm. for LDV; 14 September [Return to school so missed the bombing which followed].

Anon
They would take a cocoa tin of shrapnel to church hall and got 1d per tin. They were told it would be used to make more munitions.

Mr. Bartlett
He remembers the first night of the blitz on 24 November 1940. He was at HQ Ashton Court. The Yanks had a company at Burwalls (house just over the Suspension Bridge), and he had gone to visit them. His father was firefighting, the sirens went off. The timber yard in the docks was struck. He ran across the bridge into the pub. A policeman in a tin hat told him to go back to base. It was a terrible night – Park Street, and Castle Street got hit.

John Humphries
Used to live 8a Glendale. A German bomb struck the roof in Glendale – unexploded. His father disarmed it and they kept it for years. John planted it on the Polygon allotments and then picked it up and said 'look what I have found.'

Letter to Peggy Knowlson from Ruth Trapnell 8 May 1941
Stephen Trapnell (Ruth's son) sent me this letter. Both the Thompsons and Trapnells have their names on the plaque at the top of the zig-zag at the look-out point. Hallam and Ruth Trapnell lived at 19 Sion Hill for 50 years, with sons David and Stephen. John and Peggy Knowlson of 16 Sion Hill, went to Washington DC for much of the war – Knowlson had a job with the Admiralty. The Lewers occupied no. 16 Sion Hill while they were away. John Basil and Line Sawyer lived at 18 Sion Hill. Sonia was their daughter.

> Dearest Peggy.
> I've just come back from the Walker Dunbar hospital annual meeting and it was dreadful not to have you there. Ruth (Elliot Trapnell née Fells) was taking her mother to Clevedon after her cataract operation, so it was most old-fogieish, but such a good speech from Mr Gould of the University, on the Fall of France.
> We had another night of excitements here at 1:00am today. The usual alert was on, and Hallam and I were asleep when terrific HE (high explosives) soon got us up. We went down to compare notes with the Lewers, and were standing in your hall (at no. 16 Sion Hill), when a terrific rain of incendiaries fell on the grass, and all up Sion Hill. – even three on the Bridge. We saw two on the roof of number 21, so Hallam and Basil Sawyer flew in and pumped at them with their stirrup pump and I dashed up to our roof & found that the one labelled '19' had fallen in our garden & all seemed well with us, and no. 18. Then there were loud yells for help from 25. The Powell's had one too, but no pump. So I raced one up to their attic where I found 2 men but no water or sand. As their roof is not yet fully repaired after the last do, it was a bit tricky. So I rushed and got that bath from our hall, and with another girl, staggered all up with it, and more buckets soon arrived, while the two Miss Goodman's, who live there, dithered in the basement. The fire tried to spread on to the new roof on 24, and there was great pumpery before it was all out.
> Sonia had just got it all out, and a lot more water on us than on the flames I suspect. Meanwhile, one had gone straight through the Sawyers roof & Sonia was alone to tackle it (it was evidently in the roof space when I had been up there). Sonia and I got a spade and hacked at the bathroom ceiling till it dropped through. Bit by bit she dealt with it, while Line (her mother) ran yelping up and down stairs, cursing the 16 year old maid, 'who, with mongrel blood in her veins, sulked in the Hall' as she expressed it. Sonia had just got it all out when a deluge if water of water came thro the hole in the roof on top of her. The Lewers had performed literally alpine feats to get a bucket of water to the spot from your roof, and chucked it into the hole from the outside.. Line was by then in a fearful state of flap, & on the impulse of the moment, I gave her something between an embrace & a shake & called her an 'old silly.' However I think she has either forgiven or was too far-gone to notice, as I've had a friendly cup of tea with her this morning.
> The loudest bomb meanwhile had landed plumb on a huge gas main in Regent Street, Clifton, outside Prossers shop [a haberdashers] & by 3:00am a huge fire was raging & it had got the water main too, & one little hose was all they could work. When we were there it had spread, on the left from Liptons' [a grocer's] down round by where our Bank was, round the corner past Bromhead's, where they stored the furniture to the ceiling. They just had time to get the cars out of Windmill and Lewis [a garage] before it went up too. Then Phillips the ironmongers on the corner of Princess Victoria Street caught & the oil

spread it quickly down the hill, & Bishop's Stores [a grocer's] and the National Provincial Bank were the next to go. Then they got a pipe from the docks and it was got in hand, but all today they've been hosing, and the gas main was still blazing in the crater at lunchtime. It was fearful seeing shop after shop going and so little able to be done about it. The gas has just come on again which is a mercy, & the water was never off.

We're going to bed just after supper tonight as we're all so tired. I'm glad the boys are out at Butcombe (Butcombe Court, the venue of Clifton College Prep School during the war 1941-45), where they went last week. They are so happy there & I've had a delightful letter from David saying he's at last in the fire squad and has been given a bit of the garden. Stephen says he has found 35 bird's nests. Poor old David had his adenoids out in the holidays, as they had grown again after seven years. He was the only one at the Chesterfield for two nights, but I do really hope it will be the end of his 'stuffy noses'.

Olive our maid went to be a WAAF cook on 3 April having given us only 5 days notice. I had hoped she'd deal with Hallam, while I wasn't at Harewood (the home of my grandparents in the lane to Cadbury Camp, Clapton in Gordano) with the boys. So I tried to do both jobs- not very successfully, and now it is just the two of us here, life is easier again. Your Mother came and saw us in our living room and will have told you all about it. Olive is at Melksham being trained and she had cooked 200 jam tarts when she last wrote.

Oliver my mother's brother: a doctor in the RAMC (Royal Army Medical Corps) heard yesterday he's being moved to Yorkshire in a fortnight. A camp at Pocklington where there is miles and miles of damn all, as he puts it. I shall miss them, as they used to drop in. Mercifully the Lewers are charming & we see lots of them.

Mrs Deverson
Born in 1907, she lived at Easton in a family of thirteen. They could see planes coming up over Avonmouth. She could tell from their direction if they were heading for Centre, Filton, Cardiff or Newport. She can remember American tanks all along the Portway in 1942. They were ready to go, and had vanished the next day.

Mona Duguid
Born in 1915, Mona sent me 17 pages of her wartime memories, 1940-45 (and those of her husband who was in a POW camp). She was a local district secretary of a group from June 1940 who contacted next of kin of prisoner of war to give information about sending letters and parcels. She could send money orders for household soap or toilet soap (not both), razor blades, khaki shirts, boiler suits. She was also a volunteer three nights a week as cashier in the forces canteen set up in the new council chambers on College Green. This was used later as war progressed, by Free French, Polish and all allied forces including Americans.

Church bells were no longer rung as it would be a warning that paratroopers would be landing, no street lights, blackout frames on homes. She remembers black adult gas masks, childrens' masks were red with two bits of rubber at the side (hence 'Mickey Mouse' masks). The babies' masks were worse as the baby was placed inside and it had to be sealed. One had to pump in air so it was bad enough with ones own smell and the smell of rubber. Hand rattles were used to warn of gas, and lamp posts painted yellow which would change colour. Rationing was very severe (July 1940). One had to register with one butcher, grocer, and milkman. Bakery goods could only be obtained if available. Fish shops were selling whale meat. All shops were closed at 7:00pm (8:00pm Friday, 9:00pm Saturday) to give staff time to get home before dark. Raids usually started 6:00pm or 7:00pm and continued till 6:00pm or 7:00am. Buses and trams were off the roads by 9:00pm. Limited to a maximum of 5in of water in the bath. Coal deliveries were stopped until a rationing system could be worked out during one of the coldest winters on record. Mona became a 'female maintenance inspector' for Bristol Gas Company on 8 December 1941, when the company began to employ women. All women had to register for war work or forces, but those with young children were not compelled to work. She was issued with a steel helmet and a gas mask to be carried at all times as well as a peak cap with a Bristol Gas badge and a case of tools, and later, light blue coat overalls. She used her cycle with her tools strapped to a carrier on the back. She went to a very large house in Pembroke Road which had been taken over by government and military. There was a big reception for the American ambassador. She attended to all the gas appliances and was given a large brown paper carrier with butter, tea, sugar, oranges and even eggs.

Phyllis Farmer
In 1940, staff from the BBC moved to Clifton and had studios in Clifton Parish Hall (on the site where there is a block of flats in Merchants Road). When the blitz began, local residents used it as an air-raid shelter. I believe bombers would leave Christ Church Clifton standing because it was such a useful landmark. On moonlight lights we felt apprehension because the

River Frome was clearly visible from the air, so the docks were an easy target.

Iris Mitchell
Born 1921, she lived in Victoria Square. She worked at BAC as a welder. She was not allowed to go in the Square gardens since she was the daughter of a housekeeper. She sheltered in Litfield House (Clifton Down) cellar since her mother was caretaker. She had a pass. She left a chair for return. There was canvas over the window.

David Summers
Born 1929, he lived in Bedminster, and had experience of evacuation and Anderson Shelters.

'In November 1940, a bomb fell on the house and garden nearby and demolished the house and adjacent sweet shop. I was in an Anderson Shelter with my father, just a few yards away from the bomb crater. Our house was badly damaged, but not destroyed and the first thing I remember when going up the garden path from the air-raid shelter after the all-clear siren, was seeing my goldfish still swimming around in its bowl. This was amazing as the bowl had about an inch of window glass in small pieces at the bottom.

Like many other people, my early education was badly affected by WWII, due to frequent visits to the air-raid shelter, and later, on the 19 February 1941, being evacuated to a farm at Hartland in North Devon. The farm was a smallholding, but I was expected to help with feeding the chickens and pigs before going to school and later on milking two cows by hand. When the hay making and harvest season came around I had to help, instead of going to school!!

The trams were an excellent mode of transport before the war. The frequency was such that if you just missed one, or one was full-up, you only had to look down the road to see another one coming. Also, if you sat on the top, and it was raining at least you sat on a dry seat by folding the seat flap over.

I also remember having the yearly outing to Bristol Zoo, as a family, which included my older brother. We would walk down Duckmoor Road, through Ashton Park and across the old Swing Bridge to Cumberland Basin and along the Portway to the Clifton Rocks Railway. From the top of the railway it was a short walk to the Zoo for a ride on the elephant Rosie and a picnic on the grass.

thirteen

BBC

Operation and Memories

This chapter explains how the Clifton Rocks Railway tunnel in Bristol (as opposed to the Portway tunnel) came to be used as a backup to the BBC in London. Much of the background history in this chapter has been provided by Neil Wilson,[1] from Washford Radio Museum, John Penny Bristol historian, Gerald Daly[2] Engineer in Chief of the BBC Bristol Region, Frank Gillard,[3] Lord Asa Briggs[4] who wrote a widely acclaimed five-volume history of British broadcasting from 1922 to 1974, Max Barnes,[5] JW Godfrey's books,[6] Edward Pawley,[7] H. Bishop[8] and Richard Hope-Hawkins.[9] At the outbreak of hostilities, in the event of air attacks or invasion eliminating the London station, arrangements were made for the BBC to operate from a headquarters in Bristol. On one occasion, Broadcasting House in London suffered a direct hit, and thousands of gallons of water began pouring into the basement control room. The switch to a Bristol newsreader was made only when the whole network was in danger of being shorted out. The Germans were unaware they had hit such an important target. The news had to be switchable in this way at a moment or so's notice.[10] More of the operation is found in section 13.9 which contains memories of the BBC staff at the time. The rationale for the conversion is to be found in chapter 10.

13.1 West Region headquarters comes to Bristol

The BBC's first office in Bristol was acquired in 1931 above the Midland Bank in Queens Road, but there was no proper studio. Gerald Daly was engineer-in-charge (EiC) of Gloucester Repeater Station, which was a BBC amplifying and switching centre in the Programme Line Network (owned and maintained by the GPO) between London, Birmingham, Cardiff and Plymouth. Daly was given the task of finding larger premises in Bristol to set up offices, studios and a switching centre to replace Gloucester. Gerald Arthur Daly was born in 1918 and died in 1987 in Bournemouth. He would have been 22 in 1940 – young for a BBC chief engineer for the Western Region.

The Whiteladies Road studios were opened on 18 September 1934, with Daly as EiC and the Gloucester Repeater Station was closed. Bristol became the HQ of the West Region, a position previously held by Cardiff, which had covered both the West and Wales, but the two regions still shared a transmitter at Washford. The West and Welsh Regions were not given separate transmitters until July 1937.

In 1938, with the threat of war looming, a search was made for suitable 'country' premises to which various BBC departments from London might be evacuated in the event of an outbreak of hostilities. In the spring of 1939, the BBC purchased Wood Norton Hall and set up studios there. However, it was soon decided that in order that balanced, entertaining programmes could continue to be produced in wartime, it would be necessary to have a series of other dispersal points around the country. Arrangements were made for centres in Bristol, Bangor and Bedford to which various programme departments could be sent; there being an assumption that cities outside the London area would be immune to attack by aircraft. On 15 September 1939, the variety, religious and music programme departments were evacuated to Bristol along with listener research and some other administrative departments. Several halls around Bristol were found to help accommodate them. In April 1940 the schools' programme department was moved from Wood Norton. With the fall of France to the Germans, Bristol came within easy reach of German bombers and a search was made for a broadcasting station site in Bristol more immune to enemy attack.

13.2 BBC West Region search for a more secure broadcasting station site

Gerald Daly gave a full description of this search in his letter of 23 April 1974 to Patrick Handscombe,

217

13.1. Gerald Daly (engineer in charge) in the studio in Clifton Rocks Railway. Note the microphone and twin gramophone desk. (*Illustrated London News* 6 April 1946)

when he was researching the Railway's history. He also sent the BBC photographs which can be seen in this chapter and chapter 10. It gives a brilliant insight into the work done and use by the BBC during the war. More of Daly's comments are found in chapter 10, referring to the facilities required in 1941. Below are the comments that refer to the use of the facilities. The window-less conditions were not exactly healthy.

Gerald Daly:

> The whole project took six months to finish and I moved the main engineering staff down to the tunnel in 1941. Henceforth for the rest of the war this was the nerve centre of the BBC in the West of England. Through the Tunnel control room for the next four years passed all the programmes of the BBC to home and overseas. We went back to the prewar control room in Whiteladies Road at the end of the war. We held onto the Tunnel for nearly another two years – we only paid a peppercorn rent to the Bristol Corporation for renting it. I think it was about £5 a year. Then we removed all our equipment and the City took it over – the chambers of course remain, and it could be useful perhaps in a nuclear war. We of course made it gas proof, and with all that rock above even a nuclear explosion would have little effect I think.

> We maintained a permanent military guard over the tunnel throughout the war by means of our own BBC Home Guard Company of the 11th Gloucestershire Regiment. Well-known broadcasters were members of this Company: Sir Adrian Boult, Stuart Hibberd, Uncle Mac and many musicians and variety stars of radio of those days. (For my sins and being as I say a local dogsbody I was Captain of the Company of course.)

> No sooner had we got properly settled in the safety of the Tunnel Broadcasting Station than air raids on Bristol pretty well ceased as Hitler made his onslaught on Russia. The odd enemy aircraft bent on raids to the North used to drop a bomb or two at the Suspension Bridge but in the Tunnel we could not even hear the explosions. The main cause of complaint from the staff in the Tunnel was of the smell from the river at low tide – it was awful in spite of our gas curtains. The experts gave some relief by introducing the continuous sparking from an induction coil – the smell of the air round the spark overcoming the river smell to a certain extent. (Curiously enough it was the same sort of induction coil spark whereby Marconi had transmitted the early wireless waves – so we were on familiar ground.)

> In the darkest period of the bombing of Bristol there was talk, in the case of invasion or complete destruction of London, of the Tunnel becoming the BBC's last ditch.

Churchill had broadcast to the country that if we were driven out and conquered by the Nazis he would carry on with the British government from overseas. This, I understood from my mother-in-law who occasionally had dinner with the Churchill family, would be Canada probably. Thinking of me she said she once asked him if the BBC heads would go with him to Canada, and he said *What use would they be there. Let them make a last stand in their Bristol Tunnel – it's the best place for a last stand that I know of.* I had discussed making the Tunnel into a still more secure place with my commander in Bristol – the military commander of the Bristol area – and he had agreed that our pre-war HQ in Whiteladies Road had little chance of stopping an enemy assault, but that the Tunnel was infinitely easier to defend and make secure, and that I should so but not let anyone know, not even my second in command, who was Stuart Hibberd the famous announcer. The trouble was that Stuart, an ex soldier, was mad keen on making the old HQ in Whiteladies Road into a veritable fortress, and sandbagged and barbed wired it so that one could hardly get out or in. And I could not explain that we had no intention of trying to defend the old place and that the Tunnel was really our wartime fortress.

The work was carried out in 1940-42 and we moved in in 1941.

The planning was done by ourselves, ie. the maintenance engineering department and the building department of the BBC. The cost was £20,000. The engineers who were there permanently during the war were Q. Fisher, Douglas Gibb, and F. Dennis, Senior Engineers in charge of shift, while their staff including some girl engineers consisted of about four per shift of three shifts throughout the 24 hours [to ensure broadcasting of essential information. Shifts 9-5, 5-12, 12-9. During the night, as not much happened, one was allowed two hours to sleep].

The apparatus was dismantled by ourselves and the BBC's Equipment department in 1946-47. The equipment was stored in the stores of the BBC in London, but a great deal of it would be reused as required elsewhere. [Not all of the apparatus was dismantled since transmitting of the Third Programme continued until 1950.]

Security was maintained by the BBC Home Guard Company. [They sat at a desk between the toilets and the steps up to the canteen and insisted on seeing passes even if the person was known to them.] They had rifles, machine guns, and hand grenades, should invasion take place.

Photographs were taken by the official BBC photography department. [seven from 3 March 1942 and two from August 1945. Some were published in the *Illustrated London News* 6 April 1946 and some in Max Barnes. 'Miniature BBC in Avon Rocks' article in the *Bristol Evening World*. Clearly publicity about the secret fortress was avoided until after the war. We are lucky he sent some to Patrick Handscombe else they would have been lost forever. The BBC did not have copies.]

13.3 The Home Service and transmitter frequencies
Before the war, different frequencies had been used to broadcast the National and each of the seven different regional programmes, but in wartime, Britain was to be served by just one programme, the Home Service from London which actually began on 1 September 1939, two days before war was declared. In order to prevent enemy aircraft using the signals from BBC transmitters for direction finding, it had been decided, at the outbreak of war, to synchronise the frequencies of eight of the main regional transmitters in the country in two groups on medium waves 668 kHz (South Group A) and 767kHz (North Group B), with an additional chain of low-powered (100 watts), low-frequency transmitters (known as 'Group H') on 1474kHz appearing later.

If several transmitters used the same frequency, it would be impossible for aircraft to get a fix on any one of them until it was within 25 miles. If enemy aircraft were plotted within 25 miles of any transmitter, Fighter Command could instruct it to close down immediately. Other transmitters in its group remained in service, so a listener would be conscious of nothing more than a slight reduction in volume.

The first warning the change was to take place, was given in the 6:00pm news on 1 September 1939, and for the next two hours the announcer Robert MacDermot played records on the national programme frequencies, whilst urging listeners to tune into either of two designated frequencies. This complex frequency change was completed ahead of schedule and was launched at 8:15pm. Droitwich long wave station, and the three medium wave national transmitters then ceased transmitting. This left eight transmitters to broadcast the Home Service to the whole country though some had weak signals. Bristol lost its local transmitter at Clevedon which had been released to BBC's External Services and converted to short wave so had to make do with one at Washford.

When enemy bombing intensified in the autumn of 1940, many transmitters were often closed down at the same time making reception of the Home Service very difficult. There was also concern that if a transmitter were put out of action, in the event of invasion, essential information could not be communicated to the local population. It was decided, as a partial solution, to install low power Air Ministry T77 transmitters at main studio centres and it is possible that one was installed at Whiteladies Road before being moved to the railway. To make the programme line network less vulnerable, an emergency switching centre was set up at Falfield, near Thornbury.

As a more permanent solution to the problem of communication with the population in the event of main transmitters being out of action, it was decided to install 61 low-power, synchronised transmitters with an output of 50W-1kW in towns having a population of 50,000 or over around the country. These all broadcast the Home Service unless called upon to deliver local information and were known as Group-H. The Air Ministry agreed that provided these transmitters were accurately synchronised and would close down on a local Air Raid Warning RED, they could be freed from Fighter Command control. It was specified that these transmitters would also have to close down if local gunfire was heard, even if no air-raid warning had been received. The first consideration in finding a site was to obtain adequate accommodation adjacent to a structure capable of supporting a transmitting aerial. Few had conventional mast-supported aerials. In September as many as eight stations were commissioned in one month.

The first ten of these (including one at Whiteladies Road) were brought into service on 15 November 1940 and all used 1474 kHz (203.5 metres). Whiteladies Road was closed on 13 September 1941. Transmitting was then switched to Clifton Rocks Railway on 14 September 1941 and closed on 28 August 1945.[4] The Group H transmitter in the tunnel later broadcast the Third Programme from 29 September 1946 until 14 March 1950.

A number of imported American RCA I E transmitters were used for these earlier Group H sites but for later installations a variety of home-built equipment was used. The T77 transmitter which had already been installed in Bristol, was kept as a stand-by in the Railway. There were aerials for both kinds of transmitters in the tunnel, as shown in the transmitter room plan (plate 13.22), and the aerial plan (plates 13.23 to 13.26).

There was also a Harvey-McNamara short-wave set, and an ex-RAF medium-wave transmitter for restoring communication between the other main stations should the Post Office telephone lines be damaged by enemy action. Private lines to various Government departments, Fighter Command and RAF Headquarters were needed to receive instructions to close down various transmitters if required.

13.2. Gerald Daly (BBC Engineer in Chief) in the transmitter room with two telephones. Group H RCA (American) transmitters operating on 203.5 metres in the background, Hammarlund (American) radio communications receiver in the foreground. (*Evening World*)

13.4 Conversion Starts

Work on converting the Clifton Rocks Tunnel began on 24 February 1941. The remaining carriages were removed on 28 February. The main construction work was done in three months and described in chapter 10. After six months, the equipment installation was completed at a total cost of £20,000. In April 1941 the variety department had moved completely from Bristol, having been installed partly in Weston-super-Mare, before ending up in Bangor. The music department (including the BBC Symphony Orchestra) was transferred to Bedford, so the tunnel was never required to accommodate the largest evacuated department. There would not have been room either.

Only the lower part of the tunnel was used by the BBC and this was equipped with (starting from the bottom) a control room, a recording room, a studio and a transmitter room as well as a stand-by generator, a gas filtration system (in event of a poison gas attack) and canteen facilities above the Control Room. Everything and anything to prevent the enemy from destroying or disrupting broadcasting. The programme was fed down the line from Whiteladies Road to the tunnel control room for distribution to the SB network. It was the very great degree of close co-operation and discussion between the EiC and the chief of the presentation of the programmes which put into practice the close liaison between programme and engineering people so one could link the studios together and send the output to line and have a presentation assistant sitting alongside the engineer in doing mixing, fading and communication. Timing was crucial. The rooms are described starting from the bottom.

13.3. Sectional plan giving dimensions. Note how the 1:2 gradient allows interconnection between the rooms by going under the floor on the upper room. These rooms are all built on top of the railway lines- which were cut at the beginning of the generator room. There is a thick wall at the top and at the end of the generator room. The function of each room is given.

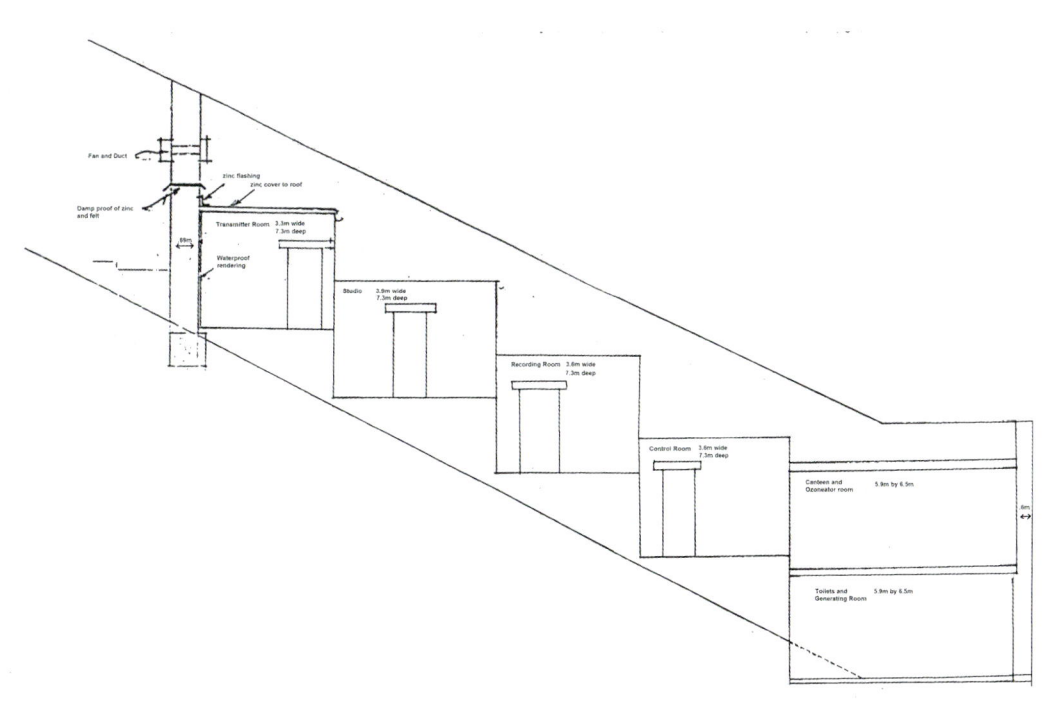

13.4. BBC stairs in 1941, BBC rooms on the left-hand side, the original tunnel wall on the right with sheeting and makeshift guttering to drain rather than drip on personnel. There is an additional telephone on the left-hand side, and a fuse box on the wall. The AC cable is on the right-hand side wall (see plate 10.24) with the telephone cable below it. The ducting provides heat. There is also ducting to provide ventilation.
(London Illustrated News)

13.5. The same view today. The staircase and ducting looks the same today. The rooms are in a state since the wooden ceilings and floors have collapsed. There are wooden strips on the steps to deaden the noise of people using them.

13.6. Ducting flow control can be shut or open or somewhere between.

13.5 Generator Room

13.8. Left, the doors in 1970. 'Danger Unauthorised Entry Prohibited'.
13.9. Above right, by 2002 the door panels had gone; by 2005, the doors had gone. The ladder leads up to the ozoneator room.

13.7. Bottom station, March 1942. The BBC have blocked off the main entrance and created a new entrance on the left-hand side. The windows have been bricked in and there is ducting to extract gas and to get fresh air. The exhaust pipe from the generator can be seen on the right, exiting horizontally. The post is a receiving aerial used for short wave communication with other BBC sites. It was probably only for emergency use in the event of failure of telephone lines. It was installed at the foot of the tunnel rather than at the top, to keep it away from the transmitting aerials which might have interfered with reception. (*London Illustrated News*)

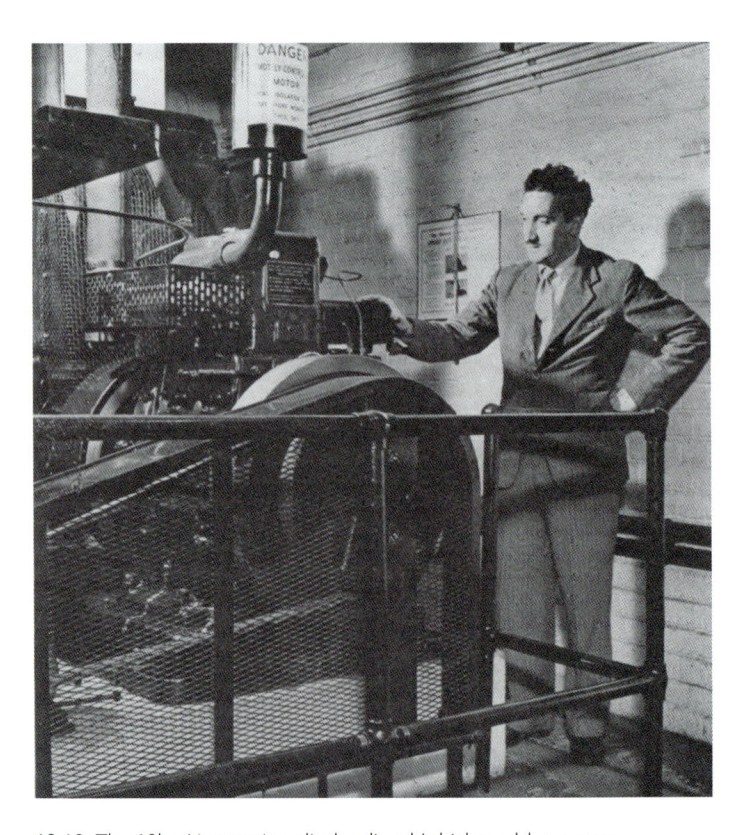

13.10. The 18hp Lister twin-cylinder diesel (which could generate 21hp at 1,200 rpm) and the belt drive for the generator (which would have been on the left). This powered a large generator needed for lighting and equipment. Note the guard and the post-war notice warning attached to the exhaust pipe that it was remotely controlled. (*London Illustrated News*)
The diesel tank was under the floor and fuel supplies were held in the adjacent cave, see section 10.18.

13.6 Air-conditioning Plant Room

13.11. Ozoneator. The tubs are for cooling and condensing. (Mike Edwards)

13.12. Extractor fan and ducting. On the right is air filtration plant and a second fan. The tunnel roof can clearly be seen. (Neil McCoubrey, Bristol Photographic Society)

Here, large intake and extractor fan motors sucked in fresh air from the outside world, which was passed through the ozoneator, and circulated around every corner of the tunnel. The air was then expelled by the extractor fans. The plant room made a constant humming sound which could even be heard by the shelterers of Shelter 3 (see chapter 12).

13.7 Cooking Area

There was a warming cupboard for the food that was delivered daily (read Molly Foss's memories of what could happen to the food in transit in section 13.19), a primus stove (paraffin-based), and an ordinary kettle (read the account when the catering officer got involved in chapter 10). There was three months' supply of food and water (and tapes) in the ozoneator room next door.

13.8 Toilet Facilities

The toilets were plumbed in, unlike in the shelters.

13.13. Molly Foss suggested there were also showers for ladies.

13.14. The BBC section had Ladies and Gentlemen and Women's Toilets and Men's in the shelter area. Gentlemen's facilities are tucked under the stairs.

13.9 Control Room (the nerve centre)

13.15. The Control Room (known as Control Room No.2). Note the all slave clock on the wall. There were 80 pairs of GPO lines linking the tunnel with the outside wall and the BBC's network of transmitters scattered throughout the UK. Douglas Barlow on the left (read more about him in Elizabeth Taylor's memories), Peggy Richie and Dougie Gibb on the switchboard. (Gerald Daly)

Transmitting speech over a distance can be done by telephone. To do the same with high-quality music and other programme material is one of the main jobs of the broadcasting engineer, for it is by 'telephone' wires that the programmes are carried from microphone to transmitter. All telephone wires used for broadcasting were rented out by the BBC from the Post Office. Thirty-nine Post Office circuits were taken up in March 1941 including 14 established between the tunnel and Whiteladies Road (to extend studio items, two for trunk line, six PBX (private branch exchange) to the tunnel for inter station calls and calls between offices in Whiteladies Road and the tunnel control room). A list of line allocations dated 31 December 1941 shows PBX tie lines from the tunnel to Redland Police, St Johns (two) and Clifton Parish Hall, which was at the end of the Fosseway (two) via West Exchange which was in Clifton. The Central Exchange was in Telephone Avenue off Baldwin Street in the centre of Bristol.[11] A report published in *The Star* on 11 January 1940 established there were six studios in Broadcasting House. It added:

> The BBC fun is produced amid surroundings of ecclesiastical gloom. Studio 9, 10, 11 are at Clifton Parish Hall – Studio 8 is the Lady Chapel of the Cathedral and is used for the Morning Services. Studio 12 and 13 are at All Saints Parish Hall (sometimes wanted for choir practice). Studio 14 is attached to St Paul's Church and is a gas-mask distributing centre. The Symphony Orchestra plays in a Co-op Hall.

The control room was manned continuously doing all the line switching until the end of the war in case the one at Broadcasting House in Whiteladies Road was put out of action by bombing. It acted as a replace-

13.16. Medium waveband 877 Kc/s (342 metres, 877 KHz) crystal drive purported to have been rescued from the tunnel. On the side is inscribed Type 63 No 106.

ment for the one at Whiteladies Road, handling all the programme lines (about 80) in and out of the West Country. Balancers and amplifiers were needed to keep the 'programme-to-noise ratio' as high as possible along the telephone line. High frequency wireless waves were then used to carry the low frequency current. To stop the frequency from varying, the transmitters had to be kept close to allocated frequencies to with extremely fine limits. A crystal drive (made to a universal pattern) was used which had to kept at a steady temperature of 40 degrees centigrade by a special device (oven) which was thermostatically controlled and never switched off. Each station had its own crystal. Emergency checking was done, one station (Tatsfield) being the master, all the others slaves. H Group was not so easy to deal with, so a schedule was produced, and a certain number checked each night.[12]

Once the threat of bombing had receded (the last raid on Bristol was on 14 May 1944), there would have been no need to man it continuously. It fed the Home Service, Forces Programme and various European and Overseas services to the main transmitting stations at Washford, Clevedon and Start Point as well as sending the Home Service to Group H transmitters at Taunton, Plymouth. Redruth, Torquay, Exeter, Weymouth and Bournemouth. It may also have handled programmes to main stations Droitwich and Rampisham, if only as a reserve route, and to 'Group H' stations Cardiff, Gloucester, Fareham, Swansea, Swindon, and Shrewsbury. Programmes from the studios at Bedford and Bangor might also have passed through the tunnel, as well as those from receiving stations on the south coast, set up at Group H sites, to receive reports from mobile transmitters in Europe after D Day. Gerald Daly:

> the Control Room in the Tunnel handled all the home and overseas programmes of the BBC during the war. The Home and Forces (now Light or Radio 2) and the various coloured networks of the propaganda and Empire and foreign services generally, i.e. Red network for programmes to the Dominions, Colonial Empire and the United States of America; the Blue network to Central and Western Europe and the Central Mediterranean; the Green network to the Near East in Arabic, Persian and Turkish, African, Hindustani and Maltese and Malaya; the Yellow network to Spain, Portugal, Scandinavia and the Balkans.
>
> Of all this massive handling of foreign languages in our time in the Tunnel, only one serious technical hitch came to my attention. A programme from the Arabic quarter in Cardiff was scheduled to go out to the Middle East. Afterwards the engineer in charge of the shift, one Arthur Fisher, came to me and said that some of the staff were slightly doubtful if the Arabic programme in question was true Arabic. One of the engineers was in Allenby's Egyptian Army in the First War, knew a few words and was doubtful. However we heard no more of it until some weeks later a Welsh speaking professor of English in Cairo wrote to say that he had heard to his surprise a programme on the BBC's Overseas Service in the Welsh language on Welsh piggeries. The professor pointed out that few Arabs in the whole of the Middle East spoke Welsh and were not all that interested in Welsh pigs, so he thought such programmes should be dropped. We were at first puzzled but it suddenly occurred to us that someone somewhere or other had got mixed up and passed this pig programme instead of the Arab programme meant.

Frank Gillard said:

> Whenever the sirens sounded in Bristol, the essential programme staff on duty at Broadcasting House were rushed down to the tunnel studio. An armoured car was available for the journey if danger was imminent. But, alert or no alert, the tunnel control room was manned by technical staff day and night from its inception until the end of the war. Every day, hundreds of broadcasts in English and in scores of foreign languages passed through those grey panels on their way from the studios in London and other parts of Britain to the transmitters in the West of England which would radiate them to listeners at home and overseas.

The tunnel, while maintaining its routine place as one of the key points in the BBC's wartime distribution system, was ready to shoulder its full responsibilities at any moment. But that moment, of course, never came. The Rocks Tunnel was never put to the supreme test. But it remains a monument to the thoroughness of the BBC's technical preparations for whatever emergency the war might bring, and as such it might well be preserved.

Programmes could be introduced by a light musical record 'Teddy Bears' Picnic'. It embraced a wide range of frequencies for musical notes so it was a good test. A more exact test was done by a frequency oscillator, which could radiate notes of all pitches throughout the audible scale for a given frequency.

The control room contained every type of electrical apparatus necessary for broadcasting: amplifiers (microphone amplifiers; controlling amplifiers; trap valve amplifiers; equaliser amplifiers); thermionic valves; telephone lines; jack fields; mains units and electrical magnetic relays; alternative current; wireless receivers; loudspeakers. The master clock for the whole station for broadcasting worked to the split second.

13.10 Recording facilities

13.17. The recording room door.

The war brought an increased demand for recording facilities and the limited number of available recording channels were put under great strain (a channel consists of a pair machines set up to allow continuous recording or playback).

In July 1940, the BBC began negotiations with Philips-Cine-Sonor to lease four Philips-Miller recording channels that they had been offered by the J. Walter Thompson Co. These had been used before the war for the recording of programmes for commercial radio stations in Europe. There were only six machines in use. Two were in the Ariel Studios at Bush House (the BBC later took over these studios) and two were at the Scala Theatre in Charlotte Street, off Tottenham Court Road. The BBC already leased a pair of Philips-Miller channels from Philips (experimental use of this system by the BBC began in 1936) which worked by cutting a variable width track into oxide-coated celluloid base film, 7mm wide, that could then be read optically. The film had a coating of gelatine and on top of that an extremely thin coating of special black paint. This was done to allow a sound track to be cut with a sapphire instead of having to record photographically, as in cinema practice. The big advantage of this is, that by using a light beam and a photo-cell, the recorded film can be monitored about a second after being recorded, instead of waiting several hours for developing and drying. Despite the system's drawbacks – the film was expensive (4 shillings per minute) and could only be used once – they did produce better quality recordings than any other type of machine then in use by the BBC. It could be stored easily for many years, and, as it could be cut with scissors and easily rejoined, was simple to edit. Problems had arisen with their use early in the war since spare parts were supplied from Holland, and film from Belgium, and supplies of both ceased after 10 May 1940 when these countries were overrun by the Nazis. However, alternative supplies of film were sourced, and ways found to cut two tracks in each film instead of just one. In November 1940, one of the Philips-Miller channels from the Scala Theatre was installed at Whiteladies Road. This channel, together with a similar one in Manchester, was used to record the final rehearsal of all major productions so that if the live performance was interrupted by enemy action, the recording could be substituted. Prior to this the only recording facilities in Bristol consisted of a mobile recording van (M53) with a pair of MSS disc recorders which was used in a static capacity.

The recording room was constructed to house the pair of Philips-Miller machines in use at Whiteladies Road, designated recording channel FB/1 (Film, Bristol /1), and these were moved to the tunnel, together with their two equipment racks (known as bays) in July 1941. The swarf (the film waste), was collected every recording and disposed of by the studio attendants— it was highly inflammable. Molly Foss in her memories stated there was a baby grand piano in here too.

Gerald Daly said:

13.18. The recording room. Note the all-important slave clock on the wall, fuse boxes on the walls, double-ender cables hanging up on the left-hand side. (Gerald Daly)

Apart from the control room operation at the Tunnel which was as I say, used throughout the war, most of the recording work was carried out there in the recording chamber. All recorded programmes were stored there too for safety's sake.

Sometime between March and August, 1942, a new American-made Presto disc recording channel was installed in Bristol (possibly at Whiteladies Road) to replace the M53 recording van. By the end of the war the BBC's Philips-Miller equipment was getting quite elderly and, with technical improvements in disc recording, was no longer pre-eminent in terms of recording quality. It was decided to surrender the machine in the tunnel to Philips and this was done on 1 July 1945. It is now in the Philips museum in Eindhoven, Holland.

13.11 Studio facilities

Frank Gillard stated there was an upright piano (to save space) in here so it seems there were pianos in both the studio and recording room. When the equipment was removed at the end of the war, it confirms that more than one piano was removed.

There was a red light operated from the control room by the announcer pressing a button. There was a white light as a signal to someone in the studio that he must go out and answer the telephone (control room calling the studio). There was also normally a green light operated from the listening room, but here the listening room was combined with the studio. That light was for the studio to go ahead with items etc, and also for the gramophone effects, or if a number of studios are taking part in the production. Note the peephole in the door, shown in plate 13.20.

13.19. The studio contained at least one BBC-Marconi Type A ribbon microphone and a TD/7 twin gramophone reproducing desk can also be seen. This is a talk studio which is too small to be a music studio. It could take a cast from ten to fifteen actors.(Gerald Daly)

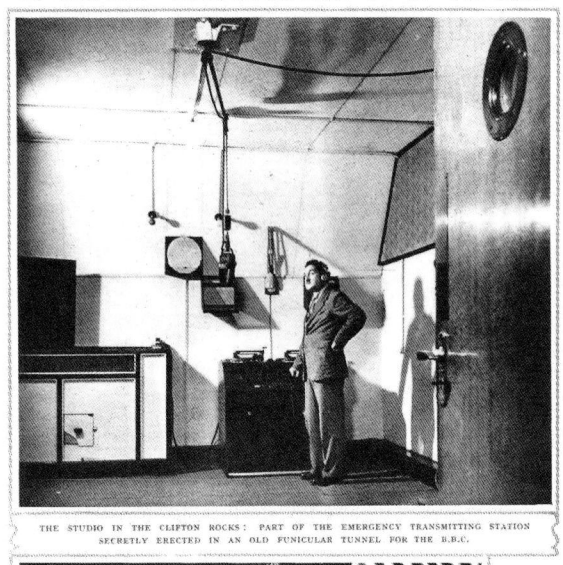

13.20. From the photographs of the studio, there were either two TD/7 (gramophone) desks or, more likely, one was moved from inside the door on the left to the opposite corner between March 1942 and August 1945. Note the lights on the back wall. (*London Illustrated News*)

Acoustical treatment carried out by Bray and Slaughter for the studio, was to fit picture rail on the walls 1in from ceiling, 6in skirting to walls and 6ft x 3ft standard, 20 picture hooks, three Cabot's Quilt curtains (from Messrs May Acoustics Ltd, Wimbledon) on each long wall, two curtains on the short walls. There was also carpeting on the wooden floor. Gerald Daly:

The studio was little used as we had only one real emergency, that was after a heavy raid when the city's water supply was blown up. The Regional Commissioner appointed two Bristol citizens whose voices would be familiar, a Mr Hindle (the City's Publicity Officer) and a Mr Wiltshire, a solicitor and musician. They broadcast warnings about the water supply on this occasion.

This shows that the studio at the tunnel was used for broadcasting important warnings to the local population. For example: it was announced that listeners should boil all water used for cooking and drinking.

13.12 Transmitting room

The transmitter room also contained a stand-by (a T77, also moved from Whiteladies Road). It was an ex-RAF medium-wave transmitter for restoring communication between the other main provincial and metropolitan broadcasting stations should the Post Office telephone lines be damaged by enemy action. A third, a short-wave, transmitter was also used for communicating with other BBC sites in the event of line failure and was an American-made Harvey MacNamara UHX25. These are known to have been used for this purpose at other BBC premises. The plan in plate 13.22 shows their locations.

Everything was grey, light grey, medium grey or dark grey. RCA senders needed copper sheet for screening fixed to the floor. We retrieved a piece 2ft 6in x 9in.

Aerials for the transmitters (plates 13.23 to 13.26)

were mounted at the top of the tunnel and a receiving aerial, for short-wave communications and possibly for receiving the Home Service in the event of line failure, was situated at the bottom. The aerials at the top were damaged by enemy bombing on at least one occasion but quickly re-erected. Frank Gillard states there was:

> a local transmitter to give a programme service to the city of Bristol, and communication transmitters which would maintain contact with other BBC centres even if all line communication failed.
>
> The aerials for these transmitters were erected up on the top of the gorge. Screened leads were brought down through the tunnel to the transmitters so far below ground. When a German bomb blew those aerials down, they were replaced within an hour.

13.13 Aerials

The Aerial Plan dated 2 January 1942 (plates 13.24 to 13.27) shows the length of RCA and T77 aerial 130ft between spreaders attached between the roof of the hotel and the roof of the first aid post with a 9ft sag. There is a 50ft mast and a 40ft mast and a hut for aerial intakes on the first aid post roof. A 1ft x $1/8$in copper tape was buried under Princes Lane by the CRR tunnel and four earth plates 4ft square were buried in the round north west of Tuffleigh House and connected by copper tapes.

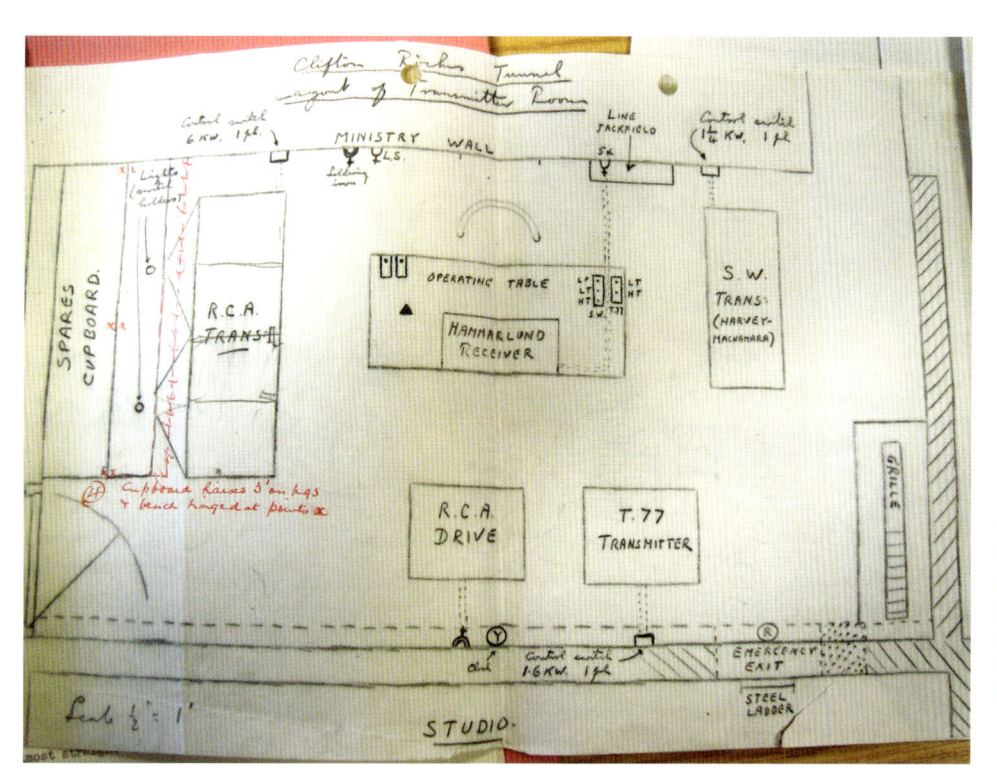

13.21 The RCA 1 E Group H transmitter shown was transferred to the tunnel's transmitter room from Whiteladies Road on 14 September to radiate the Home Service. This was the largest transmitter and was erected and operating within six days. It was brought over on lease-lend. The transmitter looks a little different than in plate 13.1 but there appears to be a hinged panel over the gauges. (Gerald Daly)

13.22. Layout of the Transmitter room. Note the location of transmitters and receivers and the emergency exit under the floor. (BBC Archives, Caversham R35/207)

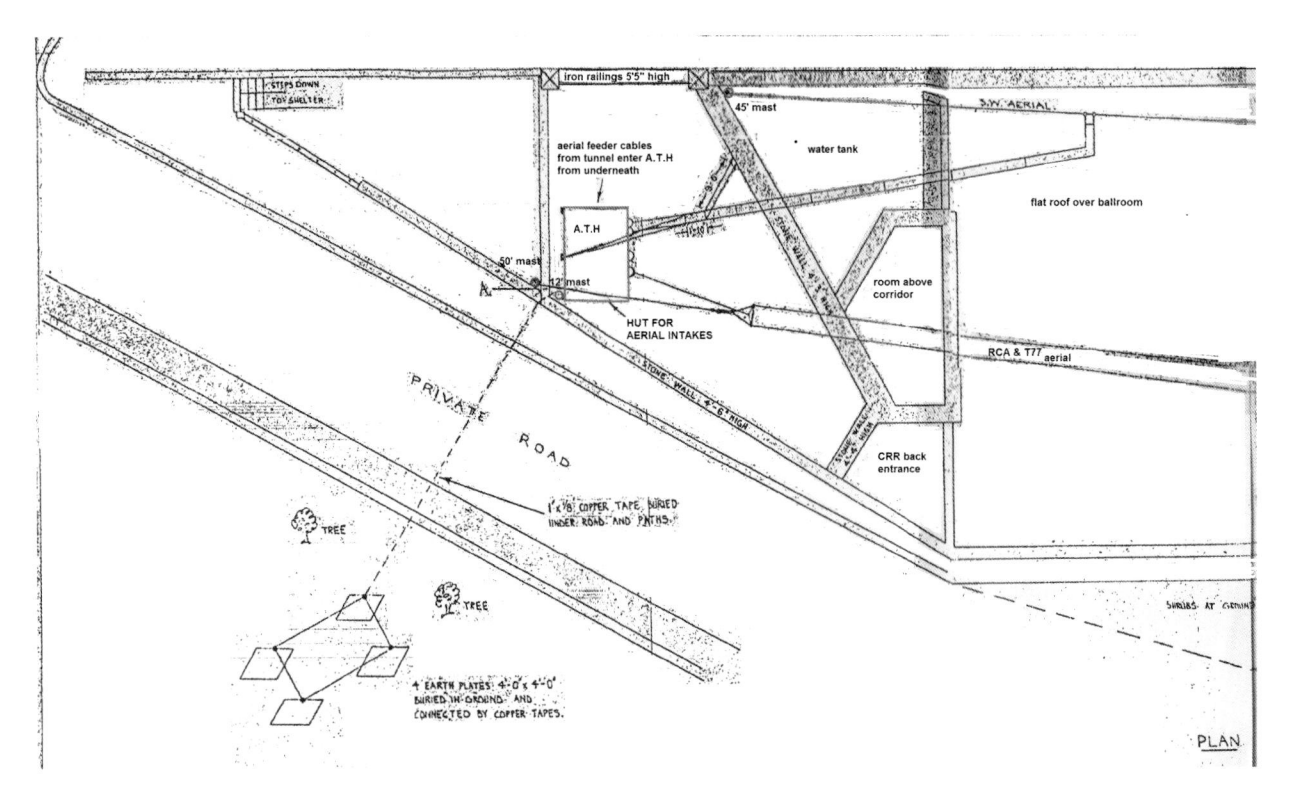

13.23. Mast, hut and earth plate on the Railway.

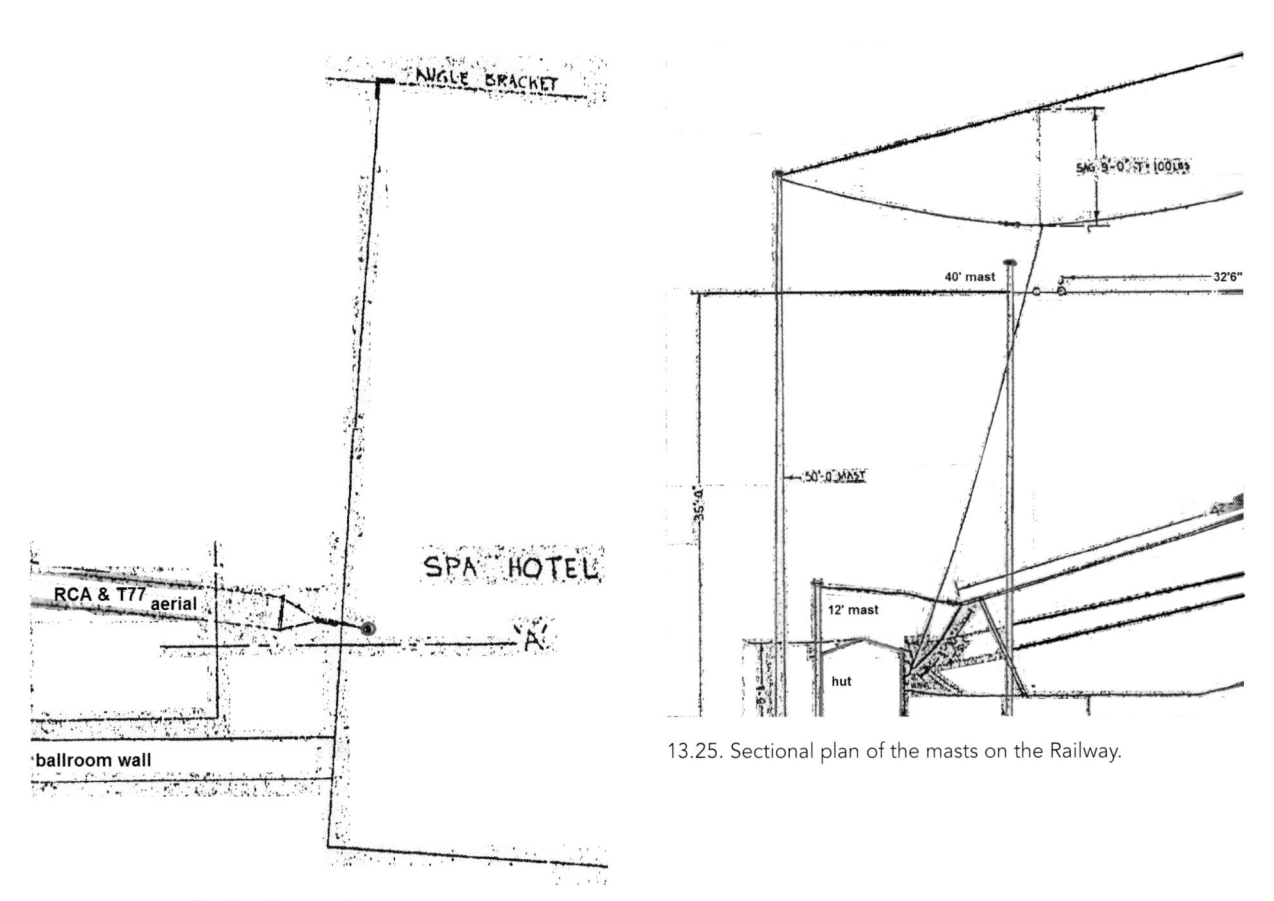

13.24. Aerial attached to the Hotel.

13.25. Sectional plan of the masts on the Railway.

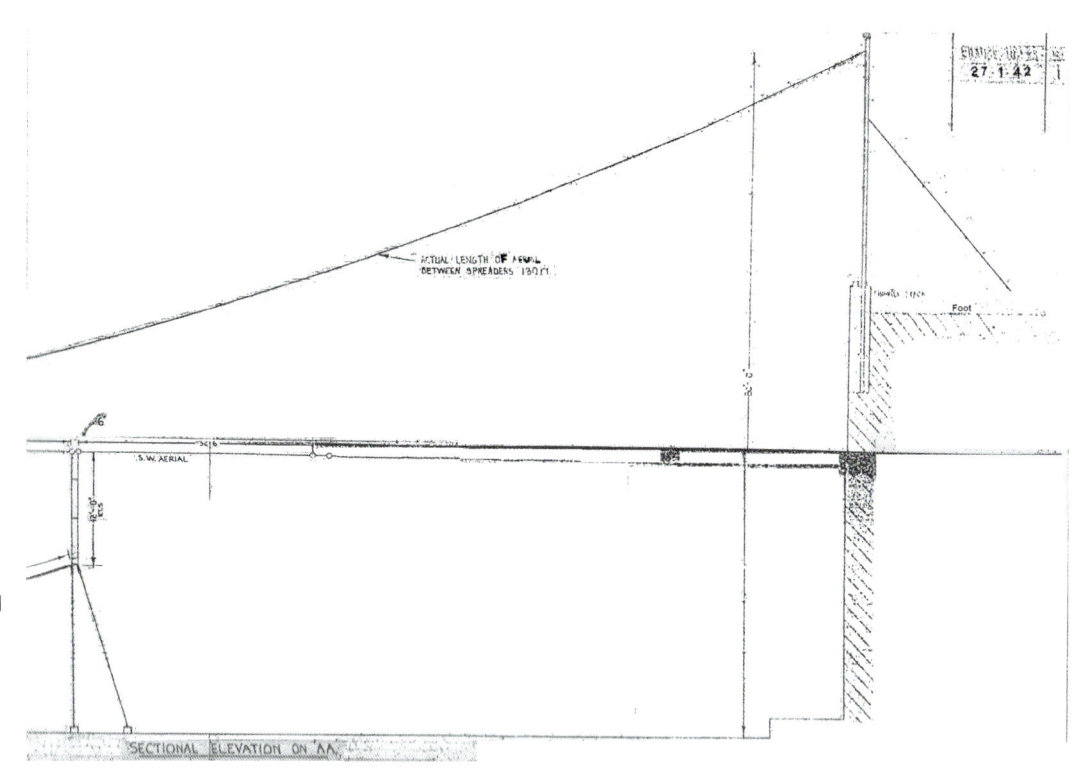

13.26. Sectional plan of the aerial attached to the Hotel.

13.27. Location of aerial hut on the Railway. The entrance steps are to the right.

13.28. Insulator on the Pump Room roof.

13.29. Location of aerial mast near the top of the entrance steps.

13.14 After the War
The Group H transmitter network was in use for the Home Service until it closed down on 29 July 1945, when the BBC implemented its peace-time broadcasting plan but the transmitters at most sites, including Bristol, were retained and some equipment in the tunnel was maintained for many years after. Line switching operations returned to the Whiteladies Road Control Room and the Studio and much of the Control Room equipment was probably dismantled. Many BBC sites set up during the war were secretly mothballed afterwards in case of future emergency.

In place of the pre-war national and regional programmes, the BBC's post-war plans were for the continuation of the Home Service on a regional basis (within 90 days of VE Day – it resumed on 20 July 1945), a programme called the Light Programme along the lines of the war-time Forces Programme and a completely new service, the Third Programme for serious literature, drama, music and intelligent discussion broadcasting from 6 to midnight. The West of England Home Service was then taken over by the 20kW transmitter at Clevedon on 1474 kHz (203.5m) and by Start Point in Devon on 583 kHz (514.6m). It was intended to merge the West and Midland Regions to release a frequency for the new Third Programme but a public outcry led to the abandoning of the plan. The final decision was taken at Cabinet level. Eventually, the Third Programme was scheduled to start on 5 May 1946 using 583 kHz, currently in use for the West Home Service. This frequency had been allocated to Tunisia and Latvia in the Lucerne Wavelength Plan of 1934 but Tunisia never used it and the low-power Latvian station had been destroyed by the Germans. France had lent the Tunisian use of it to Britain during the war. The original intention was to broadcast the Third Programme from Droitwich on 583 kHz, supplemented by 15 low-power transmitters on 1474 kHz, including Bristol, but due to problems obtaining permission to use the frequency, the start was delayed until 29 September 1946. On 25 August 1946, the Russians announced that they intended to broadcast from Latvia on 583 kHz and the station came on air on 17 September at high power. It was decided that the BBC should start on the planned date and continue to use the frequency but at much reduced power and with increased coverage from extra low-power stations on 1474 kHz.

On 26 September 1946 (a Thursday), the *WDP* announced that when the BBC wavelength changes took effect on Sunday, listeners in Bristol and most parts of West of England Region would get reception as good as, or better than the present West of England Home Service, but the new Third Programme would not be heard satisfactorily in the immediate vicinities of Bristol and Plymouth. To overcome this, at least partially, the BBC would also send out the Third Programme on 203.5 metres from about 20 low-power transmitters located in main centres of population to supplement coverage. They would have an effective range of only six or seven miles. Bristol district listeners would get good reception on 203.5 metres from the secret war-time Group H transmitter in Clifton Rocks Railway – built for emergency use if invasion came. Frank Gillard outlined the programme plans. Up to August 16 he had to plan for two lots of programmes, one if the Region was absorbed by Midland, and one if the West Region stayed. Before the war West Region programmes occupied eight or nine hours per week, now it was 12 hours and rising to 19.

The ex-Group H transmitter in the tunnel broadcast the Third Programme from 29 September 1946 until 14 March 1950 except between 10 and 26 February 1947 when, due to the severe winter, the transmitter was shut down to save power. Droitwich was also forced to work on a lower power than intended hence the chain of temporary relay stations, a number of which were re-activated wartime Group H transmitters to improve Home Service reception, particularly in coastal areas.

In March 1950, the 1948 Copenhagen frequency allocation plan for radio broadcasting was implemented in Britain, with the band extending from 530 to 1600 kHz. This allowed the Light Programme's Droitwich transmitter to increase its power. The UK was formally allocated a frequency for the Third Programme and an additional frequency for the European Programme (now part of the World Service). The frequencies that were used were within 1 or 2 kHz of the UK's current high power allocations. Radio Luxembourg's English service moved to medium wave, broadcasting only in the evening and was subject to fading. On it's new frequency, the Third could be broadcast at the full 150 kW. [13] A new main transmitter came into operation at Daventry making the Bristol relay redundant.

In spite of the closure of the Bristol medium wave relay, the BBC lease on the tunnel was not relinquished, and they continued to rent it from Bristol Corporation until 24 June 1955, at which time they began renting at £10 per annum, the first aid room on the top platform of the tunnel to house a possible temporary relay transmitter for Bristol (more details about leases are to be found in chapter 10). However when the need arose, the 10 watt VHF installation actually came to be sited on the top of the water tower on Durdham Down, Clifton and this radiated the Third Programme across the city on 94.7 MHz from 28 October 1957 until the permanent installation at Wenvoe in South Wales went on air on 1 March 1959.

The equipment in the tunnel was then retained as a deferred facility (DF no.46), a secret arrangement to keep the studio and transmitters in reserve in case of Emergency Situations (Soviet attack, cold war nuclear threat etc). Because the DFs were kept in readiness they were considered secret, and people involved with them often pretended they did not exist. The equipment was removed, the tunnel vacated and the lease determined in 1960.

Gerald Daly confirmed that the equipment in the tunnel was kept operational for a year or two after the war and then dismantled. Frank Shepherd recalls a trip made in the late 1950s (having joined the BBC in 1957), when he accompanied an engineer to the Rocks Railway to bring back to Whiteladies Road any small pieces of BBC equipment that was still there.

So some local broadcasting took place from the tunnel studio, and the Group H transmitter was used throughout the war to broadcast the Home Service and all overseas programmes.

13.15 Employment

Youth Transmitters were paid a weekly wage of £1 2s 6d plus 3s 0d cost of living bonus in 1941. In 1941, BBC representatives searched for bright young students age 16 as potential employees. These were essential to replace staff enlisting in the armed forces. Once hired, an official envelope from Broadcasting House in London would arrive. It contained a lapel badge, a multitude of employment papers, agreements, regulations, staff instructions, and a reference to secrecy. A clause in the contract stated that employees were subject to transfer at anytime and to any posting in the UK.

13.16 Training

Molly Foss's training was at Maida Vale (read her memories in section 13.19). We were given her training books: *Technical Glossary for BBC Engineers* E.L.E.

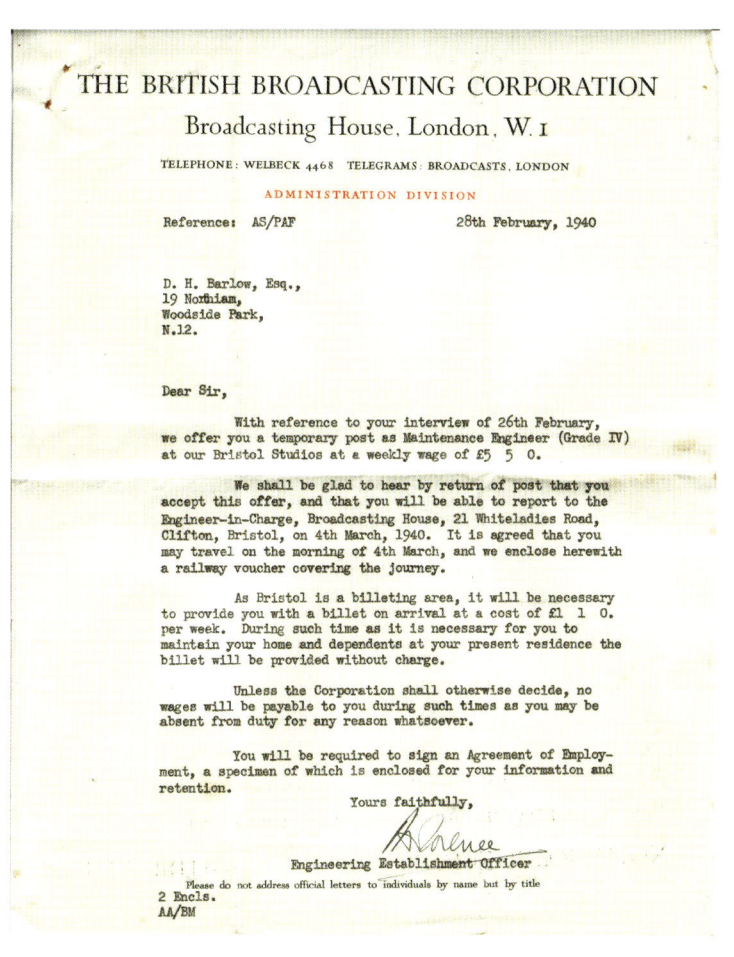

13.30. Douglas Barlow's letter of appointment as Maintenance engineer for 4 guineas a week in February 1940. Note billeting costing £1 1s per week is offered for his family. (Elizabeth Taylor)

Pawley 1941; and *Engineering Division Training Manual* 1942. Her handwritten exercise book from Maida Vale noted:

Course A Lecture 1 Organisation and Administration of the BBC; Lecture II The Wartime Organisation and Administration of the BBC; Lecture III 24 hour clock. Network System etc; Lecture IV The Fundamentals of Electricity and Magnetism; Lecture V Microphones; Lecture VI Outside Broadcasts; Lecture VII Microphones, studio lights, Relays, prefaders etc; Lecture VIII Sound Range, Acoustics and studio equipment from the point of view of a studio engineer; Lecture X Control Room Equipment; Lecture XI Recording; Lecture XII Lines etc; Lecture XIII Line Measurement; Lecture XIV Transmitters; Lecture XV Levels of programmes; Lecture XVI Shortwave and Transmission

Course B Chain of events; Method of Superimposing Time Signal; Studio signalling; Crossplugging for faulty amplifier; Plugging if trap valves faulty; Balance and control of programmes; Faults that may occur;

OFFICIAL APPOINTMENTS

THE BRITISH BROADCASTING CORPORATION.—Applications are invited for a number of appointments to the unestablished staff, for the duration of the war, as MAINTENANCE ENGINEERS. Commencing wage £5 5s. per week, with annual increments to scale, subject to satisfactory service. Candidates, who must be of British nationality and parentage, should produce evidence of technical qualifications in accordance with the following: Possession of some recognised theoretical qualifications in electrical engineering or radio communications, and some experience of electrical engineering practice, not necessarily including wireless experience; or several years' electrical engineering experience, part of which should preferably have been gained in wireless telephony transmission or speech input and low frequency amplifier practice, together with a good standard of theoretical training. Successful candidates will be posted to any of the Corporation's transmitting stations or studio headquarters in Great Britain and Northern Ireland. Future permanent employment will be determined by merit and the existence of vacancies. Further details of the conditions of employment in the service will be given to candidates who are selected for interview.—Applications, giving full details of education, qualifications and experience, should be forwarded to the Engineering Establishment Officer, Broadcasting House, London, W.1. Envelopes should be marked "Unestablished M.E." Candidates who wish to have their applications acknowledged and to be informed if their application is unsuccessful should enclose two stamped addressed envelopes.

THE BRITISH BROADCASTING CORPORATION.—Applications are invited for a number of appointments as unestablished JUNIOR MAINTENANCE ENGINEERS. Candidates, who should be not less than 17½ and not more than 19 years of age on 1st November, 1940, must be of British nationality and parentage. They should be of good education, preferably to Matriculation standard, and have a keen interest, both theoretical and practical, in electrical engineering and radio communications. Minimum commencing wage £2 5s. per week, with annual increments to scale, subject to satisfactory service. Successful candidates will be posted to any of the Corporation's Transmitting Stations or Studio Centres in Great Britain and Northern Ireland. Future permanent employment will be determined by merit and the existence of vacancies. Full details of the conditions of employment in the broadcasting service will be given to candidates selected for interview. Applications, giving full details of age, education, theoretical and practical qualifications and experience (if any), should be sent to the Engineering Establishment Officer, Broadcasting House, London, W.1, by 17th October, 1940. Envelopes should be marked J.M.E. Candidates who wish to have their applications acknowledged and to be informed if their application is unsuccessful should enclose two stamped addressed envelopes.

13.31. Maintenance Engineers with technical qualifications were paid £5 5s per week, Junior Maintenance Engineers £2 5s per week. (Elizabeth Taylor)

13.32. Silver and blue BBC lapel badge.

Method of logging origin of the Programme; Control Rooms at Regional Stations; Incoming Programme; Rehearsal Circuit; Lines Testing; Studio Testing; System of events when Broadcasting (transmitting) programme from a studio in Bristol.

13.17 Play

At some stage, obviously in a more relaxed time after the war, Gerald Daly wrote an amusing play for 14 actors *The Tunnel: The story of a BBC Wartime Fortress*. It starts in the 1890s with a narrator telling the story of the construction and opening of the Railway, with background actors including a woman who wants to ride on the railway on opening day. It then moves on to 1940 with two narrators telling the progress of the war, how to fit a broadcasting station into the tunnel, and the steps they took to prevent the enemy destroying or disrupting broadcasting should an invasion take place. They even had the ghost of an elderly woman who is said to have died of typhoid in the 1900s after drinking the waters from the Ppa.

13.18 BBC artefacts

category	type	condition	material	Height inches	Depth inches	Length inches	quantity	inscription	Where found
BBC	worktop/ shelf	broken	black glass		0.35		1		cafe
BBC electrical	6 fuse on bracket	good		11	3.5	1.7	1	T	control room
BBC electrical	cover	good	bakelite	4.5	2.7	2	1		cafe
BBC electrical	connector box	good	bakelite				1	GPO 235 BT no 6	control room
BBC	kettle				9	30.	1		cafe
BBC copper sheet								t	Transmitter room

Table 13.1

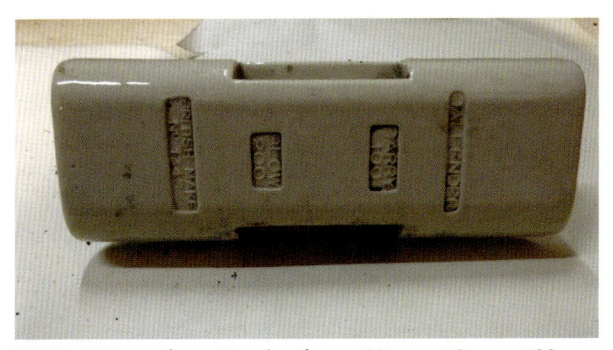

13.33. 200amp fuse. Simplex fuses, 30amp, 50amp, 100amp were found.

13.34. 200amp fuse from side.

13.35. GPO 235 BT no.6 bakelite fuse box/connector box with two banks of ten connectors in the foreground.

13.36. GPO multi-line (40 lines) connection block from control room.

13.37. Enamel lamp shade found in the studio (4.5in x 9.4in).

13.38. Blue enamel lampshade, 4.5in x 9.4in. We retrieved one from the studio and one from the recording room. There would have been one per room. There is still one in the Ladies' toilet.

13.39. Generator fuseboxes, and a light switch in the foreground.

13.40. There are fuseboxes everywhere.

13.41. A trip

13.19 Memories

I was lucky enough to speak with many of the engineers who worked in the tunnel. This helped understanding of their work, and what it was really like in the tunnel. There was also a programme *Secret Underground* recorded in 2005 with Chris Serle as the presenter and shown on ITV1 West. Cliff Voice, Molly Foss and David Pearce were interviewed and a transcription is given in section 13.19.1. These people were interviewed again later, so more information from each appears in section 13.19.2. Some of their memories conflict with each other, but it is fascinating to read.

13.19.1 Secret Underground: BBC, What does it feel like now?
Cliff and Molly

'Looking at the control room photo (plate 13.15) – Douglas Barlow was the engineer, and a girl not long in Bristol (Peggy Richie – she was the kind of girl everyone would remember) and Martin? a junior engineer. Very memorable. The transmitter in the photo (plate 13.21) was American. There was two transmitters in top room. The recording room (plate 13.18) was Cliff's space. Notice the speakers looking like beer cubes, the 6-footer double-ender cables. There were holes through the floor in each room so a person could go to each room through the floor. You could get to the recording room from the control room. Mr Daly is in the photo. A great man. Cliff had a strange interview with him. He was asked about red and blue terminals on a battery and if he could make good tea. He was told you would be a good engineer if you were good at making tea. Later on David Jacobs was on a programme 20 years later with Daly, and Cliff was told to make the tea rather than his wife, because he was an engineer so made good tea. Very likeable Irish man. The boss.

Shifts were 9-5, 5-midnight and midnight-9. Not uncomfortable. During night as there was not much happened, one was allowed 2 hours to sleep. Mattresses were by the transmitter and in studio. There was no contact with the shelterers. There was a big door. She once looked through and saw people sitting with water dripping. It was horrible. She was warm and comfortable. Air from river stank. There was a corrugated roof, duct and handrail. The first door was the door to the generator, a canteen, and toilets. You went through the bottom gate, down the corridor. There was a uniformed Commissionaire at a reception desk by the toilet – you had to show your card.'

They remember Sergeant Waite who inspected the card. He had his gloves on his shoulder. He knew the driver.

They felt important – it was not just a job. They felt jolly safe. On the corner of Tyndalls Park Road was gramophone Library which got hit and disappeared – only 20 yards away from Broadcasting House. The A building on the end also got hit. Cliff felt safer in the tunnel. He did not mind coming here to work.

Queen Mary came to visit the BBC. Molly was in Broadcasting House, sitting not 20 yards from her. A wonderful woman. She was given a photo of her. Molly was asked who she was listening to – Ivy Benson and the girl band. Cliff was in tunnel. The boss had warned them she was coming. She went straight up to the top to the transmitting room. The door opened and she marched in with people trailing behind her. Cliff took her hand as he was not sure what to do. He had a BBC badge in his pocket she had seen. He was told later if he had given it to her that she would have given him a silver pencil. She apparently took a liking to things and expected them to be sent to her. She was tough, not young. She had a reputation for visiting sites. At Broadcasting House the cleaners spent hours and days cleaning one toilet for her but she did not use it.

> Gerald Daly reminisced in 1974: A memory which still hangs around the back of my mind is the time when we were asked if Queen Mary, who had been evacuated to nearby Badminton, could see over it. It had been kept such a secret that we were surprised that she could ever have heard of it. Anyway she came along, but as she was getting on we thought that she would not want to climb the hundreds of steps to see into each chamber, so we arranged that we would tell her about it in the entrance hall. When she came however she said she wanted to see it all and started up the steps. She climbed to the very top apparently without losing her breath, while we men panted behind her very much out of breath.

Molly remembers going in the Railway when she was about six, her grandfather had taken her up. They then got the bus or tram home. It was lovely. She sat down and up they went. She saw the other car coming down at the same time, just like in the picture. Not scary. She remembers it well indeed.

The Railway was secret. There was no discussion. That applied to most things. No one knew what they were talking about if they talked about the tunnel.

Cliff said it was a shock to see the rooms in the state now – no longer spick and span. Most are just a pit. No floors – they had just rotted away. Unbelievable after 50 years to disappear. It was warm and dry. Just like home. Unbelievable to disappear. Molly said she could not bear to see the rooms now. There was a carpet on the studio floor, lino on the floors of the other rooms. two transmitters. They could go through floor/roof to get to the other rooms avoiding steps.

Due to river pong through the fans there was a machine to make everything smell nice. It was so revolting she preferred the river smell.

When the Americans came they put their transmitters everywhere. Douglas Barlow, Peggy and Martin are with the transmitters in the photo (plate 13.15). The other photo (plate 13.18) shows the Philips-Miller film-based machine in the recording room. Everything was grey – light grey, medium grey or dark grey.

In the other photo Daly (plate 13.20) is standing by TD7. It plays two 78 discs so one could do continuous programmes [that is, switching of programmes by engineers]. The arm goes straight across. At the end, the record lined up for overlap. Often a disastrous result.

They could not remember the man in the trap door (plate 10.23) [It was Norman Morse]. They did not like going under the floor. Molly remembers having two hours to sleep on the night shift. Someone for fun, put her mattress over the trapdoor hole and she was woken up by someone trying to get in from underneath.

Dave Pearce thought it was exciting to recreate daily life 60 years ago. He was there from 1943 as a junior engineer. They used a different main entrance in the war to get into the railway. There was a three position front seat in a Dodge car. Its roof had particular armour flat plate welded to the roof, and a dome filled with sand, so the car was top heavy and rolled. Sand was put there to protect the occupants from shrapnel and buildings falling. He had a uniformed driver to take him to the tunnel. They were not allowed to drive. No street lights or they were dim, and headlights dimmed. Very little traffic, but a lot of people walking around. There were a lot of soldiers in uniform particularly Americans. Bristol had a high population of servicemen

The studio had a table, four chairs, and a microphone (plate 13.19). It was a talk studio, too small for music. The recording studio had Philips-Miller recording equipment. The transmitter was 1KWatt or 1 rat power (engineers term). The main control room had racks of equipment.

Mr Daly was fantastic, incredible. He was the Engineer in Charge. In 1936 he started the BBC in Bristol-he had seen the empty building in Whiteladies Road and thought it would be suitable for a new broadcasting house for Bristol. They were in a hurry so he went in via a coal delivery chute and opened the door to another engineer, and that is how it all started.

Ernie Mace is the person in the trap door photo. He was an engineer. All the rooms were linked under the floor by vertical iron ladder

There was a big diesel generator to power the place (plate 13.10). No electric starter, so one had to swing it. It was brutal. He was only 16. He tried to swing it and it would bounce back several times. There was a decompression lever to operate, then one would swing like fury, switch the compression lever back and it went 'thwump, thwump'. It was a single cylinder diesel.

The rooms are still as cold, drips dripping. He remembers the cabling and ducts. Fantastic. He was here from 1943 on. He could not hear the traffic as there was a big steel door. It was only on night shift they came down. First thing he did was he started the diesel. In the night shift they came in an armoured car. There was a dish on top filled with sand to absorb shrapnel. It wallowed like a pig. The Railway was an emergency back up. There was a routine to check everything was working.

There was a canteen, food and water emergency store and a primus stove. They had to minimise power used from generator, so everything else was wartime emergency stuff.

Outside ducting was for the air-conditioning plant. They used a Commer van to bring people down if changing equipment.

There was an equipment bay in the recording machine room under the clock. It was a slave clock. There were clocks everywhere since timing was critical. Programmes started at an exact time. The clock was accurate to the nearest minute and reset twice a day.

He remembers the staircase – same as now. Remarkable.

The Philips recoding equipment had a 20 minutes film tape, so much better for talks. The other system – the wire system – could only record for four minutes and was magnetic recording. Audio signal was good for talks. One cut the tape with a short stylus and showed a light on film. This was extensively used equipment, and used if disc was not suitable.

Fun? Yes – he was young and enjoyed it so did not realise the implications. He worked long hours, sometimes 36-hour days. Although on changeover a shift was 9:00am-5:00pm on a Saturday he had to return at 10:00pm and did not finish until 9:00am on Sunday so he worked a 19-hour day – long for a 16-year-old boy. There were great people to work with.

13.19.2 Individual Memories
Arthur Bradley
Born in Brislington in 1925. He worked for the BBC twice. January to May 1940 (as office boy) and January 1942 to January 1943 (in the control room).

'I left school at the age of 14 and first joined BBC as an office boy in January 1940. I worked for Mr Page in the post room. Duties included opening of post and delivering to the appropriate departments after vetting by the registry girls. The BBC had taken over a variety of premises in Clifton including church parish halls which had been transformed into studios. Items often needed delivering to these locations, and this was accomplished by us lads on tanks (heavy bicycles with a front basket). Although my stay with the BBC at this stage was for only a few months I spent time in defence stores with Sidney Day, in the buying office, and at the music department, 91/93 Pembroke Road (the music and variety department had moved from London to Bristol). Duties here included being doorkeeper, and checking that people entering were staff and could display a BBC pass. Sir Adrian Boult was not excused from this requirement. I seem to remember that the passes were much the same colour and size of a packet of Gold Flake cigarettes. Light Entertainment Department had also been evacuated to Bristol and there was a concentration of well known celebrities of the time, some of whom would make the translation from radio to television, and some not. One weekly show was Garrison Theatre starring Jack Warner – its catchphrase was 'mind my bike'. Garrison Theatre was probably the studio in All Saints' parish hall. It was soon realised that Bristol was not very safe, and music department moved to Bedford and light entertainment to Bangor. Jack Warner, Elsie and Doris Waters were there and maybe Vera Lynn. There was quite a lot of hate mail for the stars, which I would throw away.

I watched the planes bomb Filton from the steps of the BBC. After the air raids on Bristol in the spring of 1940, my mother decided to move to Edinburgh with myself and my younger brother George since my mother thought it would be safer, and the air raids had unsettled her, so I was forced to hand in my notice. My father who had served in the army in France in WW II had joined the RAFVR in the 1930s and was stationed at a balloon barrage site by the Forth Bridge. Hence the choice of Edinburgh. George went to a local school, learnt the language, and complained about my Bristol accent. I got a job as an office boy with a firm of solicitors. There was also a language problem there but we managed somehow. I suspect my teenage grumbles at home was a factor in our next move. By late 1941 we were back south in Uphill, Weston-super-Mare. This was a bad move as we caught the full force of the air raids on Weston at the end of June 1942. Our flat was made uninhabitable, and in the confusion we were split up. I moved to my grandparents house in Bristol where I was able to continue going to work. My father got compassionate leave and located his wife and George at a reception centre. He was later able to obtain a more local posting. My mother had a break down and went for a spell in hospital. The family eventually re-established in North Bristol.

By now I had acquired an elementary knowledge of radio, and could send and read morse code at a modest speed. The Labour Exchange (there were a lot of jobs then, and lots of men from Geordie land) at Weston-super-Mare sent me to Bristol BBC Whiteladies Road where there were a number of vacancies for Youths-in-Training in their engineering department. Arthur was dereserved, which meant he did not have to be called up until the age of 18. You could get called up at 17. I was interviewed by Chief Engineer Gerald Daly who features greatly in the establishing and use of the Rocks railway tunnel for wartime purposes. He asked me some simple questions and was interested in the fact of my previous employment with 'auntie' although I had no qualifications. I was taken on at £1 2s 6d a week, and started work for the second time for the BBC on 5 January 1942, getting to work by rail and bus. Employing lads so young was a scheme started in an attempt to replace the established engineers who were in demand elsewhere. We were recruited at many stations around the country and spent about a year getting to know the equipment, and how it was applied. Learning electrical and wireless theory was also quite intense. After this we were selected and sent on a course of some 14 weeks, and then posted to a transmitting station, studio centre or wherever, when our job title became TA2 (technical assistant grade 2) and our pay £1 7s 6d per week plus a few shillings 'cost-of-living allowance'. We had a one-year deferment from call-up so that the corporation got about a year's work from us after training. Training was two weeks at Maida Vale, London, and twelve weeks at Evesham. Women had also been recruited to do the same work from Autumn 1941. Ladies taken on followed a different training route I think. Some of the write-up on women is a bit cringing – skirts v. trousers ...etc. Before the war, evening dress should be worn if there was occasion to visit a studio after.

The working day was divided into three shifts to cover the 24 hours. There used to be four or five on shift over 16 days: four mornings 9:00am-4:00pm; four afternoons 4:00pm-12.00 midnight; four nights 12:00 midnight-12:00 noon; four days off and I would then go home. After reporting for work at Whiteladies Road the people who were to work at the tunnel were transported by car or van to the entrance on the Hotwell Road. I remember working night shifts at both Broadcasting House and the tunnel.

At Bristol there were many men from the Merchant Navy including Mr Lamont who I worked with. I can remember the control room in the railway. I sat on the right hand side in front of the switchboard doing SB (simultaneous broadcasting). There was a schedule of tests of the emergency communications between BBC centres. On one occasion I was the only morse code literate person on shift since I had learnt it at the ATC (Army Training Corps). The receiver was tuned in, the transmitter was run up, and the schedule time came and went – a complete failure. Perhaps others had better results. An engineer named Lamont made or perhaps repaired an aerial changeover switch using sixpence pieces for contacts. This sounds a bit unlikely now but that is my memory of the emergency system. I think it was never used for real. There was a network of low power transmitters dotted around the country carrying the Home Service on 203m. One of these was in the tunnel. To avoid giving navigational help to enemy aircraft, these transmitters could be switched off on local information. The switching off of high power transmitters required a special message from London which should be word perfect; 'Urgent priority close [name of station/s] SCRE London'. SCRE stood for Senior Control Room Engineer'. I was only involved in this procedure on one occasion when Washford 1 and 2 were the named transmitters.

If the message deviated I knew it was the enemy and the system was closing down transmitters on an air raid. There was a short wave transmitter and there was a schedule of contacting all stations by morse. The hour was split into four: 15 minutes for English, then 15 for German, 15 for French and 15 for other languages). I would send out messages and press knobs.

I would monitor programmes. If the quality was wrong I wrote it down in a log book. The lines were rented from the Post Office. The lines were all numbered PWSW with a number. The best lines were inside the cables. There were 100 pairs – the BBC had those. The outside lines were for telephones.

I could not remember much of the tunnel, only really the control room, but I should have remembered something of the transmitting room. I remember cleaning the instrument knobs when I had nothing to do.

I can remember Douglas Barlow (who thought Bristol was a dump as he had hated being uprooted), and Molly Foss who was quite a lot older than me.

I was at Bristol for about 12 months before I was sent on a course. My time for assessment came along in 1943, and Mr Lindsey arrived from London to interview us. He decided I was not ready to be promoted. However, this injustice was overturned behind the scenes by local senior engineers, and off I went to London to the training regime detailed above. This marks the end of my connection with BBC Bristol and the Railway. I was firstly billeted at Grosvenor Square for a few weeks and then went to Evesham for three months.

Evesham was a big listening centre with lots of special aerials. Everyone was very clever and they all spoke 17 languages. The famous people did not mix with the others. Everyone was posted to different places at the end of the training course, and I was sent to Bush House in London to work for European Services – one just went where one was sent. I had digs in Clapham but was doodle-bugged (VI), so moved to Finsbury Park. In 1945 I was called up and was in the Army Royal Signals, and met lots of BBC colleagues. After training as a radio mechanic I was sent to India where I was posted in turn to Mhow, Bangalore and Poona. After Indian Independence (August 1947) and shipment back to England, I had a short spell recording for the London Olympic Games in 1948 before resuming at Bush House. Next stop was Television Centre at Alexandra Palace before moving to Lime Grove Studios where I summoned up enough courage to hand in my notice in 1951 or 1952. This was the end of my BBC connection.

After the war the BBC was obliged to employ me again, just like all other employers had to employ those who had come back from war. This was hard on those who had taken on their jobs during the war as they lost their jobs even though they were often better qualified.'

Molly Foss
Molly was born on 27 October 1918 and died in October 2008. She lived in Clouds Hill House, Summerhill Road, St Georges Park, Bristol all her life. There were six children in the family. She worked in the BBC area of the Clifton Rocks Railway between 1942 and 1945. Molly gave us her wartime training books from Maida Vale.

'I worked for the BBC Engineering Department between 22 September 1941 and 31 December 1975, when I retired. I got my job in Broadcasting House, Whiteladies Road by answering an advert in the *Telegraph*. My wage was an average amount. I started at the age of 19 as a technical assistant (I was a physiotherapist before), which entailed receiving programmes and passing them to transmitters noise-free. I always worked as a transmitting engineer.

Engineer Mr Smith sold me a car costing £100 (Austin soft top) so I could visit my nephew aged 7 at boarding school, and he could visit during holidays. His parents lived in Singapore. My favourite band was Ivy Benson and her girls band.

I received my training in London over three weeks, I thought it should have been longer. This was in Cosham House, Clapham Common. I went to Broadcasting House with Pat Whyman whose father was the vicar of Knowle. Pat went to Washford. At the end of the war I came to Bristol.

When I joined the BBC in 1941, as a woman operator in the Engineering Division, the Control Room, in which I worked was on the third floor, and overlooked a garage and two large houses with steps up to a pillared entrance, which at the time were flats. The BBC had acquired 19, 23, 25 and 27 Whiteladies Road by then. Opposite no. 27, the Territorial Army barracks, was as lively now as it was in wartime. Outside the entrance to the BBC Whiteladies Road, were sandbags and men with guns in khaki guarding. Nearly all the chaps inside belonged to the Home Guard, and the senior men mostly wore their uniform. Mr Daly, the Senior Engineer and others wore it. As I began to be aware of what went on around me and what I was supposed to be doing, I noticed that the men in the Control Room who had finished work at 5:00pm, and those off duty gathered together, putting on dungarees or scruffy clothes, and all departed, so I was told, for the Emergency Control Room at the tunnel. Gradually it dawned on me that this was the original Clifton Rocks Railway tunnel.

The BBC in Bristol had for sometime been the place where London's tragedies had sent main departments – music, drama, schools, Children's Hour. It was a wonderful place for people who knew broadcast programmes well. Here was Uncle Mac, Auntie May, masses of well-known actresses and actors, all to be met in the Canteen, as well as announcers, and producers, but as I didn't know any of them, I was bewildered. Frank Gillard was just preparing for his job as a War Reporter and gave us a lecture about the job of a BBC reporter, and his recording engineer too. Sometime in these wartime days, they had an exercise for the home guard, and my job consisted of working and staying for about 36 to 48 hours in the BBC's studio control room at Clifton Parish Hall [between the end of the Fosseway and start of Merchant's Road], which fortunately was not bombed when Clifton Parish Church was demolished. Another woman operator, a senior engineer and I stayed all this time in the Parish Hall, waiting for troops to come in at the end of operations. At other times I did my turn in firewatching from the roof. For sometime Ivy Benson and her Girls Band broadcast from there regularly as the BBC's band (women replacing men was the policy). We stayed the night in church hall if we could not get home due to bombing and listened to the noise. We could have a bed anywhere, even in the studios.

Once during this time Queen Mary came to visit the BBC and the control room at Clifton Parish Hall, and I was sitting next to her, while I was monitoring Ivy Benson and her band. She asked me what I was listening to and I told her. An amazing lady, she later toured the Tunnel broadcasting area, climbing up the many steps to see the transmitter room and studio. Then she had tea in the canteen with all the staff at the parish hall and had her photograph taken with them in studio 1.

I should mention that the tunnel control room etc was in regular use for about two years. The drill was that four or five engineering staff manned the tunnel 24 hours a day, on a shift basis, 9:00am-5:00pm, 5:00pm-midnight, midnight-9:00am. Two people on shift were left at Broadcasting House to deal with operations there. Staff manning the tunnel were picked up in an armoured car at about 8:30am and driven to the tunnel. I regret to say that several times I did not arrive on time to catch the car and then had to make my way by bus to Clifton, find my way to the zig zag, and run down it as fast as I could, high heels and all. Inside the tunnel on the ground floor was a large Lister diesel engine, used if the mains electricity failed; cloakrooms and a shower for staff. On top of these was a canteen with all facilities. However, this was not used, as food was sent from Broadcasting House, in a large boxlike container. Sometimes things slipped and one got custard on the meat course, or gravy on the pud. About 16 steps up was the Control Room, and the same distance above, the Recording Room. Above that a studio, hardly ever used, and above that the Transmitter Room. All this took about half the tunnel. There was electric lighting in the whole of the BBC area.

We used to get a phone call to go to the tunnel. A car would take us to the tunnel from Whiteladies Road. Elsie Otley, an announcer lived at no. 1 Colonnade also worked in tunnel. I had a pass but took nothing with me since the BBC canteen delivered hot food every day. In the refreshment area was a kettle to make tea. There was a cupboard, cups, teapot, and warming cupboard to warm up the food that had been sent. There were shelves on either side. I remember showers and a bath by the ladies toilet. Working conditions were warm and comfortable. I do not remember it being wet in the BBC section.

There were three people in the control room, the rest in Broadcasting House. They took it in turns. There were five or six BBC personnel in there on a shift. I worked mostly in the Control room. I answered phones, listened to programmes, tested lines. All lines in the tunnel were from London. They would transmit Home Service, the Light Programme. The programme was fed to the transmitter. Programmes went through a series of apparatus, including an equaliser for level of sound to be same for all notes. All notes had to line up on the same level into the amplifier to bring the level back up again.

Transmitters went through Washford, Stockport, Clevedon and Cardiff. Cardiff had the same transmitters. The Transmitter in the Railway was American (RCA), which was easy to use. For faults, sometimes faint speech, I rang London to get another line for the programme, so then sent that programme to Cardiff. I had to test the incoming lines for interference (I tested the insulation of each leg and resistance of the entire loop) and the amplifier in case it squeaked. There was a set number of frequencies.

We transmitted children's programmes. Uncle Mac worked in Broadcasting House, he did birthdays, but not in the tunnel – only engineers were in the tunnel. The studio was never used.

They also transmitted schools recordings, and Childrens' Hour. There was a lot of recording for schools.

Orchestras came from London including Adrian Boult. He did no recording in the tunnel even though there was a recording room – mostly live in Studio 1 in Whiteladies Road. There was a kids orchestra who played very well. The Recording room was used very little. Cliff Voice played repeats. There was nothing in the recording studio except tapes to be played if necessary. I do not recall many people using the recording machine (that was Cliff Voice's job). The boss sat in the Control room in room below, never in the studio. I worked with Alan Urwick, Douglas Barlow, David Pearce, Ernie Fletcher – other transmitting engineers.

There was a baby grand piano in the studio, table and chair and a loudspeaker in each room – huge and marvelous (type LSU10). There was a TD7 gramophone (78 rpm disc).

They did not have many visitors. Some producers came occasionally, announcers came – Hugh Shirreff was the main one, Douglas Swann, Elsie Otley. The boss of engineering was Gerald Daly. They were in uniform all the time. Churchill (Chancellor of University) did not visit even though he often came to Bristol. I saw him at the University.

The BBC never held concerts in the refuge area. The blast door at top of the BBC steps was kept locked which divided us from the top half. Refuge people came in from the top. I got a key one day and looked in the shelter to see rows of benches in a very dark cave, dripping with water, and lots of people sitting there for the night – poor things – it was absolutely dreadful. These days, wartime memories seem to think London and Coventry were the only places which were bombed. Most of the heart of Bristol was completely demolished. Only people of my age can remember Bristol as it was.

There was a hole in the floor in each room to be able to get out. Nothing was stored in the tunnel beneath the floor. Only the boss went in there. I remember putting my mattress over a hole by mistake (we were allowed two hours off during night shift to sleep) and someone trying to get out of the hole.

I never went into the bottom turnstile room – I did not know it was there.

There was a big vent for the air (oxygenator). Thing going round drew air in, stuff went out in the river, the river stank. There was a machine to switch allocation. It added a nice smell.

Frank Gillard at 6:00am would play a recording of announcements made the night before about the weather. One day I was late due to the bus (catching them was awful), ran up stairs, and fell flat on my face going into studio, since a woman had washed the floor. I had to put a record on and knew it would not finish in time. Since I was 10 seconds late, near the end I turned up the speed so it would finish at right time, but I overdid it. The announcer put it on again and said 'that's not me' several times. Frank told me I had to apologise – I didn't want to but he was very nice about it. He told me I had made a mess of it. I had to admit that it wasn't the first time I had done it.

I worked all day in Broadcasting House, then put on working suits and went to the tunnel. Four years of hell. There was nothing to do in the control room unless the phone rang, so just chatted. I remember once we sat in a circle and shut our eyes imagining the lights had gone out. When we opened our eyes it had. It was black and we had to fumble around and go and start the generator (by the handle).

The transmitter had a pole outside on top by the side of hotel. I went up with my grand father to see it.'

She bemoaned that no one knows what engineering is nowadays. She kept being asked what she did all day – she told them she got programmes and sent them. They know nothing about amplifiers – not a clue.

David Pearce
Born 1928, David lived in Lilymead Avenue, Knowle. He usually walked to work since the convoys of tanks gathered in the Cattlemarket and the docks meant the buses were held up, but by walking he could nip between the tanks.

David Pearce's recollections of the Tunnel were all BBC and he had never been in the top part of the tunnel before. He joined the BBC at the age of 15 in December 1943 as a technical assistant to work in the Whiteladies Road control room. David confirmed the picture of the Railway control room was not that at Whiteladies Road. His job had been to allocate studios – making sure that people were in the right place at the right time. He was never a transmitting engineer. He was trained on the job. He only got training at Maida Vale later, after he had come out of the services.

'The Railway was a wartime installation, no broadcasting (only if an emergency), only transmitting using a 1kw transmitter. It was controlled from BBC House. It measured the transmitting signal to check if the correct frequency and to high standard. Tattesfield was the receiving station (established 1929), and used for wartime monitoring of propaganda sites and reported on jamming. Caversham was for listening to world transmitters to see if there was vital military stuff. Maida Vale was the centre of the BBC News operation during World War II, and the centre for specialist recordings.

My first day reminiscence was meeting Gerald Daly face down on the bed having his trousers stitched up (they often had beds in rooms in case they could not get home). I had a very small BBC pass for 1944. No photo, just had D.G. Pearce and my staff number 37203, so not exactly a secure means of identification. I worked for the BBC for 43 years.

I was with Gerald Daly (quite something – he took an interest in me and looked after me) who was the EiC controller of the Western Region, and with John Chantrill (senior maintenance engineer) who was also very good. There was one senior engineer in Whiteladies Control Room for each shift. Daly said that TV would never catch on- we will carry on with radio the same as now. It was a ridiculous time but enjoyable.

It was in 1944 when I first went to the Railway. No one worked in the railway as it was for emergencies only at that time if the Germans had captured London. I did not know anyone who worked there pre-1944. Molly was there from the beginning so it may well be true that the Railway had been in fulltime use. The lines were switched back to Whiteladies Road in 1945. I stopped going there at the end of the war and went to do National Service in 1946 at the age of 18.

From 1944 as a 16-year-old, I went to the tunnel regularly at night at 1:00am after my routine jobs in Whiteladies Road, to check that equipment was in a working condition, and run it up. I helped to get the equipment ready in case of a quick transfer from Whiteladies Road. I did not go in every night just every two to three nights, when there was time, as we were very busy. It took time to get there with all the army stuff around, and the night shift was long. The big threat was over, still a bit of bombing. Two of us went in a wallowing Dodge car (we were not allowed to drive), with a roof reinforced with a metal plate with a curved top filled with sand to safeguard from shrapnel. We only went in at the bottom, in a private entrance with a big solid steel door which was very heavy and strong. This was to stop the Germans, if they had successfully invaded, from getting in too easily. It worked well. I had keys. I never sheltered in there. I would shelter at Berkeley Road or Woodland Road. Dormouse (Miss Dormer Technical Assistant Female) came with us. She was funny and good looking. Right at the bottom was a diesel generator. It was installed early in the war, and electric starters were unknown, so you had to start it with a starting handle. This handle engaged a dog on the crankshaft and you tried to turn it round and round as fast as you could until the engine fired. It was a brute of a thing for us to start. It provided power for the installation. Dormouse would start the generator. I marvelled that she could do it when I had such problems starting it, due to its kick back. I realised she was doing something on the generator (I realised later she was decompressing the cylinder so she could wind it fast), put it back into gear and then it fired. We switched everything on and did engineering checks on all facilities.

The mast at the bottom was a receiving aerial used for short wave.

As you know the BBC had an emergency control room there. The person in plate 13.15 nearest to the camera was Dougie Gibb. The equipment on the extreme right, the panels, had lines fed in, incoming and outgoing lines for routing transmitters to studios.

There were four or five rooms at the bottom each about 12 feet square with a roof height of roughly 8 feet. They were not proper rooms. They all leaked. The roof leaked as the rock above leaked and water got into the studios. It could be a problem with equipment. There was an odd bucket strategically positioned as ceilings leaked too. It used to be drier down at the bottom. You could go through to the bottom under the floor in case the stairs got damaged.

But I get ahead of myself, the BBC worked 24 hours a day 365 days a year so we had three shifts. The day shift from 9:00am until 5:00pm, an evening one from 5:00pm until 10:00pm, and then the night shift from 10:00pm until 9:00am the next day. There were shifts of seven, two upstairs in the control room in broadcasting house, the others in the tunnel.

Starting at the bottom, there was the control room with all the landlines coming in along with the amplifiers and gear to operate the studio, Next was the room, which contained a very modern and scarce recording machine. This machine could record a programme, which lasted 30 minutes, a very long time for that period. The normal recording method was onto 12-inch disks, which lasted for three to four minutes. I think there were very few of the machines we had in the country at this time. It was a Philips-Miller film machine but operated in a unique way, a specially shaped sapphire cut a groove in the film according to the sound fed to the recording head. To hear the sound on playback a light was shone through the film, which travelled at about a foot a second, and that was converted to an electrical signal. It was very advanced for its time. Next was a Talks Studio, which contained a table with a microphone on a stand and four chairs for announcers etc. It was not suitable for music or anything else as it was meant for issuing messages only. We then come to the transmitter room, which housed a one-kilowatt sound transmitter. It was not powerful compared with the big boys, but capable of covering a large area of the country with sensitive receivers. Lastly there was a small stock of spares and a water and food store with a rudimentary stove. To my knowledge it was never used in anger for broadcasting and I think we stopped checking it certainly just before the war ended in mid 1945 when Germans surrendered.

I would check everything was working and the generator was still working. All except the recorder which was peculiar and large and used tape and if the tape broke they evacuated as soon as possible as it was a fire hazard. Tape editing was part of the recording department under Mr Fenelow, who Cliff Voice worked for. The generator was interesting. They would spend 1 to 1 1/2 hours in there. It was cold as there was no electricity unless the generator was on. In the control room I would check the lines. We only checked the control room.

Change-over for the three-shift system was on a Saturday, so the Saturday shift was 5 hours off, I then did night work, so worked 19 hours on that day. One day I fell asleep just before midnight, waiting to shift lines from the Home service to Norwegian service, so they did not switch lines to Norway Secret Service at the right time. 'Calling Liage to London'. The underground was in Norway. I woke 5 minutes later to find John in Norway had sent a message to London due to a 'technical fault'. The pips did not go out. I was worried that they were doing an operation that day. I managed to get more pips at 12:15. I worried for some time, but there were no deaths due to the pips not going out. Daly had covered for him. I had to explain why the message had not got through in time.'

David looked at Molly's mathematical training books. He had the same Engineers book that Molly had. He was involved in working from the start. He only did his engineering training in Bristol two years later as he missed the course at Maida Vale since it was a busy course. He did his course in two houses in Arlington Villas off St Pauls Road. They made a funny school. Mr Brownjohn, an engineer, taught him. When he came out of the services he received no training for 12 years as he was too busy.

In 1944, BBC closed at 10:00pm. They were controlled by an elaborate master clock in Whiteladies Road, Bristol. It had a new hand>10 sec break which could not be varied. Bow bells played until the signal came back. Timing was everything. We had to be accurate to within a couple of seconds to allow for programme switching, and sender exchange between stations. It had a slave system to all studios to within one second. The master clock was accurate to one second within six months and would send an electrical pulse to the slaves. It checked twice a day. RCA senders needed a copper sheet for screening fixed to the floor.

Gerald Daly had seen the Whiteladies Road premises when they were buying petrol across the road. He had gone into the coal cellar with Tubby Myers. They had been looking for suitable premises in Bristol in 1934, and the coal cellar was the only way in to view it. Tubby was very large and took up two seats. Bristol was very busy during the war. All the big programmes including ITMA (Its that man again) were produced there. Ivy Benson was in studio A, and Hugh Shirreff. He cannot remember any local news. All the interest was in London. In the war, news was broadcast all the time. He did not take much notice as he was only 16 or 17.

David remembers the generator room on the top floor of Broadcasting House, which charged the batteries to power the control room equipment. Six systems ran all 300v, 20v, and 6v equipment. It was a major thing. It was long – 10ft long and 2ft 6in in diameter. You had to evacuate through the windows, as the hydrogen coming off the batteries knocked you backwards as it exploded, so they were kept open.

After the war David moved to London several times. He went into the army in 1946 at the age of 18. He was deferred for 18 months (deferred service not National Service as that was for hostilities only), where he served from 1946-49. When he came back there were no jobs in Bristol. He moved to London and went back into radio. There was one recording channel in Waldorf Street (he had never done recording before- all had been done live). He worked at Maida Vale in specialist recordings, and Oxford Street where the Americans were, and did a tremendous amount. He was sent a memo for an appointment near London. There were five chaps at a table: 'Why do you want to go to TV?'. He asked what a TV was as he had only just come back from North Africa. They thought that ideal and sent him to work for outside broadcasting at Alexandra Palace (Palace Arts). Wembley did not exist until 18 months later. Mobile camera would deliver pictures to Ally Pally for transmission. There were three scanners. There were long hours since it was all live. He would rehearse the recording method and repair equipment. In 1946 BBC television transmissions were sent once more from Alexandra Palace in London. Ally Pally ceased transmission during World War II, when it was needed for other purposes.

Frank Shepherd
'I'm afraid that my BBC recollections are confined to one trip made in the late 1950s. I joined the BBC in 1957 and shortly afterwards I was asked to accompany an engineer to the Rocks Railway and bring back to Whiteladies Road any small pieces of BBC equipment

that was still there. Unfortunately my memory now fails me. I only remember loading a pile of BBC grey metal boxes into the van, what they were I do not know. Despite spending another 34 years with the BBC, that was my only visit.'

Clare Springhall

Claire was born in Cumbria in 1922. Her brother, Joe Curran, was a trumpeter in the BBC TV orchestra in Alexandra Palace until the war, when no more television was transmitted. He was born in 1905 and lived until he was 99. He had started working in a colliery band at the age of 12 and worked in the pits. He progressed to playing in the BBC orchestra and to being a professional musician. He worked with Eric Wilde (another trumpeter who had a group called the Teatimers) from Canada who was an arranger of Henry Halls band. The whole band had moved from London to Bristol with other musicians who formed the BBC orchestra under Sir Adrian Boult. They tried the acoustics in the Portway tunnel. The whole orchestra would practise in the Clifton Rocks Railway tunnel during the day. When it became too dangerous to continue in Bristol because of air raids after a few months, the orchestra moved to Bangor in a chapel in Bethesda.

Claire worked with variable area recording machines. These were used for long music recordings and was the only one in Bristol – so all music was recorded from the tunnel. There was a grand piano in the recording room.

Elizabeth Taylor (daughter of Douglas Barlow born 1907, transmitting engineer)

She was born in 1941 in Duncan House, next to the Zoo and had one sister. She remembers BBC Whiteladies children's parties in studio one in Whiteladies Road for employees' children. Thora Hird and Jeanette Scott were there. They showed films such as *Snow White*, and *Popeye* and tried to dance the hokey-cokey on a sloping stage floor without falling over. They took part in BBC performances in costumes made from parachute material. Elizabeth provided pictures of her father at controls at the BBC in the control room in Whiteladies Road.

Her father's job was a transmitting engineer (his letter dated 28 February 1940 can be seen in plate 13.30 offering employment, having had an interview. He was a Maintenance Engineer Grade IV at a weekly wage of £5 5s 0d. Billeting was an extra £1 1s 0d with his present residence maintained free of charge). For training he got 3 weeks at Evesham.

The furthest person along in plate 13.15 in the tunnel control room is her father Douglas Barlow.

The man in the trapdoor (plate 10.23) is Norman Mace, who was great friends with Douglas. Their immediate boss was Arthur Fisher who had a long career with the BBC. Douglas joined the BBC in 1938/9 in London. He had studied Electrical Engineering at Wood Norton, Evesham. He was transferred to Bristol because of the war when the BBC left London. His job was transmitting various BBC programmes from the tunnel – described as

13.42. Douglas Barlow working at a switchboard in BBC Whiteladies Road (he can also be seen in plate 13.15). (Elizabeth Taylor)

'rather like running a telephone exchange'. He worked shifts on a three-week rota: 5-midnight, 9-5. He was always on time. There were two generators which pumped air through the vents (which are still in place) – very necessary because of all the heat put out by the transmitting equipment.

At the time the family lived in a bungalow in Failand (she went to La Retraite for the first term, and then to Failand school). She lived opposite a German prisoner of war camp. Douglas came to work by car. The children needed to be brought to school so they moved to near Redland station. Douglas could then walk or cycle to the Railway. Their mother missed the countryside, so the family moved to Wraxhall and bought a soft top Lanchester. During the war transmissions were from the tunnel. The programmes were retransmitted by Washford (near Minehead) and Rampisham (Dorset). When Elizabeth was five she walked to the tunnel each day after school to await her father and get a lift home. She waited in the end room upstairs. There was usually someone using the mike while she sat there. This was otherwise used by the renowned Uncle Mac. Despite being the children radio mainstay, he was unfriendly and did not like children. Douglas's hobby was his work. He had a radio set as a teenager (radio had only been going commercially since 1922).

Ernie Fletcher
BBC transmitting engineer. He was employed as a trainee engineer for two to three years from 1941 as part of his apprenticeship. He then joined the army.

He remembers wet walls, and five aerials at the top. Confinement in the tunnel was a problem, the door at the top of the BBC area was normally kept closed unless he had to check the aerials. The smell from the river was very bad. He worked in the control room, and used the studio to relax and as a rest room. He remembers the recording machine, and the huge power supply needed.

They did seven-hour shifts round the clock taking it in turns. This was in case they had to take over from London. They transmitted the Home service, closing at midnight. Programmes were passed onto the transmitter. There were no machine programmes. They also rerouted programmes if it was news or in a different language.

Roy Hayward
Born in Hastings in 1925. He had 42-years service in the BBC. As an ex-employee of the Beeb in Bristol, he had some memories of the tunnel. There are still a few old BBC engineers in Bristol who remember it fairly well. Pictures are as rare as horse feathers, but you never can tell.

He was employed as a youth transmitter (under service) on 18 October 1941, aged 16 and straight from school. His job interview at Hastings was brief – he was shown the transmitters and was asked if he knew Ohms Law. He did, and was recommended to London as a successful candidate. 'Dear Hayward. I have pleasure in offering you the post of Youth (Transmitters) (Under Training) at the weekly wage of £1/7/6d plus 3/- Cost of living Bonus...' offered by P.A. Florence, Head of Engineering, Broadcasting House.

The people who helped in the training were older men who had been employees of Marconi, wireless operators on ships, from Cable and Wireless etc. They kept an eye on the trainees and were nice people determined to help the youngsters. They taught them in quiet times at night learning about communications theory.

After six months he was sent with about 12 other youths to receive training at Bristol under Mr Smith (training engineer) for three to four weeks. He worked hard during the day and learnt about microphones and receivers and control work. There were no girls on the course. He thinks this is because the boys were so much younger, being straight from school whereas the ladies were working as part of 'women at war work' and tended to be about 23-40. He was billeted in Cotham Hill with an elderly couple who took on youngsters as their part of the war effort. He could see over the city and could see the bombing going on at night. They would go to the BBC Club bar (corner of Tyndall Park Road on the second floor). One evening they went to the cinema to see *Hellzapoppin'* – an idiotic comedy, and they all sat in the front row. Their enjoyment was so infectious that they led the whole audience in appreciation.

Later the club moved to the ground floor of Arlington Villas (Pembroke Road end), purchased by a Trust set up by BBC staff. There was rented accommodation above let out to lady secretaries.

Later Roy was transferred to work in London Control Rom at Broadcasting House, Portland Place. The war was on and all staff had to do night duty. Senior staff would often go to the control room and chat so one would meet senior staff including Lyndsey Wellington who was very tall and who was at that time 'Director of the Spoken Word' – a title that was later changed.

One morning the class was taken to Clifton Rocks

Railway. He remembers the stairs on the left. They were sent there to do testing in the control room and also remembers the studio. The loudspeakers were LSU10. He identified the microphone type in plate 13.20 as ribbon-type AXB (invented by the BBC's Mr Alexander who was in charge of microphone development and acoustics of all BBC studios and also lectured at Maida Vale training station on sound range and acoustics), and the turnstile as TD7. At the very bottom of the tunnel was a Lister Diesel generator. When Roy got back to Whiteladies Road Control Room an engineer in the control asked him how many generators there were – one or two? He was then sent on a bus back to the railway to check again and found that there were two. There was an identical one as a stand by. He thought it a mean trick for him, as a 16-year-old, to be teased. And waste so much time.

He recalls a musty smell at the tunnel. The control staff did shift work, and the senior maintenance engineer of each shift had access to the tunnel's key. There was a strange switchboard using an input switching system to connect amplifiers A, B to another station.

The transmitter in the tunnel had frequency 1465 (he thought), so it was quite low down the scale. Medium wave needs two masts and one aerial as it is not vertically polarised (they were on top station, see plates 13.24 to 13.27). It was an RCA transmitter (Radio Corporation of America) with 1 KW or 500Watt output (2 transmitters together, so depending on whether they were in series or parallel).

In the recording section one would plug in jacks (which could get dirty) to tie the contribution. Separate thin lines to test incoming lines for interference. One would test the amplifier in case it squeaked. There were a set number of frequencies to test reaction in case a valve needed replacing. The operator had to switch from the trap valve over to the listen jack.

The BBC wanted music in the Railway studio and they had a spare Baby Challen grand piano in Whiteladies Road. They thought size would be a problem getting it up the stairs so they decided to make a mock up out of three-ply to precise dimensions and try it out first. Unfortunately the whole process was brought to a halt, as they could not get it out of the workshop in the basement of 23 Whiteladies Road. (Molly Foss confirmed that there was a grand in the tunnel so they must have managed it later.)

There was a dress code. The engineer in charge on the eighth floor in London was called Mr Bottle and he wore a dark suit, starch stiff collar, dark tie, bowler hat – a requirement of Lord Reith's days. Clothes were on ration and Roy remembers wearing ginger cords. One tended to wear collar and tie, polished shoes and flappy trousers with turn ups, in spite of clothes rationing.

One did not tend to use cameras since film was so expensive. The BBC did not tend to do publicity photos during the war to any great extent.

Bristol's Clifton Parish Hall – basement crypt was a drama studio. Lance Seiveking was the first man to devise radio drama – a madman but delightful – he wanted 'blue noise' from his actors (see *Fun Factory* by Peter Opie).

Gerald Daly the Engineer in Charge at Bristol was a lovely man. He discovered Whiteladies Road site in 1934. He was looking for a suitable home for the West of England Home Service and forced a window open in the empty building and climbed through. On the strength of that the Home Service was born. He would do annual reports on his staff for them to get annual increments. A memo would summon one to arrive in his office at 2:00pm. When he went to the gents at lunchtime, there was Daly who asked him if he had any troubles- when he replied he had none, that was the end of the interview and he got his increment. Others had the same experience of very informal unexpected interviews. He was very relaxed and very kind.

Uncle Mac was a funny man – very self-opinionated and super confident. He once insisted the programme engineer came down from Plymouth to Truro while he was on holiday to do his Saturday morning record programme and play the gramophone records.

He recalls working with Cliff Voice (a recording engineer who worked in the Railway – his memories are also included in this section) at Plymouth Studios – using a mobile recorder in a recording car (Humber Super Snipe). It was cheaper to do that, than go to London to do a radio newsreel. One would get a land line to London and send a disc recording to London within two hours.

John Bull from Shirehampton was a lines engineer in Bristol Broadcasting House. He and his boss were responsible for all the lines carrying programme and phone lines based in Bristol. Only two of them for years doing this massive work in conjunction with the Post Office who owned the lines.

Elsie Otley was one of Bristol's West of England Home Service announcers. She also worked in Clifton Rocks Railway and lived in the Colonnade. Roy once walked back to the office to find an uproar. He had chosen a morning when there were no broadcasts, to check the styluses of all the gramophones in the building. One had to take the stylus out and examine a magnified picture to see if it was worn. One had a crib sheet, so it was easier to take all the needles out and shadow graph them. There had been an addition of an extra programme but the line broke, so Elsie wanted to play a record for continuity, but there was no needle in the studio gramophones, since he had removed them all. She had had to resort to reading everything in the Radio Times even adverts for something to say. Roy was called in by Frank Gillard the Head of Programmes who thought it funny, but he had never been forgiven by Elsie who did not stand nonsense.

The West of England Service was very good. There were nice pieces with extracts to trail. There was a 'programme parade – a programme each morning that trailed the major programmes of the day. They were added after the 7:00am news and 8:00am news. In two minutes the announcer would tell the listeners what was in store. Probably six or seven 30 to 45 second extracts to illustrate the drama had to be played, from about six separate discs. Very complicated indeed especially at the early hour of the morning. The announcer would have to switch everything. The programme engineers took it in turns and some of them would oversleep, and be seen running down Whiteladies Road in pyjamas and carpet slippers and getting there just in time. There was a red light in the control room operated by the announcer, a green light in the listening room, a white light to answer the telephone.

Roy never speeded up tapes to get the timing right but sometimes the old UHER recorders ran slowly when the batteries went down, so the tapes went a bit slow and sounded a bit silly. He was very good at timing. One needed enough for a trail and announcement, so programmes were never longer than 2 minutes 23 seconds.

There were lots of people producing a lot of programmes and the basic Home Service. Even Cranbourne Town Band. A variable delight. In Roy's day, between 1949 and 1957, Plymouth had five producers of radio programmes and was a very busy place indeed before television came.

He was then sent to Maida Vale, London on another course which was an extension of the Bristol course. The billeting officer allocated him a given address so he had to go and find it on his own with his suitcase and brown paper bag full of clothes (no rucksacks in those days). It was in a Victorian terraced house but he was kept awake all night by soldiers tramping up the stairs. He found that he had been billeted in a brothel so asked to be moved. He was transferred to a hostel near Euston station.

He was then directed to work in London Control Room. He was in a sub-basement in Broadcasting House and there he stayed until he was called up. He worked in the Control Room. He did not work shifts in Bristol since he was only a trainee but worked a 4 shift pattern in London.

H group transmitters were set up by the BBC – if there was an invasion then one could broadcast from that station to a limited area of 5-8 miles covered by 100 watt transmitters, so one could broadcast to the community without alarming/ upsetting the rest of the country. They were put in strategic places throughout England – Tunbridge, Exeter, Redruth, Plymouth, Brighton, Cheltenham.

It was Frank Gillard's idea, as controller for the West Region, to set up local radio. He went to America on holiday – (a restless soul and great mentor). As a war correspondent (Roy recalls Frank covering the Rhine crossing), the department got the report but static was bad and the signal left for two hours to test again. His end was perfect but the other end could not hear – it made him laugh later). He had listened to community radio in the US and was impressed with the way of local people talking to local people directly with a common interest in locality. It was a new way of broadcasting. Bill Coysh hammered out a system of setting up local radio stations in Bristol studios. Instead of the test programmes being broadcast by a transmitter, the output was recorded for later examination for feedback and the elements of the operation like staff members, shift patterns, technical requirements, costs etc were carefully considered to see how it would sound and operate before Local Radio could be started. Bill Salisbury – (later station master of Radio Bristol) with Bill Coysh was programme manager for this first experiment and very good at the job.

Roy did not work on crystals as an H receiver. They were never touched, as they determined the frequency. Under the jack plug a flag came down and so telephone communication was made.

One morning while working in London, he answered the phone from H-Group transmitter at Tonbridge Wells to hear a girl crying. She had been left all alone all night. A fellow engineer had taken ill and went home. She had forgotten how to get the transmitter up and was in a total flap. Since it was the same set up as he was used to in his earlier days in Bristol he was able to direct her how to get the transmitter up and running. Her crystal drive was also messed up. There was equipment which measured frequency, and checked frequency and strength. It was more difficult during frequency modularization as it used line of sight.

Regarding the Philips Recording machine which was in the Railway (which used celluloid film) – he only had experience of the blattnerphone and was frightened of it. It was unwieldy almost lethal. It was expensive and if the steel tape broke it had to be mended with a welding gun. Not much editing was done with it. Making gramophone records took at least 12 hours to make which is why it was essential in broadcasting to have a form of recording where one could play back the result virtually immediately. The Philips equipment cost £800 per year plus £1250 royalty and the running costs would have been £2.50 per hour. It first went into service in 1938 and there were very few in the country. It was easier to edit because the tape could be cut with scissors and spliced.

During the air raids, many of the powerful transmitters in England were closed down to prevent their signal being a help to German air craft. Locations were divided relating to German bombing targets. The locations were back ups so that if they closed, the central H transmitters they could not be of service to the enemy. They were on medium wave. Droitwich on long wave was closed down often. The telephone in the Control Room would ring. It had a special note so they knew there was trouble- either a complaint or close down. There was a very strict procedure which had to be closely followed since no slip-ups could happen. No variation was allowed when the order to close down was telephoned to the transmitter stations, and this helped a disciplined approach.

Bombings in London were difficult. War was irrelevant – it was just there. Roy had a tin helmet. As a schoolboy he had been a bicycle messenger for ARP doing night shifts. He was not scared – he had no fear. Problems were always at night. In London he walked to work with bombs dropping around. The VI doodle bugs were the most scary. A friend and colleague had sent his wife to Bath to get away and Roy was asked to be his paying lodger in his top floor flat in Battersea overlooking the park. There was a V1 doodle bug bomb two doors down. The ceilings of the flat came down, but the ceiling board protected him, so they went out to see what they could do to help. It was very upsetting to see an elderly couple in bed open to the street. Since his flat was in a mess he moved to Chelsea. Rehousing worked well. Wardens were aware of their community and their street and who was ill.

After the D-day landings there followed a special programme each day as a news requirement after the 9:00pm news called 'War Report' at 9:15pm. Malcom Frost was in charge in Broadcasting House in London and correspondents were sent out to Europe (like Wynford Vaughan Thomas (Wynnie), Frank Gillard, Guy Byam). Journalists were attached to advancing troops and send their reports by radio from a mobile van – a moving studio with transmitter. Transmissions were received at Tatsfield (a BBC listening station near Croydon) and at Broadcasting House where a 'shack' had been set up to be in touch with the people in Europe. It was equipped a shortwave receiver/transmitter and a morse key (he had been taught morse code at Holloway). It became one of 4 operations locations in touch with mobile units. The BBC units were called:

Mike Charlie Mike MCM (Broadcasting House, London)
Mike Charlie Oboe MCO
Mike Charlie Peter MCP
Mike Charlie Nan MCN

The American stations were called:

Jig easy sugar peter JESP
Jig easy sugar queen JESQ

Messages were in morse and one would talk into a microphone in ordinary speech.

There was one terrible moment for one shift to find the whole operation closed down. During the day there had been a senior engineer responsible for maintenance of the mobile units and would sort out all problems so that they could be 100% efficient. He was called Windsor (Winny for short). He came into the shack and asked for a message to be sent and dropped the news that he was going to Europe to check out one of the units. A colleague operator on MCO heard that Winny was coming out to see him, so look out. At that moment several dark cars arrived at London Broadcasting House, the men wearing dark suits, glasses and clip boards and seized him and took him to the Director General's Office. A senior establishment officer had to be called in to clear Winny of any disloyalty. By odd coincidence we learned later that Winston Churchill was going to the same place and it had been asked how did Winny know that. He was let off the hook but there was quite a scare. The day went ahead normally after that but there was a whole pile of work and day full of frantic messages as a result.

Roy was then called up to Fleet Air Arm in May 1944 at the age of 19. He served until July 1946. There was a requirement by law that if one was called up, then those called up had their jobs reinstated on return if one so wished. He went back to the BBC as a junior maintenance engineer in Aldenham House, near Edgeware, Hertfordshire. This country mansion had been converted to studios and was the administration studios of Arabic, Latin American service and Central America. He then transferred from engineering maintenance to become a programme operator. He came back to Bristol later as a radio producer and retired as deputy Radio Editor in 1982 after 40 years of service.

Alan Urwick
Born in 1925 (he knew Molly Foss, Cliff Voice, David Pearce, Arthur Heale, Arthur Markford).

Alan came to BBC Bristol at the beginning of 1944 after they had ceased staffing the tunnel. He worked in Whiteladies Road facility as a transmitting engineer. He was in the recording unit for a time.

The transmitter was kept in the tunnel and every day someone would go and maintain it on a shift basis to make sure it was functioning properly and to check for faults. Signals were sent back to the control room. He went there once a week. After a while it was possible to monitor it automatically from Whiteladies Road. The transmitter in the tunnel was closed down in the early 1950s.

The studio recording equipment was still there and a small control room. It was dry and well air conditioned. There was an air conditioning plant that had an ozone air purifying device to stop the smells from the river of the sewage when the tide went out. There used to be one mast at the top.

After he left school at 15, his father arranged for him to be an apprentice at a big electrical engineering factory, but because he was under 16 he could not start, so his father took him to get a temporary job at the Unemployment Centre. There was a BBC advert needing 4 youths on the Youth in Training scheme at £1 7s 6d per week (including a 3s cost-of-living bonus) to man the transmitter at Churchdown. Alan joined the BBC at the end of 1941 and went to Churchdown Hill, Cheltenham. There were already 4 engineers there, and he was so happy he did not want to leave to take the apprenticeship, and worked for the BBC for the next 44 years. The Home Guard were stationed at the top of the hill. The transmitting station was at the top of the reservoir in a small hut. This was shut down in July 1943 when invasion risk was over. He then went to Cardiff and then Bristol.

He went to Maida Vale for a month's training, then Wood Norton, Evesham for two months. He went on further education courses to Evesham later on when converting to television.

There was only small low-powered transmitting equipment all over the country. The BBC had been asked to hastily set up 60 transmitting stations all over the country due to the risk of invasion, so they could broadcast the announcement over the radio. In Cheltenham, it would have been the local sheriff who would come in time of invasion and make an emergency announcement. Post Office lines were used and sometimes underground cables became faulty. One had to establish other lines and reroute if the lines were faulty. They would receive transmissions from Bristol, Southampton, Cardiff and Plymouth, which came in via Bristol.

He used Morse code on the night shift, when off the radio, to practice speed. He was not billeted at Bristol because he knew people in Cranbrook Road, Bristol. He used a push bike to get around. He went to Bush House for a couple of years but then returned to Bristol until he retired.

Cliff Voice
'I cannot really add much to what I said on the *Secret Underground* film but I will do my best, although memories change over more than sixty years.

The bottom half of the tunnel was used by the BBC and built up to make five levels. At the road level was of course the entrance together with toilets, a very small area for making tea etc. and a generator in case of mains failure.

On the first floor was the control room with all necessary equipment to take over the running of the West Region should Whiteladies Road be put out of action. On the second floor was a recording channel to be used for either recording or transmitting recorded programmes on film. This was a different method used for several years and devised, I think, by Philips Miller. Recording was on a clear base with a

13.43. Cliff Voice with recording equipment in 1948.

very fine top coat of Black and about a quarter inch wide. The cutter was V-shaped and dug into the black coating removing it as it cut, with an air suction device. According to the width and depth it cut, so it varied the frequency and sound level, basically a very simple method but extremely clever to work when you realise that the black coating was only a few thousandths of an inch thick. The swarf, that is the area cut off, was collected in a bin, emptied after every recording, and taken away by the Studio attendants to dispose of as it was highly inflammable. Unfortunately one of them decided that the best thing was to burn it which they proceeded to do, just by the entrance where the intake fans were for fresh air in the tunnel. It went off with a terrific swiiiish and all the smoke and terrible, terrible smell was immediately sucked in and all the staff evacuated in record time.

My only other memory was the day HM Queen Mary came to visit. A friend and I were transmitting some programme when the recording room door suddenly was thrown open and this elderly lady advanced towards me with arm outstretched – I am sure I said, 'Your Majesty' but I can really remember nothing clearly after that.

The next floor was a small studio with, I thought, a grand piano. Maybe Molly could be the final arbiter [Molly confirmed there was a baby grand piano].

The floor above that was where two transmitters were housed in case of emergency.

Above that of course, was the top end of the tunnel which was used as an air-raid shelter.

Tudor Gwilliam-Rees
'I used to run a business in Bristol – dealing in old radio and valve equipment. I bought a lot of valves and other items from a famous radio shop called Roy Pitts in Picton Street Bristol – many, many valves and bits, all from the old WWII radio transmitter in the Rocks Railway.

I regret I have given you the impression that I still have these valves – SORRY – long sold, it was in the 1970s.. One has to sell items to make a living.. Had a successful business in Bristol – from home in Mangotsfield – then a shop at 64 Broad Street Staple Hill Bristol – then from a converted church in Mangotsfield – exporting vintage valve Hi-Fi to Japan mainly. Still trading – but only just. So nothing to photograph really. I remember the valves – huge triodes with green BBC test labels stuck to them.

I am very interested in the Rocks Railway and its history. I climbed over the fence once and ventured down with a torch – all those steps. I was offered once, in my Broad Street days, the entire BBC collection, it was dumped on the City Museum by the BBC – who were just not interested. I just could not accommodate it. I viewed so much equipment – mostly in the old Fishponds tramways centre – with rain dripping through the roof onto same – terrible. So, the point is that maybe the BBC or the City Museum might have some info on the wartime transmitter. One further lead. Before I started my strange business I worked at the HH Wills Physics Lab – they had many valves in the lofts there, much development work being perfected there.'

The next chapter describes the sad years after the BBC moved out, and the site was left to go derelict.

fourteen

Life after the BBC 1960-2004
Planning Problems and Dereliction

The strata of the Avon Gorge in the area of Clifton Rocks Railway is characterised by seams of stiff clay sandwiched between carboniferous limestone. This had caused problems in 1891 when the tunnel was constructed (see chapter 3, 3.12). The area between Princes Lane and Hotwell Road had been terraced, and masonry walls built, to retain the earth on the various terraces from 1720 onwards.

14.1 The stability of the bottom station

At the end of 1956, it was reported[1] that cracks had been noticed between the masonry of the tunnel façade in Hotwell Road and the face of the limestone cliffs. A crack 4in wide had also opened up in the brick lining of the tunnel immediately behind the façade. Scaffolding was used to shore up the masonry façade, and glass tell-tales placed to indicate any further movement.

In August 1957, Dr Skempton and Dr Henkel from Imperial College, London were approached by the Bristol city engineer to advise on measures that should be taken to ensure the stability of the tunnel façade and cliffs, all the way up to Hotel, and to consider the movements of the concrete wharf further upstream.

In April 1959, they produced a report entitled 'The Stability of the cliffs of the Avon Gorge in the Vicinity of the Clifton Rocks Railway'. From a survey, it was discovered that the masonry retaining wall, and the masonry lower down the cliff, had moved forward. The problems were of a local instability rather than the general instability of the whole site. Five feet of clay sandwiched between the strata and loose strata was removed. Four exploratory boreholes were put down including one that penetrated the tunnel and can be seen in plate 3.12 of chapter 3. It was shown that some of the clay had a high water content and less cohesion, thus causing some sliding. Adequate drainage was needed, and supports designed to withstand a normal thrust of 200 tons to ensure the stability of the cliff face, the tunnel portal, steps, and the wall on Hotwell Road. Twelve buttresses were tied back into the limestone below the clay layer with steel tie rods. Six drainage boreholes were also put down through the clay layer and backfilled with gravel. The exposed faces of the clay layer were covered with concrete to prevent further softening.

14.1. 1956. Note the entrance to the shelter is still in place, and the vents are still in place.

14.2 The man with a head for heights is Mr Frank Marks, seen at work high up on the side of the Avon Gorge demolishing an unsafe wall over the old Rocks Railway, 20 February 1957.

14.3. Buttress of scaffolding on the Hotwell Road, and the wooden chute for removing the rubble. The steps to the side have gone. 1,000 tons of material was removed. Compare this plate with 13.7. All the rock and wall above the highest left hand ledge (by the shorter ventilation shafts) has gone. The masonry was 2in out of plumb at the top west corner. 20 February 1957. (BRO 37167/97)

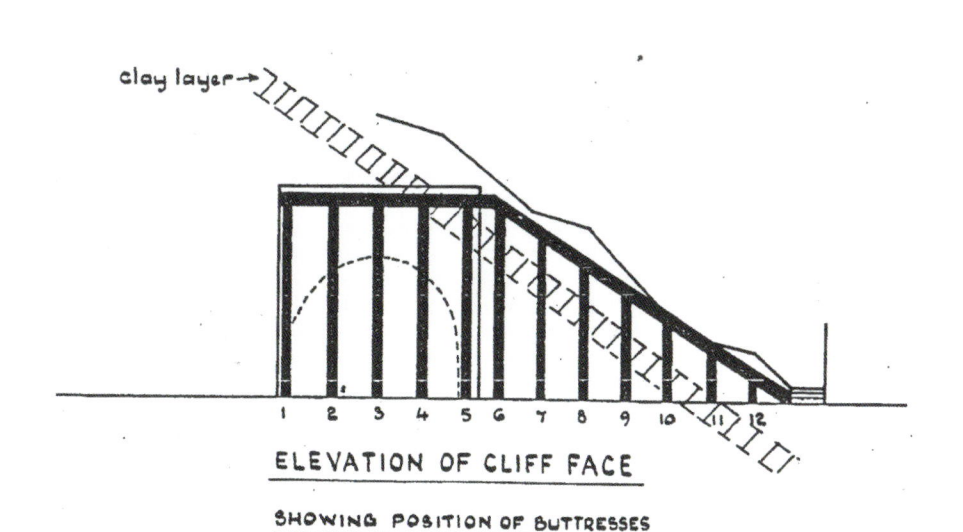

14.4. Clay layer shown in relationship to the buttresses. (Skempton)

14.2 Regular break-ins at the bottom station

14.5. 1975, two people trying to get in. The first-floor windows were bricked in when the vents were removed. The buttresses are in place. The shelter entrance porch is still there. (David Vousden)

14.7. 1981, the door has been broken down so anyone can get in.

14.6. 1975, a crowd waiting to break in. Boulders are not in place yet to stop parking. (David Vousden)

14.8. 1989, it looks like the door had been removed for some time. There appears to be a gap above the upstairs windows. Boulders to stop parking were put up in 1998.

14.9. The bottom station as it is today. Compare this image with the bottom station image when the BBC occupied it and the ventilation ducting is visible (plate 13.7, chapter 13). Note that the windows are blocked up, the middle entrance has been removed, and the wall above has been removed. We have removed the foliage from the slope. The supports are suffering from concrete rot and have cut into the façade. The porch for the bottom shelter has now gone. Note that right of the main façade and some pennant blocks, there are old bricks possibly from the rear of the old spa building. The landing stage is rotten too. Very sad.

14.3 Deterioration and undergrowth

14.10. 1950s: the signs are complete but glass is missing from the ornamental sign (Davey collection)

14.11. 1964: note the light on the bracket, and corrugated tin.

14.12. In 1986, when the hotel had been renamed Avon Gorge Hotel. There are now hoardings around the railway, and a false roof of disintegrated fibreboard placed on top, supported on timber beams.

14.13. In 1986, weeds. The hoarding has come off showing the disintegrating roof. (C.G. Maggs)

14.14. In 1986, a shocking jungle. (C.G. Maggs)

14.15. In 1999, lots of undergrowth (this was removed by the Hotel in 2001). (Junior Chamber of Commerce)

14.16. In 2005, at least some of the railing can be seen and it looks better than 1986. Hoarding prevents one from looking over. (Ross Floyd)

14.17. In 2005, hoarding over the railings. Peeping through the slot, one can see lots of debris. (Ross Floyd)

14.18. In 2005, there was a huge heap of debris on top of the railway lines. The staircases on each side to get down to the shelters during WWII had been built on top of two lines. The water-pipe leading into the top reservoir is on the left. Although there was some glass in the undergrowth, very little of the original glass roof remained (see plate 4.1 chapter 4). We found many railway artefacts in the debris (see chapter 7).

In 1962, a planning application[2] was granted to use a portion of the Railway as a bar extension to the ballroom.

On 27 January 1965, the *Western Daily Press* and *Bristol Evening Post* announced that Nuthalls (Caterers) Ltd wanted to buy the tunnel from Bristol Corporation. They stated that the tunnel ran under the Grand Spa Hotel [it does not], and the top of the tunnel was near the ballroom. They wanted to tidy up the top end of the tunnel that had become a rubbish tip and an eyesore. Negotiations had first started nearly four years before, but there was a covenant that the purchasers could not do anything to prejudice the use of the tunnel for Civil Defence purposes in the event of war. The Bristol Civil Defence Committee agreed that the tunnel would not now be needed as an emergency shelter.

On 27 September 1965, Bristol Corporation sold 95% of the Railway to the Grand Spa Hotel (owned by the Bristol-based Mount Charlotte Investment Group). The bottom 5% was kept by the Corporation to be used as a ransom patch.

Interestingly enough, Nuthalls bought the tunnel on 7 March 1985, and according to the Land Registry still owned it in March 1996. Plates 14.13 to 14.15 show little evidence of tidying. It was bought in August 2002, by Peel Hotels.

14.4 Contentious planning application

On 6 January 1971, a planning application was submitted to build an eight-storey building on the top of the Gorge[3] above the Clifton Rocks Railway tunnel and fairly close to the retaining walls of the bottom façade, entailing demolition of Tuffleigh House. This would house banqueting facilities for 250 people, 126 double rooms, 12 single rooms and parking space for 203 cars. This would have been eligible for a £126,000 tourism grant. The planning officer at the planning committee recommended granting outline approval, despite many objecting letters. The Clifton and Hotwells Improvement Society created a group of interested people to start a campaign. This hit the national press when 1,200 objectors wrote to the Secretary of State who announced a public inquiry. The developers submitted detailed plans and began to build foundations. The National Trust, the Georgian Group, Civic Society, and even Sir John Betjeman, all submitted expert evidence – in addition to the local groups and individuals. The inspector recommended that the building should not be approved. The fallout was messy. The company hit the Council with a compensation claim for £251,000, while the Council said the Department of the Environment should face at least some of the liability. An application for a smaller building followed next following, but was abandoned.[4]

In 1974, a report[5] published by geological surveyors came to the conclusion (having drilled more boreholes) that parts of the site were in, or near, a condition of 'limiting equilibrium', that is, on the point of sliding. The imposition of massive building loads would aggravate this situation and high-level foundations would not be suitable – the building would slide down the hill on the mudstone layer and along a clay-filled plane layer. A massive concrete raft on large diameter piers up, to a maximum of 140ft though the clay layers to sound limestone, would be required, and

259

which could have passed through the tunnel. The piers would straddle different slip surfaces; rock bolting was not an option. It is just as well planning permission was overturned.

14.5 Bristol Junior Chamber feasibility study

In March 1983, the Bristol Junior Chamber of Commerce initiated a feasibility study on the Railway.[6] The perception at the time was that the hotel management would like to divest itself of any financial liability for maintaining the railway tunnel. It was suggested, perhaps tongue-in-cheek, that the tunnel might be used as a vertical subterranean extension to the hotel's residential accommodation (which hardly respects its history), or as a nuclear fall-out shelter (for which it was too small). The report conceded that the 320 sq ft of land occupied by the top station would not make a significant contribution to the proposals. The Junior Chamber considered it had little potential as a tourist attraction: there were no scenic views; the Hotwell Road was too close to the bottom entrance; there was no parking facility at top or bottom stations; and there was not enough of the original railway to refurbish it to run again to British Standard Specifications for lifts. The options considered were to refurbish the complete railway (and remove all wartime brickwork), or to refurbish the middle two lines only, either of which would cost between £300,000 and £500,000. Four blast walls would need to be removed and 25,000 sq feet of tunnel arch brickwork repointed. The potential returns would be insufficient to justify examining the matter further, funding would be limited as the site was not listed since it was not 'listed', and it would be hard to obtain approval to operate as a water-balanced railway again, and comply with safety standards. An electrical winding system would be required, which would require a specially excavated pit; and the bottom station would need to be rebuilt.

14.6 Permission to demolish the Railway and Pump Room

On 22 April 1985, the Council refused outline planning permission to build 67 new bedrooms, bar, club, sports and conference facilities and 73 parking spaces. This would have entailed the demolition of the mainly derelict Clifton Rocks Railway, the Turkish baths, and the Pump Room (which could be re-erected somewhere; the marble pillars were to be used in a swimming pool). This was unfortunately overturned at appeal by the Secretary of State, in April 1986.

On 20 March 1989, there was an article in the *Evening Post* stating that the loss of the Railway and the Pump Room would be a sad loss, wanting them to be restored and for the spa waters to be tapped once more. (There was also a comment that English water failed to pass the European Community standards, so a natural, good water source must be of benefit.) It stressed how popular the Pump Room had been as a ballroom after 1928 and how in the 1950s, Peter Sellars, Shirley Bassey, Laurel and Hardy, and Cary Grant had all been clientele. The 15-piece Grand Spa Band was the first to bring Latin American rhythms to Bristol. In the 1960s, the marble columns and plaster mouldings were sprayed spinach green and the place became more popular (I danced there in 1965-8 when I was at college – there were different nights for different types of dancing). The windows were boarded up in the 1970s. The Evening Post wanted the Railway to be incorporated as a small branch of the new metro proposals, bringing commuters into Clifton from Portishead without going through via the city centre.

In May 1989, the Pump Room was listed as Grade II, but not the Railway because the street frontage was no longer complete, and the interior was much damaged. In June 1990, the Pump Room was quietly de-listed after an appeal by owners Mount Charlotte Investment. Ministers decided it was not of sufficient special architectural or historic interest..

On 18 August 1990 a letter was published in the *Evening Post* from someone who had toured the Railway (organised by the Junior Chamber). They felt it to be impossible for the Railway to ever return, but that the entrance and tunnel would make an excellent museum or tourist attraction.

On 22 November 1990, the *Evening Post* stated that time was running out for a unique piece of Bristol History. The upper station and the Pump Room would soon be demolished as neither were listed; the ironwork railings of the Railway had been boarded up; a breezeblock wall had been built in front of the Pump Room.

Campaigners from several groups (Clifton and Hotwells Improvement Society, Bristol Conservation Group, Clifton Spa Pump Room and Rocks Railway Refurbishment group), started to step up the pressure to stop demolition. On 19 December 1990, legal action was taken to stop the continued demolition of the Pump Room that had been started by Mount Charlotte Investment. Outline planning permission would expire on 30 April, but may or may not be granted again. The Council granted a temporary injunction against the owners.

On 21 December 1990, the Department of the Environment decided to uphold the de-listing.

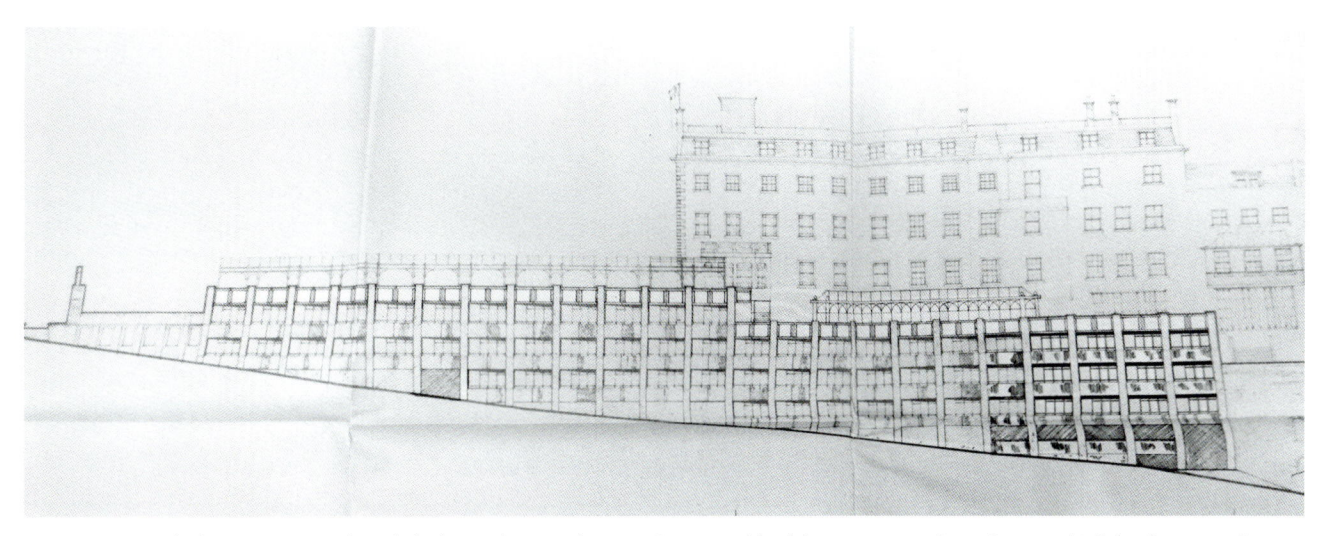

14.19. Proposed plan in 1985 to demolish the Railway and Pump Room and build a grotesque 2- to 4-storey high bedroom and conference block. The fenestration was objected to.

On 8 January 1991, several campaigning groups declared their intention to take direct action – they would obstruct the bulldozers to prevent the demolition. A 250-signature petition was presented to Mount Charlotte who asserted that restoration was uneconomic.

On 10 January 1991, protesters celebrated a reprieve of the condemned Pump Room. From 7 January to at least 13 January, there was a sit-in at the Hotel, and campaigners circled the Railway day and night to stop the bulldozers. The Hotel owners agreed to negotiate on the future of the two buildings; campaigners agreed to remove the blockade of the Hotel car park. A County Court hearing was held on 14 January 1991, to decide whether to lift the injunction. Barry Taylor of the Council suggested Mount Charlotte should radically rethink their plans; Bristol West MP William Waldegrave, telephoned Baroness Blatch, minister in charge of Monuments at the Environment Department, asking her to spot-list the building (which did not appear to happen). Jerry Hicks, vice-chair of Bristol Civic Society asked for revised plans that might co-incide with the renewed sense of heritage and preservation in Bristol.

On 14 January 1991, Clifton Suspension Bridge Trustees denied that they had put in a £1 million bid to buy the Pump Room. (Their ambitious plans to build a museum near the bridge had been thwarted in 1986.) At the County Court, Mount Charlotte agreed to reconsider the plans and, the next day, agreed that demolition would be postponed for a year. It was also considered that the whole of Clifton should be listed as a European Heritage Area [this has never happened, but it keeps on being raised].

14.20. 13 January 1991 photograph shows a crowd of protesters on Princes Lane (Jane Jenard)

14.21. 13 January 1991 shows the scaffolding lorry and makeshift fabric roof over the Railway. (Jane Jenard)

On 21 February 1991, according to the *Evening Post*, it was agreed by the planners that the best approach was to stand back and rely on the company's goodwill not to flatten the buildings before consent expired at the end of April.

On 7 March 1991, the *Evening Post* reported that jubilant protesters were celebrating after the owners confirmed the demolition had been shelved. The Pump Room would be retained as part of a new conference and banqueting development. It was a fitting date – the hundredth anniversary of work beginning on the Railway. The anniversary was marked with a candle-lit procession of over 100 people, from the upper station to the bottom station, and a firework to commemorate the first blast. Actor Tony Robinson, who lived locally at the time, led the procession dressed as Brunel. The author drove the Lord Mayor and Lady Mayoress down to the bottom station in a 1924 Lanchester tourer.

14.22. A wooden plaque was placed on the bottom façade.

A steering group was created to organise action and focus attention on the Railway. This was part of the Clifton Spa Pump Room and Rocks Railway Group (no relation to the Clifton Rocks Railway Trust). The group thought they could provide a link with the proposed Advanced Transport for Avon metro system (which was bankrupted in 1992). It was a campaigning group and without access to either the Railway or the Pump Room and, since it had no ownership of either, could not apply for funding for its grandiose plans to restore them as one unique, combined structure, with income from the Pump Room used to fund the Railway. The war-time structure would be removed. Clearly Mr Peel for the owners had other ideas for the Pump Room too. In 1998, the Council suggested that having two groups with competing ideas for the site (the other group being the Bristol Junior Chamber) would cause them to fail, and that they needed a proper constitution.

On 18 April 1991, the breeze blocks that screened the Pump Room were removed. Suddenly there was something to look at that gave substance to the campaign. Later that year *Secret Underground Bristol* by Sally Watson was published by the Bristol Junior Chamber. This covered the railway as well as seven other sites and was aimed at lifting its profile.

14.7 Business plan by Bristol Junior Chamber
I have added comments about how the current Clifton Rocks Railway Trust's experiences compared with the Junior Chamber experiences.

On 27 February 1996, industrial archaeologist Tim Rew (who had worked for the Hotel group), teamed up with thirteen members of the Bristol Junior Chamber to make the bid to clean up the Railway and, within months, enable visitors to enter at the top and walk down the staircase. Rew reckoned debris could be cleared and the railway opened up with the help of volunteers for as little as £10,000 (in fact in 2005 when the Clifton Rocks Railway Trust started work, we spent only £2,459 on skips, hoists, buckets. We cleared rubbish and retrieved artefacts over 1,637 working days). He said:

> This work is unique in Britain, but it is deteriorating so something has to be done. We want to co-operate with the hotel. This could be an asset to the hotel and to the local community.

The 'Rocks Group' drew up a business plan, and hoped to form a company to run the scheme, initially for two years. After that time, they could continue or cease. They agreed that it was not realistic to get the railway working again, despite various groups having put forward plans in the past. The reasons were sound: the lower station is inches from the pavement after a blind bend, and there is no space for a pavement – if passengers could not get off and out at the bottom, the sole attraction would be a 40-second underground journey; the railway cars had been scrapped and, for a scheme that had never made a profit, 4,000 feet of new rails would be required, as well as major work in demolishing the WWII structures. Instead, it was hoped that making the tunnel accessible to guided tours would help focus attention on the tunnel and possibly inspire its redevelopment as a heritage attraction.

The Junior Chamber reckoned a hard-hat tour would last 30 minutes, with a maximum number of 15 at £2.50 per person (Clifton Rocks Railway Trust tours

	1995	2005-
Financial requirements	£14,750 to set up a company, skips, marketing, lighting, handrails (15 working days)	Trust and company cost nothing, skips £600, lighting and insurance provided by Hotel, handrails £100, initial marketing from a £1,500 lottery grant (1,637 working days spent in 2005)
Running costs over 12 months	£15,450 (includes two guides plus admin assistant £5 each per trip £4,500)	All voluntary. Running costs excluding work are £500-£600 (major projects excluded)
Expected receipts	four parties on two weekdays and Saturday and Sunday £30,000. In practise, Tim took groups on request rather than regularly.	800 trips difficult to achieve unless you have enough volunteers. Achieve £5,000 for about 85 trips/year, donations £2,000, merchandise £1,000, friends £1,000. Groups on request and on specified days.

Table 14.1. Comparing plans of the Junior Chamber of Commerce with that of the CRR Trust.

take 2-2 1/2 hours or sometimes longer, with a maximum of 10 people at £5 each). It is interesting to compare their proposals with what our group achieved in practice (table 14.1 above).

No electricity source was available. This made it impossible for the Junior Chamber to do any work.

Tim Rew mostly worked with Sir George White from 1994 to 2000, rather than the Chamber of Commerce. Visitors went in at the bottom station, and were not allowed to go into the top station.

In March 1996, Sir George White wrote to Tim. He considered that the Railway would never pay for itself and few visitors would visit it more than once if it was no more than an impressive 'hole in the ground'. The difficulties comprised the traffic at lower station; removal of a vast tonnage of rubble; safe delivery of construction materials, rolling stock etc., and safe delivery of passengers to the bottom station. One needed a business plan to demonstrate that the project was a 'runner', before anyone would take it seriously and in particular to convince the owners that the project was viable – otherwise, they would concrete over it for good.

In October 1998, the Council's transport planning team confirmed that the railway:

> In terms of transport, could not form part of an integrated transport network since there was little demand for a link between the Portway and Clifton. An integrated system is more about rationalising existing movement than creating the need for additional trips.

A signal-controlled pedestrian crossing on the Hotwell Road at the foot of the railway could not be considered because visibility was not good, and traffic speeds high. It would be a tourist attraction, and necessarily a private initiative.

In March 1999, an engineering viability report was prepared for the Chamber of Commerce. It confirmed that removing the wartime walls was not straightforward since it was in a confined space in which dust, noise and fumes would have serious consequences to health if undertaken without the required expertise. Access to remove rubble from the base of the tunnel was difficult since it was unlikely that permission to use the road for construction was unlikely to be given, so a conveyor belt system would have to erected over the Hotwell Road. This would be difficult since the upper windows of the façade through which the debris would flow would have insufficient clearance over the road. A barge would be used to remove the debris on a tidal basis. Prices would vary dependent on tide, availability of barges, unloading jetties and tip sites. Extract ducts would be needed to discharge the dust laden extract air without impairing road conditions. Leaks in the tunnel could not be sealed otherwise the resultant loss of drainage would potentially destabilise the slope. Demolition vibrations could cause damage to the tunnel lining causing destabilisation to the slopes. It was envisaged that simple hand driven tools would have to be used. An access route was also considered from the Floating Dock but a 'road train' was considered unviable. A foot trail could be established, but a safe crossing was needed. Another expert, in assessing Heritage Lottery Funding, warned that the scheme would have to be tempered with considerable care, so as not to overreach financially. They concluded that the cost of conversion far outweighed possible returns, and described it as an eccentric, exciting, white elephant.

14.8 Change of ownership and listing

Whoever would buy the Hotel would acquire 95% of the railway. The Council own the bottom 5%.

In February 1998, Peel Hotels PLC was created by Robert Peel, one time chief executive of Thistle. On 4 September 1998, Thistle Hotels PLC, after a failed attempt to sell the company, agreed to sell 30 of its 87 suburban hotels (including the Avon Gorge Hotel) to Lehman Brothers Holdings. As a part of the deal, Lehman appointed Peel to take over management of Thistle's smaller regional hotels. Under the deal with Lehman, Peel had the option to buy two of the hotels. In February 1999, the hotel directory of the Peel Hotels PLC showed the Avon Gorge Hotel as having a management contract with the company.

Finally, on 7 January 2000, the Pump Room was once more Grade II listed, but not the Railway.

> Since there were 6 funicular railways: Lynton and Lynmouth; Bridgnorth; Hastings East Cliff; Hastings West Cliff; Folkestone; The Leas; and Saltburn-on-Sea built by the same promoter which continue in use and are listed. The architecture of both top and bottom stations were very damaged and was entirely filled up by the BBC wartime structures of brick and concrete and were of no interest architecturally. There was also another cliff railway built in a tunnel at Hastings West Cliff. English Heritage concluded that the building at Rocks Railway did not possess sufficient architectural or historic interest to merit listing because of the damaged nature of the buildings and the alterations within the tunnel. The BBC wartime use was not considered to be of special interest.

As the Railway is not listed, it makes it harder to qualify for grant aid to assist restoration.

On 29 September 2001, 70 people, mostly near-by residents, were invited to view the Pump Room and Clifton Rocks Railway at their own risk with no hard hats and no lighting save for torches. This was my first visit inside. I asked if I could help restore it, and make it more accessible, but was told it was the work of professionals – health and safety regulations made it difficult to use volunteers.

In June 2002, Peel Hotels PLC bought the Avon Gorge Hotel.

In 2004, a new Clifton Rocks Railway Group was formed, as a result of Robert Peel talking to James Tonkin, and the rest is history. Read what we have achieved in chapter 15.

14.9 Memories post-war
Michelle Howe (1960s or 1970s)
'We befriended a tramp who slept in the tunnel called Len and his small dog, Toby. Apparently they used to eat pretty well courtesy of the hotel staff – plus the stuff we used to take to him and his dog. [We found some tins of Tiny Tim food for poodles in the tunnel when looking for artefacts.] He told us he was from Malta, originally. Tramps seemed to be quite a feature in those days – there was a whole ghetto of them on the opposite side of the Avon on the old Portishead line, Trampstown we called it. I was in total awe of anyone who dared to spend a night in that horrible place, that it was. I can still smell it and I would dearly love to go down there again but even better to see it restored to its former glory.'

1985: Six-minute music video by the Yakometties[7]

14.23. Yakometties playing in Shelter 2

Garry Smout reminisced:
'The intro has a very impressive entrance way, through the pump room but I believe this has nothing to do with the actual station and was part of the Avon Gorge Hotel. I have no idea how we got keys or permission, but the hotel was very supportive and offered changing rooms and hot beverages.

Power came from a generator they provided at the bottom of the tunnel. It was very difficult to get it there and I believe the police had to slow traffic down so we could get safe access. A second generator was at the top, but the fumes from it nearly killed those in the tunnel so it was abandoned.

All in all about 30 to 50 people were in the tunnel at some point. The room was, I think, the second one down as it was the cleanest. There were lights and music but not really a party – you can see the people coming down the tunnel carry hardly any drinks (no six packs etc.) and any glasses were props from the hotel – very strange. The band's gear, drums, keyboards etc. and the film gear – lights, tripods etc. all, obviously, had to be carried up and down those steps. All were numb with exhaustion at the shoot's end. In one shot looking down on the band, you can get an idea how steep it is – how did the BBC work there? The intro and the African dance were the only sequences not shot in the tunnel. The video was shot and edited on VHS which explains the poor quality. I think this is possibly the only version surviving.[7] The music is a bit of a dirge but I hope you find the shots, and use, of the tunnel of interest. Rich Monday was keyboards and backing vocals, Paul Dowding on drums, Dog on bass, Beef on guitar, Simon Fraser vocals.

Adam King

An Inspector Calls (J.B. Priestley) was staged in the tunnel (Shelter 2) in 2001 with eight actors. It was based in 1912, focusing on a prosperous upper-middle-class family.

We did three performances over three nights in summer 2001. As it was all a bit dodgy, the audience was by invitation only – but we managed to get about 70 (very scared) people in each night. At the time it was possible to climb over the railings at the top with relative ease, so that's generally how we would get in for rehearsals, and to get the tables and chairs of the set in there. But obviously you can't get 70 people to climb over a dangerous railing before it was properly dark so we used the bottom entrance by the river on the night, leading people down the windy path from the observation point at the top. What was slightly weird was that a couple of days before the performance – the chain always having been padlocked thus far – we went down with bolt cutters (yes, naughty I know, but we were planning to replace the padlock at the end and send a new key to the Avon Gorge hotel), and found that the padlock had gone and the chain was just wrapped around. I suppose it's not impossible that one of our number had secretly removed it beforehand to freak everyone out, but still pretty weird. We hung a black sheet up by the bottom of the steps so the light wouldn't show onto the road, and lit the stairway and auditorium with candles. It was amazingly good luck that it all went smoothly. It was pretty cold, but we told people to bring coats and hats.

Two other weird things happened. During one of the performances, a couple of members of the cast backstage could DEFINITELY hear someone moving around at the bottom of the tunnel, even though the

14.24 Lots of candles since there would not have been any electricity, and other props.

14.25. All dressed correctly in dinner jackets.

bottom door was padlocked. We went to investigate, but saw no one.

And some months previously I was down there with a girl on a slightly unconventional date, in the spa bit by the hotel, very late at night, when someone appeared at the entrance (i.e. on the steps of the tunnel) and shone a torch at us, and then disappeared without a sound. Pretty scary at the time.

It's such an amazing place. I particularly liked the changing rooms for the spa – where the rusty old coat hooks and tiles were still in place.

fifteen

Life from 2004

This chapter describes the huge amount of refurbishment the Railway required and the success of the volunteers who have achieved so much to put the Railway 'on the map'.

In December 2003 seven people met to discuss the future of the Railway: James Tonkin, joiner and North Somerset District Councillor; Peter Scott, engineer; Peter Davey, tram expert; Mike Rowland, Clifton Suspension Bridge; Tim Rew, who had helped the Chamber of Commerce; Malcolm Frith, BBC business correspondent; and Delphine Lydall, Clifton business community), Robert Peel, owner of the Hotel and Railway from 2002; and Roger Howard (his engineer and project manager). From this on 6 February 2004, a new 'Clifton Rocks Railway Refurbishment' group was formed to restore Clifton Rocks Railway to as much of its original glory as was technically and economically possible.

Roger Howard had worked on the refurbishment of Mr Peel's listed cottage in late 2003; Peel liked the unobtrusive repairs Roger made, and the way he ran the building contract and asked Roger to assist him with his plans to restore and refurbish the buildings, terraces and features that comprised the Avon Gorge Hotel. Believe it or not, his exact brief was 'to make things happen'. In a chance meeting with Peel, James Tonkin had asked him what he was going to do with the Railway and with similar brevity, Peel turned the question round to ask Tonkin what *he* was going to do with it.

This time the new group had the support of Mr Peel, who wanted to improve the image of the hotel and have it upgraded to four-star accommodation. In 1985 and 1990, Peel had wanted to demolish the Pump Room and Railway, and groups were formed to save them (see chapter 14). He recognised that the earlier events with Mount Charlotte Thistle could have been handled better, and he was keen that, under his own banner of Peel Hotels, he would consult more with local residents and involve them in his plans to restore the Avon Gorge Hotel. In support of that end, he attended a meeting of the new group in February 2004, when he outlined his plans for the Hotel and his desire to see the Railway restored (in some form) at the same time, declaring his intention to organise a commitment dinner to formally announce his plans. He would invite representatives from the council, local residents' associations and other conservation/interest groups. This time the group had a better footing from the beginning.

Mr Peel's wish was that the Railway should be developed, and the top station formed into a museum area. He saw it as a community project, with the group providing maintenance responsibilities. Once restored, it must become self-financing. This was a special interest group, but with no official constitution. The chairman was Peter Davey, the vice-chair was James Tonkin and David Bell secretary. 2004 was spent in discussions. January to July saw the completion of the first action plan. Feasibility of the project needed to be accurately determined. A study was needed to determine what type of attraction would bring the most appropriate financial flows. At some stage, an organisation was needed to take the project forward. Its aim was to advance the education of the public through restoration of Clifton Rocks Railway. This would happen when expected income for the feasibility stage was received. The group was very action focussed; key stakeholders were brought together. They foresaw a feasibility phase (which would cost about £120,000 and which would not be funded by Heritage Lottery Funding), a restoration phase and an operational phase. Large estimates (circa. £105,000) were obtained for appraisal studies. (I was later advised that feasibility studies have short lives and should be treated with circumspection – others were done by me, having been given helpful advice on what was required]. The original vision was to

restore the Clifton Rocks Railway (alongside the restora-

15.1. Peter Davey, Robert Peel, Tim Rew, Malcolm Frith, Mike Rowland, James Tonkin, Roger Howard, and Terry Prater stand where the Pump Room meets the top station, in June 2004.

tion and refurbishment of the Avon Gorge Hotel) as an integrated part of Bristol's Victorian heritage and to operate it as a sustainable regional attraction.

To be sustainable it had to be integrated with other attractions: part of a Clifton Trail, Victorian Heritage Trail, or an Engineering Trail. It was advised that it was wise to move forward with one clear message about what the aims and objectives were. This is all familiar territory to the author, having now read the work of the previous groups, and her own experience.

Mr Peel hosted the commitment dinner for stakeholders in June 2004. The problem of car parking for visitors was a major issue. There was a good understanding that this part of Clifton needed improving, but there were a variety of views about the preferred nature of improvements. There was a positive response to the Railway and the Hotel actively engaging local people in proposals prior to producing definite plans and formal consultation. Improving the area would need to be done in sections and required careful planning. Heartened by the goodwill shown towards him and his plans, Mr Peel asked Roger Howard to liaise with the group and find ways that the Hotel could assist with the group's endeavours – organising/improving power and lighting in the tunnel was an obvious item, as was arranging insurance cover and providing a room for the group meetings.

In August 2004, the next action plan was agreed: fundraising for the feasibility study phase. It was envisaged that full restoration would cost £10-15 million so it would be a major project. In September, the action plan for establishing the organisation, and communications was agreed. In November, it was decided to start the fund raising campaign to tie in with the transport heritage weekend on 20/21 May 2005, and to restore the railings.

Mr Peel paid for scaffolding to paint the railings outside the Railway, and for a shot blaster to prepare the railings ready for painting. He also paid the costs of setting up the domain of a new website.

15.1 My involvement

I got in touch with the group in the middle of February 2005, offering help in any way, and being able to provide extra publicity via the Bristol Industrial Archaeological Society and the Clifton and Hotwells Improvement Society (at the time I was webmaster for both). I heard they were going to repaint a Bristol bus in Bristol Tramways colours and I had vintage paint charts. The charts could also help with working out the colour to paint the railings. I was welcomed into the group, asked to post photographs on the Bristol Industrial Archaeological Society website to aid publicity, and bring some of my vintage cars to the first open day, and get my hands dirty. I took many photographs to document progress.

15.2 Restoration highlights

This section shows just how quickly we competed gratifying tasks. The public took us to their hearts since there was so much visible progress to a site that had remained secret and inaccessible to the public for so long. We were not just a talking shop, we were very enthusiastic volunteers, keen to work and learn about this neglected site. A detailed diary is shown on the website[1] together with some of the *Bristol Evening Post* articles. The first year was amazing. There is only enough room in this book to describe the highlights of our work. Peel Hotels recognised the importance of not being perceived as having a negative attitude towards these aspirations, and during the period of their ownership, official bodies like the BBC were able to update their archive footage on the Railway as well as include it in various official city events.

15.3 Exciting times in 2005, something new every month

In 2005, we spent 344 man-hours on four open days, 30 man-hours on nine tunnel trips and 1275 man-hours at work on the railway, let alone hours of work done at home. The site quickly looked less neglected. We gathered history. Many were keen to tell us their story and we were ken to hear them. Everyone wanted to see and hear about this secret underground site that had been neglected for so long. We were featured in newspapers and magazines at least 24 times. All of the comments we have ever received confirm that the people, the businesses, and the institutions of Bristol have taken the Railway to their hearts and all have expressed a desire that it be included in the city's trading and industrial history.

15.3.1 Railings

In March, we removed the hoardings, exposed some beautiful railings and painted them – starting on Good Friday, 25 March we worked the Easter weekend). We had tangible results to show for our efforts and, as a result, lots of publicity; offers of help came pouring Being in a high-profile spot in view of the Suspension Bridge and adjacent to the Avon Gorge Hotel with its popular terrace bar brought curious passers-by who asked us what we were doing.
James started off as our project co-ordinator.

15.2. The front boards were removed and the ironwork which held the signs was welded up to take new ones. (Author)

15.3. Cobwebs on ornate railings after being hidden for so long. (Author)

15.4. Handrail and supports were missing so we replaced them. Weeds were growing. (Author)

15.5. The top gates were found in the top station complete with padlock. These were refurbished. (Author)

15.6. Man from Mars using his portable grit blaster. (Author)

15.7. Progress after being grit blasted. Roger Howard and Mike Taylor getting on with the painting. (Author)

15.8. A very pleasing result for one month's work and a big change from plate 14.12. The Suspension Bridge can be seen in the background, and means many people pass by. (Author)

15.9. Both cable wheels needed uncovering. (Author)

15.10. The cables and fittings were exposed. (Author)

15.11 We found the railway lines under the rubble. (Author)

15.12. The lines could now be seen from the pavement. The new hole in the wall was cut so that visitors could view them, and for access. This was later enlarged. (Author)

15.3.2 Railway track and cable wheels

In April, we started to clear the railway lines having discovered they were under the rubble (plate 14.18 shows very clearly why we did not know the lines were still there). We cut a hole in the wall to make it easier to move buckets of dirt via a chain of people to skips in Princes Lane. Our aim was to enable passers-by to see the rails from the pavement. Festoon lighting was installed in the top station. The website was set up (www.cliftonrocksrailway.org.uk), 1,000 leaflets were produced to give away (not enough) and 6,000 postcards to sell (funded by a lottery grant of £1,200). Bristol Blue Glass made limited numbers of commemorative paperweights and gifted them to us to sell to raise funds. Founder-members were registered on payment of £10 (including me). We had a double page spread in the *Bristol Post*.[2] There was even a folk record produced by Rob the Rich called 'A Ha-penny Down, A Penny Up'.

15.3.3 First open day

In May, after only two months' work, we opened for the first time to the public since before the war. Sir George White did the honours. He said his grandfather had reopened it, his father had closed it, and he was very privileged to come and re-open it. An estimated 3,500 visitors descended on us. I brought four cars to display outside for something for visitors to look at while waiting. Eighteen display boards were erected relating to the history and restoration. A jazz band

15.13. Sir George White cutting the ribbon, James Tonkin looking on. (Author)

15.14. The author's 1929 Peugeot, 1925 Talbot, 1924 Lanchester, and 1985 Moss Monaco out on display for the transport rally. (Author)

played, and there were various events in the village including dance, marching bands, music, vintage buses, balloons, streamers etc. We raised £3,788.20 by donations and sales. The link with the heritage transport weekend was very successful as the buses provided free transport from the Harbourside (and doubled our attendance), and lessened parking problems. We resolved to improve signage, organise bunting, work on the website and produce another set of postcards.

An impressive light-bracket was returned by someone moving house. It was fantastic that in August we found a gas light that fitted the bracket when we retrieved artefacts from the turnstile section. We found a rim for the light in December. The light is proudly hung up every open day (see chapter 7, plate 7.54).

In June, it was decided that the group should be established with interim trustees of six sub-groups (communications, finance, fund raising, membership, operations and restoration), and then be registered as a charity. I was appointed chairman of the restoration sub-group (which also meant I was in charge of feasibility specification and contracts). The *Bristol Post* referred to me as the 'leader of the heavy gang' – which I thought was great. We also resolved to do more work in the tunnel; in fact I worked virtually every weekend, sometimes both days until December. Sometimes there was just the two of us, sometimes as many as 14 if something momentous was happening. I would draw up a list of jobs to be done. Sometimes the volunteers wanted to do other jobs, but it all was done very democratically, and everyone had a job to do. I personally clocked up 248 hours that year, despite having a full-time job. The team clocked up 1,807 hours in total.

We also went to the salerooms and bought a first-day medallion presented to the Munro family (see chapter 2 plate 2.6), a standard medallion, and a limestone paperweight made from the first rock that was cut (see chapter 3 plate 3.24). Peter Davey recorded a video about the railway which was available for sale.[3]

15.3.4 Clearing the rest of the railway lines
Tom and Ed Scammell came along to help clear the

15.15. The hoist on scaffolding in Princes Lane. Another rope was used to stop the wheelbarrow catching against the tunnel roof. It was a high lift. (Author)

15.16. Josh Townsend on the guide rope. The wheelbarrow had to be hoisted about 36ft, so it was good that the hoist was electrically powered. (Author)

15.19 Nearly there. We could not lift the pillar cap as it was too heavy. The original access to the first chamber can be seen at the top of the wall. (Author)

15.20. Mike Taylor, Ed Scammell, the author, Josh Townend, Alan Griffiths, Mike Edwards, and Lyn Coleman celebrating the end of a month's work.

Left: 15.17. Brothers Tom Scammell (our secretary) and Ed Scammell (our treasurer) digging. Many interesting artefacts were found. (Author)

Left: 15.18. The author digging carefully round buffer springs (see plate 6.5).

273

rest of the railway lines and called themselves 'The Power Workers'. Tom contacted me 'I am very fit, strong, healthy and practical and can turn my hand to pretty much anything, no matter how dirty, dusty or monotonous it may be.' I can vouch for that. They also joined the committee.

In July, we hired skips, scaffolding and a hoist to clear the rest of the rubble (about 16ft by 16ft and 2ft deep) from the lines visible from top station. This meant we could winch the rubble in buckets in a wheelbarrow over the Princes Lane wall and use chutes to put it straight into the skip. Three people were on the scaffold, and usually four below. Two wheelbarrows, trugs and many buckets were used. Nineteen volunteers were involved, an average of seven per session over six days. Cost was £1,600. We also took the National Trust, and our MP, for guided tours.

The BBC asked us to be on the Restoration Show to compete for £1 million, but we felt we were not yet in a position to proceed. We were very pleased to be asked, and that they recognised the worthiness of our project such a short time after beginning work.

15.3.5 Retrieving a Turnstile

In August, we removed foliage from the façade of the bottom station, refurbished various artefacts, and

15.21. First we had to cut a 4ft square hole in the 14in wall. Chris Bull doing the honours. It took 10 hours with the power saw. (Author)

Right: 15.24. Bob Steadman, Pete Luckhurst and Mike Taylor lifting the turnstile out of the hole. It was a tight fit. (Mike Edwards)

15.22. We chose the turnstile with a counter, and belayed both Tom Scammell and Dave Hewgill, and the turnstile, to stop then rolling down the 45% rubble slope. We were hooked onto the wall too. Pete Luckhurst (health and safety) is standing by the hole watching progress. (Mike Edwards)

15.23. Mike Taylor at the top of the ramp ready to grab the turnstile to get it through the hole. Tom Scammell and Dave Hewgill discussing the next movement. (Mike Edwards)

15.25. Dave Hewgill and Tom Scammell guiding the turnstile up the steps. (Mike Edwards)

15.26. Dave Strawford winched up the turnstile up the stairs while the others relaxed. (Mike Edwards)

15.27. We had a big tea break to celebrate the turnstile coming back to its original location. It only just went through the doorway. (Mike Edwards)

275

15.28. Queues of people waiting patiently to get in. (Author)

brought the turnstile and railings up from the first chamber. Chapter 7 gives more details and photographs.

In September, we opened for Doors Open Day for the first time. We had 3,000 visitors. In October, we spent much time in the top station when raining, so we could work out the source of the leaks and try to fix them. In November, we were filmed for the *Secret Underground* series, which was shown on HTV West on 2 March 2006. This included interviews of people who used the tunnel during WW II. Their memories are given in chapter 13 and chapter 12. We also set up a ten pin bowling team to play against the ss *Great Britain* to raise funds for Children in Need. We met some solicitors about creating a trust.

In December, more research, more patching up leaks and more tidying up. We also found lots of empty tins of Tiny Tim poodle food.

15.4 2006: a year of tidying up, media interest and a feasibility study

In 2006, we still worked most weekends (697 hours work over 32 days, but now started to do more trips too (41), as well as the open days. More research was done about the artefacts we had found.

In January, we continued tidying up and started writing part of the feasibility study. I was concerned that the group kept on quoting large figures such as £15 or £20 million, even £25-£30 million., when no costings had been done. From a survey I made of running funiculars in the UK, I found insurance would cost around £10,000; the cost of track for two cars was estimated to be around £170,000.

There was a superb 'heavy gang' two-page article in the *Clifton Chronicle* with lots of photographs. We took our first Lord Mayor down the tunnel. Over the years, six Lord Mayors and one Mayor have visited.

In February, we submitted a competition to design a footbridge over Hotwell Road as part of the Brunel 200 festivities. In March, I wrote an article about Clifton Rocks Railway that was published in the *Bristol Industrial Archaeological Society Journal*.[4] We started to make a profile of the back of a car to put on a pair of rails because visitors were confused as to which rails were a pair (only six are visible as two are covered WWII stairs). A horse-drawn car was measured in L Shed (next to M shed) at Wapping Wharf. I talked to Eura Conservation who had worked with ss *Great Britain*, and other companies about feasibility studies and conservation plans. They were all incredibly helpful and gave me good advice and useful examples. They told us what we as a group could do for ourselves. They advised us to think of which story had the most historic significance, and as a historic destination, what would bring revenue. This should be done before feasibility stories. As a result we designed our own survey, the results of which are in chapter 16.

In April, there was a one-hour recording by Radio Bristol which was broadcast as a series of programmes. We were in the top 50 Brunel 200 Ideas. We started to write a newsletter, *The Cable*, to send to the Friends of the Trust.

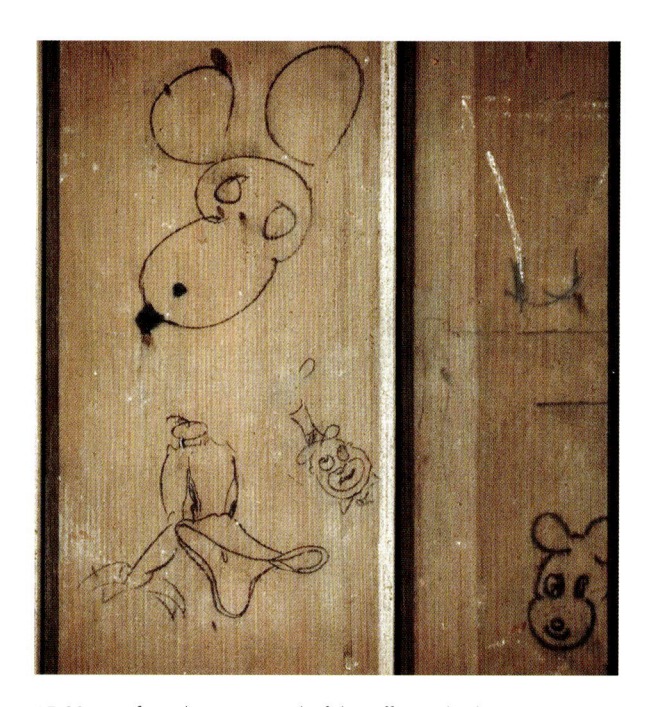

15.29. We found some wonderful graffiti in the bottom station. Pre-war Mickey Mouse and Donald Duck (born 1932), a clown, a wineglass and a bear. There are also some curvaceous nudes with 1930s' hairstyles hidden away on the other side of this panel. (Author)

15.30. Johny Hall and helper starting to make a hole in a WW II wall in an original alcove. We found a new tunnel (see plate 1.12). (Author)

15.4.1 Discovering a new tunnel under Sion Hill

In May, there was an article about Clifton Rocks Railway published in *Bristol Magazine, Clifton Chronicle* and the *Evening Post*. We appeared on HTV news in regard to our new Sion Hill tunnel (described in chapter 1 and shown in plate 1.12). Open Day attracted over 1,100 visitors and a bus specially painted in Bristol Tramways and Carriage Company colours was used to ferry visitors round. The Association of Independent Museums had a conference in Bristol and they joined us for a trip; we received lots of advice from their delegates. In view of all the people I had spoken to, I thought it was important to discuss all the alternative plans for the railway with the committee. In July, we cleared the top water reservoir, and supported one of its walls with acro props.

In September, we had our fourth Doors Open Day. We were still a top attraction with 1,665 visitors.

In October, we cleared rubble from the bottom station and searched under the ledges for artefacts. The *Evening Post* picked up the story about future plans.[5]

In November, we had our first AGM and held an election for our officers. Ed Scammell became our new treasurer; Ed was an accountant and had worked well with the restoration team.

In December, Dave Hewgill and I started making models. I made a scale model (1in to 1ft), which is very useful to show visitors the proportional sizes of all the sections, and to point out the conundrum of having to look at each section to decide which is the most important story: wartime or railway. Dave made a working model of two cars moving up and down. I managed to get the committee to change the aims from:

> To advance the education of the public through the restoration of the Clifton Rocks Railway as a sustainable visitor attraction and transport system, which is fully integrated with other visitor attractions, transport systems and the Victorian heritage in the area.

to:

> To advance the education of the public through the preservation and restoration of the Clifton Rocks Railway and its wartime history as a sustainable visitor attraction, which is fully integrated with other visitor attractions, transport systems and the Victorian heritage in the area.

This change was to reflect the importance of the wartime development, and also why 'preservation' was added. This was as a result of me discussing the project with many experts, and having spent a great deal of time in the tunnel with the restoration team. This

51.31. Bus painted in BTCC colours especially for us. (Author)

15.32. Alan Griffiths and the author standing by the car profile they have made.

was a big U-turn for the group, but at least we now had a committee singing from the same hymn sheet and working together.

2007 We still worked most weekends (377 hours work over 24 days but now started to do more trips too (48), in addition to the open days. Because we were closed by the new Hotel owners from September, the number of tours were fewer. We had 3,212 visitors on four open days and a 'Blitz' night.

In January, we carried on with constitutional matters. Stating that a site was to be used as a transport link could cause a problem if setting up a charitable trust, because a transport link is not normally recognised as being for educational purposes. We had gained a new volunteer, Rod Barnes who was an accountant with experience of company law who carried on with this task. Tom Scammell became our new secretary because David Bell had moved away.

In March, an article was printed in the Spring issue of *Pints West* (CAMRA Real Ale journal) about the beer bottles found in the refuge areas. Wilf Watters came

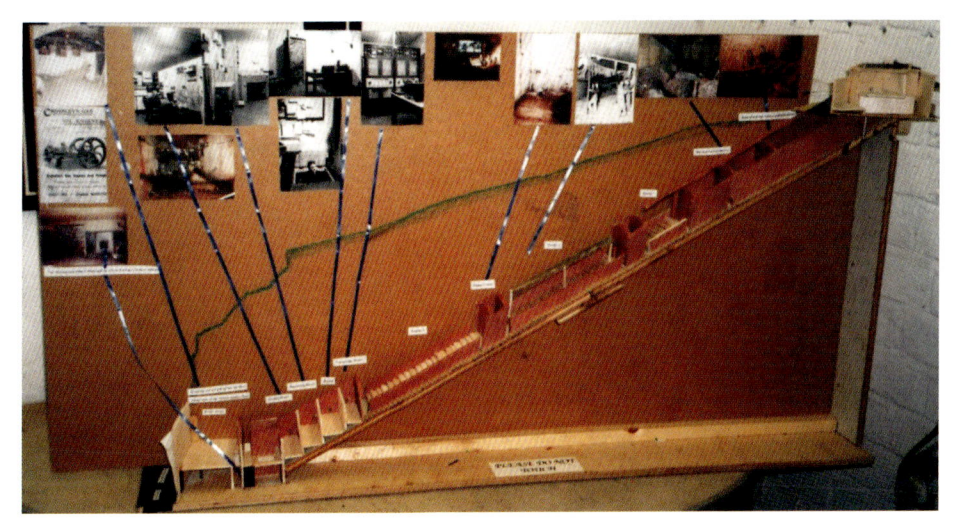

15.33. My sectional model 1in:1ft. (Author)

15.34. Much detail went into my model. The top station took ages to get the levels right. (Author)

15.35. John Perkin's model. (Author)

15.36. Dave Hewgill's model. (Author)

15.37. A visitor showed us his Meccano model. (Author)

to film the railway as part of a video he was making about 'Cliff Railways of Great Britain'.[6] His video used archive footage to show older cars in service as well as current railways.

In May, our models were completed, and a booklet about barrage balloons in Clifton printed. We had our fifth open day in conjunction with Museums and Galleries Month, and the Harbourside bus rally. This included Blitz night as part of Museums and Galleries at night event. We had a Green Goddess, a hand-driven siren, and a spotlight on the lookout point, and we all dressed up. Seeing the spotlight, the police came to check on us.

In June, we became a limited company by guarantee, the Clifton Rocks Railway Company with charitable status and directors Ed Scammell, Rod Barnes and James Tonkin. It was confirmed that we saw ourselves as a museum rather than a railway, otherwise we would have become a community-based company. The Hotel confirmed they did not mind whether we were a museum or railway. The conservation plan was now virtually ready, having been written by the author (helped by my two sons who are both archaeologists) and Rod. It was far too early to create a business plan. Having the project owned by both the Council and the Hotel could cause difficulties. For example, the Council could be challenged if it undertook work at public expense that benefited the capital portfolio of the company owning most of it. There were a number of potential conflicts of interest. How could any claim to ownership, be justified by any transfer of ownership, by any legal authority?

In July, we rescued a clack valve from the bottom water reservoir (see plate 7.15, chapter 7) to put acro props under the floor. It was a challenging task. Another valve is still in situ, since there were two independent systems.

In August, there were many more finds in the refuge shelters. Extreme weeding on the top station windows and bottom façade was carried out by a volunteer – a proficient climber who belayed himself down to the buddleia. We were given permission to paint the bottom gate but nothing else, by Bristol City Council. They wanted a business plan before they would consider a transfer of property.

In September, we held our sixth open day in conjunction with Doors Open Day. The University of Bristol Archaeological Department offered to help us record the BBC rooms and barrage balloon section with the help of MA students in Historical Archaeology. The Avon Gorge Hotel changed hands on 3 September just before Doors Open Day, and the Railway was closed by the new owners (the International Swire Group) on 24 September to sort out insurance.[7] Their insurance company did not like railway tunnels. All our trips had to be postponed. They were amazed at the commitment shown by the volunteers. Trips were now a bigger source of income than open days.

2008 Due to work being stopped, we worked only 36 hours tidying up for open days, and did 46 trips (from the end of May). We had 1,627 visitors on four open days and a Blitz night. I helped several students with their heritage projects.

In January, the Hotel asked us what official status we had, and if we had prepared annual accounts. They assured us they were making positive progress with insurance, and that they were as frustrated as we were at the delay. One of the groups waiting to go on a trip which was an insurance company, even offered to run it on their insurance – but to no avail. We obtained a list of possible insurers to help. We were advised that as a spin-off from all our endeavours, we now had good public support and good media support. Working together with the hotel raised their public profile too, so it was a mutual benefit for them to let us continue our activities. They allowed us to advertise our open days.

We were locked out until May. When we were allowed back in we were not allowed to do any work (though we carried on weeding and sweeping and small maintenance to keep the site safe for visitors). We were allowed to continue doing trips and open days. We were not allowed to do the archaeology surveys with the University, or do engineering surveys, so there was no point in progressing with a feasibility plan, as we could not highlight the principle areas of further work required and thus get funding. Rod had already highlighted that best practice was to ensure the charity did not become funder led rather than mission driven. We were so relieved that we had done so much work before the Hotel changed hands. This meant we now had more time to show visitors the fruits of our labour.

In March, another article about Clifton

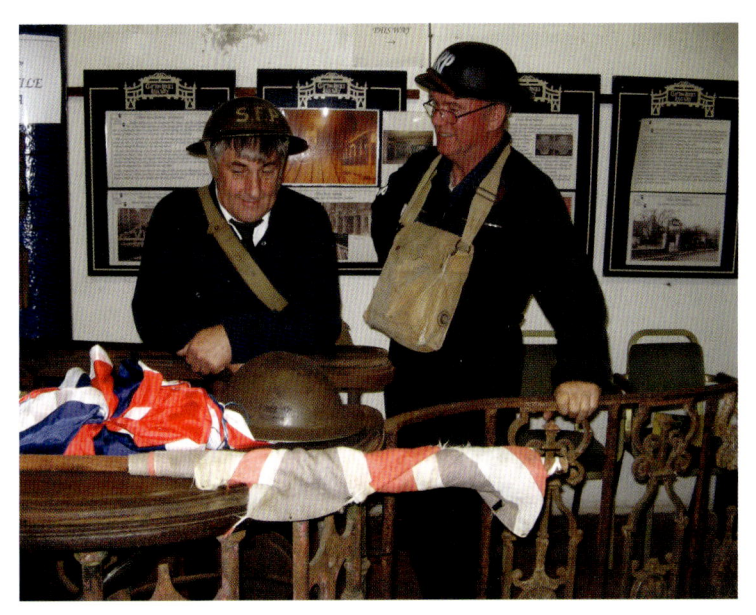

15.38. Alan Griffiths and Mike Taylor as wardens in our Blitz experience. (Author)

15.39. Wartime spotlight on the Suspension Bridge. (Author)

281

Rocks Railway was published in the BIAS *Journal*.[8] I attended courses to learn about accreditation and collections' care.

The Hotel spent £15,000 to satisfy the insurance company. This included an extra set of lighting in the tunnel and more emergency lighting (£9,000), putting barriers down the stairs (and putting several signs up to 'mind the step'). They employed a structural engineer (£5,000) to carry out an initial assessment of the structural integrity and condition of the tunnel for insurance purposes (who acknowledged that our maintenance had helped save the site), and provided reflective jackets. The insurance company needed to be reassured that the tunnel was safe to enter. We sent our risk assessments to them (we were lucky that Pete Luckhurst who had been in the project from the start, together with his wife Donna who had handled the design work of the information boards, was a health and safety expert). The Hotel wanted the railway to be a community heritage project.

In May, we were open again (but we were only allowed in the day before open day, so we had a huge amount of cleaning up to do, and no time to do the weeding. We had started to wonder whether we would ever open again). This was our seventh open day in conjunction with Museums and Galleries month, the Harbourside bus rally (Bristol Buses were celebrating 100 years of running on Bristol streets), and Blitz night as part of Museums and Galleries at night event. The latest Lord Mayor came to visit us which was very welcome and an announcement in the *Evening Post* stated we were back in.[9] Our numbers were down, and we were told we had to escort all our visitors through. This actually worked very well since people enjoyed having things pointed out to them by enthusiastic volunteers. Children under 14 were not allowed – which caused some problems. We had a big backlog of group trips down the tunnel to catch up with. Mark Passmore, General Manager of the Hotel said

> This dedicated group of volunteers has done a fantastic job in restoring this important piece of Bristol's history and we are delighted that the Clifton Rocks Railway is again open for business.

The Architecture Centre, Geographical Association, and Friends of Museums confirmed our views about the future use as preserving the WWII history and keeping the site as a museum, with a railway running possibly part of the way.

> WWII is on the school curriculum and will attract funding. It is totally unrealistic to consider that the Railway would ever run as a transport system. The WWII history is as important as the Railway. The WWII history belongs to all of us.

In September, we had a lovely sunny Doors Open Day. We managed to keep up with the queues despite escorting people through, but it was hard work. At least we had more control over the time we spent with each group of visitors depending on the size of the queue. Lots of nice comments, applause, laughter and publicity. Numbers were up.

In October, we were on BBC2 Newsnight. Peter Davey took the BBC down to relate life in the tunnel to the credit crunch.

In December, Nicola Williams, one of our volunteers who was a keen photographer, arranged for the Bristol Photographic Society to go down the tunnel. They took some marvellous photographs that they let us use, some of which we display in the top station, and some of which (by Neil McCoubery) appear in this book. I wrote an article for the Bristol Civic Society newsletter, stating again why a railway would not be running again, and how important the war time story was.

2009 We made 59 trips, and saw 1,417 visitors on three open days. The trips included the Civic Society, Executive of Bristol City Council, Antiques Road Show, and the Conservation Department of the Council with their Chief Archaeologist. They all considered it should stay as a museum. Railfreight thought our tunnel was in better condition than many of theirs. We still had fun even if we could not work. The Antiques Road Show group described their visit beautifully;

> I wasn't sure what to expect when we started the tour – but as our guide peeled back the layers of history inside the Clifton Rocks railway I felt like I was attending an architectural autopsy. The building and tunnels would have been interesting in their own right, but the use of the building as an air-raid shelter and later as a BBC Radio base during WWII has made an impression on the fabric of the building which was fascinating. The narrow stairs that lead down into the tunnels are deceptive, doubling the impact of the wide high ceilinged sections that hang onto the side of the hill following the path of the tracks towards the Portway. As we descended, I could imagine what it must have been like, huddled in these dark tunnels during WWII air raids. Slightly more difficult to imagine was the arrival of a grand piano, and BBC entertainment presenters in dinner jackets. The Radio studios are long

gone, but with some imagination you can easily conjure up yet another period in the railway's colourful history. We enjoyed the tour very much.

I helped with some more student heritage projects. A conservation architect gave us some useful help. He admired the different styles of brick layering (this came in useful when I wrote chapter 3). We also had an electrical engineer who could appreciate the antiquity of the fuse boxes and other bits and pieces left on the walls.

In January, it was 14° F (–10° C) outside and 38° F (3° C) in the shelters (the coldest we had ever recorded), with icicles in the BBC area where it is draughtiest. We took Norman Thomas (see chapter 12) down to see what he could remember from sheltering in the wartime at the age of 11. He sang 'Home on the Range' – very poignant as he had sung it in the tunnel for a Christmas concert in 1942. We spent a lovely three hours with him and learnt a huge amount from him. A BBC cameraman recorded it all. We also went to see the Temple Meads wartime shelters.

In March, I went to see the grandson of George George, the gangmaster of Clifton Rocks Railway (chapter 2 and chapter 7). This led to other members of the George family getting in contact with me as a result of seeing Harold's photograph on the website and re-uniting family members who had lost touch with each other.

In May, we set up an exhibition of Bristol Photographic Society pictures in the Hotel as part of a very wet Open Day. This was very successful and we received many positive comments. The visitors could talk more to our volunteers and it showed support from the Hotel. Once again there were free bus trips from the bus rally at Lloyds Amphitheatre. I also took 60 University staff round the top station as part of Positive Work Environment week, ably assisted by Mark Horton of the Archaeological Department who walked us from the Water Tower to the Railway via the lead mines on the Downs via the Observatory. I was glad we did not have to go up and down the steps at the Railway too.

In July, we helped Robin James of Heritage Railways with a book about curious and unusual railways.[10] He really enjoyed his visit. The book was published in November and contained a four-page article on Clifton Rocks Railway using some of our photographs. In August, I helped Pete Williams with a photography project for opening Colston Hall. He wanted to show images of the railway projected onto the building.

In September, over 1200 people were shown round on Doors Open Day. It was beautiful weather, and free vintage buses ran on Saturday again. The Hotel again let us use a room to show off the Bristol Photographic Society pictures. The Rag Morris Mummers presented *The Nine Lives of Isambard Kingdom Brunel* by the side of the Railway to celebrate 150th anniversary of his death. It is lovely to be a part of so many events. GWE Business West held *An Evening with the Avon Gorge Hotel and Clifton Rocks Railway*. This showed collaboration with the Hotel.

Clifton Rocks Railway was featured in *Somerset Life*. I took photographs of the Railway for the new M Shed museum to show one of the unique aspects of Clifton and Hotwells. They are still on display. They also show a film with people talking about the Railway.

In December, we bought a clock at auction made in 1893 to commemorate the opening ceremony. Plates 2.7, 2.8 show the clock. We are lucky to have been able to acquire so many unique important artefacts.

2010 We did a record 70 trips down the tunnel, and saw 2,250 visitors on four open days. We held a 10-day exhibition in the Photographique gallery in Baldwin Street, which was very successful. It included new contributions and additional archive images. Stephen Williams MP came and said

> I really enjoyed the exhibition and was amazed by the transformation since I visited the tunnel in 2005. The huge volunteer effort has been really worthwhile and I look forward to another visit to the site.

I gave a talk at the NAMHO (National Association of Mining History Organisations) conference on the construction of the tunnel, and did two trips down the tunnel for the delegates. I added an image of the curfew sherry bottle (plate 12.11) from the Railway wartime shelters to the BBC History of the World Project. I couldn't put the whole site in, but thought it suitable to think of someone sitting in the shelter enjoying a glass of sherry with such a suitable name.

I got in touch with the Yakometties – a pop group who had performed in Clifton Rocks Railway in the 1985 (see chapter 14) and made a short video. I wrote an article about our Bristol beer bottle with a Guinness Label (see plate 12.20, 12.21 in chapter 12) for *Pints West*.

We started to get requests from paranormal groups wanting to investigate the tunnel. As they all wanted to be there at midnight, expected us to stay up all night, and there were no facilities for them, we declined them all. If they had found anything they may have wanted to come back again too, and encourage more groups to come. We never knew there was so many groups, or that they had so much investigative equipment. There is a limit as to what unpaid volunteers want to do, and losing sleep is not one of them. We would rather tell visitors the history, and point out the features, and enjoy the atmosphere.

We took Dr Brian Hawkins, the eminent engineering geologist down the tunnel to hear what he had to say. His particular research interests included the Quaternary geology of the Bristol region, slope stability problems in soils and rocks, and the influence of ground chemistry on construction. He said that where the tunnel passes through the mudstone horizons, down-dip water flow through the limestones is also inhibited, such that water pressures may accumulate. As the tunnel is only shallow and close to the free face of the gorge, some water will move southwestwards, reducing the head of water abutting the tunnel. The evidence from a fairly quick inspection following relatively dry weather suggested seepage into the tunnel is mainly some 50m from the Hotwell Road. This is consistent with Skempton's 1961 paper (see chapter 14): In the wetter area the water seeps/drips through the side and crown of the tunnel over a zone of some 33ft.

We also had visits from various engineers who had worked on funicular railways. The Stationary Engine Society told us about a write up about our Crossley engine in the Institute of Civil Engineers Proceedings of 1896, and a report in a book written in 1906. Having found them in the University of Bristol library, this encouraged us to look for studs on the floor. It is great that we have received so much free advice and visits from all these eminent people.

The organiser of Doors Open Day contacted us:

> What is it about going underground? I sometimes think really all I need to open on DOD is yourselves, Redcliffe Caves and Temple Meads and all Bristol is happy.

I became a trustee of Clifton Rocks Railway Company with the role of secretary. Unfortunately Rod had changed jobs. James Tonkin and Ed Scammell were the other trustees.

2011 2,019 visitors came on open day and there were 57 trips down the tunnel. We were visited by Mr Woolley, whose grandfather had worked on the Crossley engines in the bottom station and whose father had done the engineering work for the Railway. I was given his daybooks for 1899-1901 and 1904-7 (see chapter 6 for his work in the Railway). These provide very important engineering history at a time when so much was being developed. I wrote a paper for Bristol Industrial Archaeological Society about his work. It was lovely to pick up a buffer spring (plate 6.5) and know that Mr Woolley had made it.

The BBC ran a feature in the tunnel to celebrate the 70th anniversary of their transmitting from the tunnel. I went to Caversham to look at the BBC archives. As you can see in chapter 10 this was a fruitful visit.

We were asked to provide a display for the Pride of Place event to be held in the Museum of Bristol by the Neighbourhood Partnership. We were given the local Clifton banners afterwards, which have proved to be very useful. We raffled an 1893 opening-day medallion for £323. We helped more students with their heritage and cultural projects.

2012 We had 2065 visitors on open day, 1669 to the exhibition and 71 trips down the tunnel. Trips down the tunnel included the Worshipful Company of Engineers in their dinner suits. They invited us to lunch. The Lord Mayor was very impressed when he came to see us on an Open Day. We were pleased that some events came from the Hotel due to conferences, weddings, and birthday celebrations. Chris Rayner, an architect, who was doing a project for English Heritage researching public air raids (including workplace and institutional ones) all over the country, also came to visit us. He first got interested when carrying out some conservation work on a site that included them. He said

> Each one has its own story to tell once you are able to read the clues, and they've invariably also had a post-war life of abandonment, sealing, partial collapse and vandalism. They are usually very dirty, wet and of course dark, and if you don't enjoy the film *Arachnophobia* they are probably places to avoid. But I find them the most moving and eloquent of the survivals from the Second World War (and occasionally First World War) covering some of the lowest and highest points in peoples' lives at the time.

I received a huge amount of information about WWII from a 94-year old lady in Australia who had worked as a gas fitter in Bristol during the war (see Mona Duguid in chapters 11 and 12). We also met more members of the George George family (see chapter 2). I was also in correspondence with the great-nephew of Croydon Marks (see chapter 2), who was very helpful.

I was awarded the British Empire Medal for Services to Clifton. This also brought more publicity for the Railway.

We held an exhibition in Bristol Records Office for three months, part of the time overlapping Doors Open Day in September. This benefited both of us, and they let us have the images they had used from their collection which we now use in our exhibitions. There were several interviews on Radio Bristol and a nice feature in the *Evening Post*. There were many positive comments.

We nominated Roger Howard (vice-chairman) and Dominic Hewitt (liaison officer) to be trustees of the Clifton Rocks Railway Company. Peter Davey continues to be the chairman of Clifton Rocks Railway restoration group which looks after the day to day running of the site.

James bought another unique Munro 1893 opening day medallion in Nailsea (see plate 2.6).

2013 1,614 visitors arrived on open day, with 71 trips down the tunnel including 10 trips for English Heritage. Michael Portillo filmed an episode for 'Great British Railway Journeys'. We also took down two groups of heritage students with Mark Horton of the University of Bristol Archaeology department looking at various issues of conservation, volunteers etc. (We regularly take down graduate engineers too. There is a push to get more young people involved with engineering and this is an ideal site to show historic engineering features close up.)

We helped a student from the Czech Republic with her heritage project. Part of her project was to have work experience on heritage sites, so she helped us on open day, tunnel trips and clear up sessions. She now has a career in heritage.

The University of Bristol hosted the prestigious Fulbright Commission's Summer Institute[11] established to explore the culture, heritage and history of the UK. Ten undergraduates from the US, had been awarded scholarships in recognition of their academic achievements, spending four weeks in Bristol and taking part in various meetings, workshops, seminars and field trips; looking at the role Bristol has played in American history, and exploring the city's rich heritage. We were privileged to take the students round the Railway where they helped us clean artefacts. It was such a success that we have done it every year since. We certainly are fulfilling our educational aim.

We also took two groups down to celebrate 'Love Architecture' organised by the Royal Institute of British Architects. One group that we did not accommodate, was a company that wanted to use the tunnel for combat simulation and training exercises using replica weapons – we thought it was not respectful of the history and might cause damage.

Another Lord Mayor came to visit us on an open day in April. He was able to visit our photographic exhibition in the Hotel too. George Ferguson, the first elected Mayor attended Doors Open Day – we were so busy I told him he had to go round in one of the groups. He was as good as gold and did what I told him to do.

Gordon Young of the Bristol Film & Video Society came to film in the tunnel as part of a film about the Avon Gorge: 'Bristol's spectacular route to the sea', incorporating Clifton Rocks Railway, peregrine falcons, trees, plants, rock climbing, goats, the source of the gorge and the Observatory.[12] It won a couple of awards. *Clifton Matters*, a new magazine, had a four-page spread about the Railway. Penny Mellor wrote a book about 20 years of Bristol Doors Open Day, having organised it from 1994 to 2013. She selected 31 venues to illustrate the diversity of the sites, which included Clifton Rocks Railway.[13]

Several companies contacted us to put information about Clifton Rocks Railway on their own tourist, and specialist websites – this all helps lift the profile.

Someone moving house returned a gatepost from the top station. It was hard to work out where it had come from because war conversions had destroyed its original location (see plate 7.60, chapter 7). We had an amnesty for items to be returned – very successful, it brought back lots of useful objects.

I gave a short talk in the Wesley Chapel for Bristol's Heritage Buildings: Relics or Legacies event. So many novel experiences.

15.5 Work begins again 2014

We had 1,590 visitors on four open days and 78 trips down tunnel including six trips for English Heritage, so trip numbers were increasing. 121 hours spent working – better than 78 hours spent weeding and sweeping the previous year.

In January, we heard from the Hotel that the local head office of the Swire Group had closed, so we would deal directly with the Hotel, and that from now on we could do more work, just so long as we gave them visuals before work began. Anything to tidy up Princes Lane would be welcomed. We told them we would like to replace the back station windows, and the Newnes' arch over the junction of Princes Lane and Sion Hill. We would be allowed to do more filming so long as we told them in advance, that it was factual, and respected the history. This would give a commercial benefit to the Hotel. This was all welcome news. We had amassed quite a lot in donations since we had not done anything major since 2008, so we could afford it. The Hotel was still owned by Swire Pacific. Swire UK would be closed. We were now in a progressive state.

The Ministry of Entertainment, presented a play *Normal Service will be Resumed* set in Clifton Rocks Railway, at the Brewery Theatre in Bedminster. We helped with images and attended in force.

The *Evening Post* ran an April Fool's joke[14] based on the Railway (books were to be moved from the lower floors of Bristol Central Library to make way for the Cathedral Primary School. It was suggested they could be moved into the tunnel. The bookshelves could be installed on the rails to make them moveable).

We still managed to park vintage cars outside on open days to attract the visitors, and have successful exhibitions in the Avon Gorge Hotel. A jazz band played. Another Lord Mayor came to visit us and praised our work. People had come from all over the country (and from overseas). Our Facebook page (looked after by Ed) attracted a lot of interest as well as our website so these helped to advertise the event.

The granddaughter of Philip Munro the architect of the Railway came to visit us to see the clock and medallions presented to her family (see plate 2.9) – we made sure we had an image of her grandfather on display.

We participated in a short film presented by John Craven celebrating 80 years of BBC Bristol, as part of Inside Out West. He was rather puffed by the time he got up the stairs from the BBC rooms to meet David Pearce who had worked here for a short time during the war (see chapter 13).

15.5.1 New railings for the top station

In October, we replaced the hoarding at the apex of Railway with new railings made by Mike Taylor. This was a vast improvement to the rotten wood tied up with rope. Mike carried the railings from home on an old GWR truck, which came in very handy.

15.5.2 New windows for the top station

In November, two new sash windows for the back station made by James. It is unbelievably light in there now. We then had to paint the walls and ceiling as they looked so dreadful.

In December, Mike Taylor did a short film for Made in Bristol – a local television station serving Bristol and surrounding areas.

2015 We made a record 91 trips down the tunnel, including six trips for English Heritage. We had 2,166 visitors on four open days. Numbers had increased due to publicity about the new signage. We spent 83 hours working. The Lord Mayor attended open day.

We looked after the telecommunications engineers when they had to repair a fault in a live 1967 telephone cable holding 300 lines running down the tunnel. It took them a week to knit the lines together. The original 1941 cable is still in place at the top, but must have been sealed off when the 1967 cable was installed.

15.5.3 Port and Pier drinking fountain

We were offered a Victorian, Ham Baker of Westminster drinking fountain found by Ian Webb, from the Port and Pier Terminus just beyond the Suspension Bridge. Ham Baker was founded in 1884. They started to specialise in equipment for the water and waste industries in 1904. We thought it would be a good reminder of the spring industry so we got it home, refurbished it, and put it in the top station. This can be seen in plate 15.46.

15.5.4 Replacement of George Newnes' sign

We commissioned Dorothea Restoration to make us a new sign based on the sign on an early postcard, and put George Newnes' initials on to show that it was over the original exit. This was very successful and attracted a lot of attention and praise. Before and after images are in plates 15.47 to 15.49.

James resigned from the Clifton Rocks Railway group due to lack of time, and I became vice-chair as well as restoration officer. James is still chair of the Trust.

I took Edward Stourton, the celebrated BBC broadcaster, down the tunnel to help him with his research for a book about the BBC during the war. This was published in November 2017.[15] Ed added fridge magnets to the merchandise selection, which sell well. Postcards and booklets are still selling well too. In September, we appeared on Antiques Road Trip with Charles Hanson.

In December, Swire UK Hotels was bought by Malmaison and Hotel du Vin – new owners of the Railway again.

2016 Eighty-six trips down the tunnel and four open days with 1,776 visitors. Numbers were down due to the closing of the Portway in May, and the closing of many local roads due to the Tour of Britain in September. An impressive 258 work hours was spent (excluding work done in my garage), partly due to two big projects – the stairs and another sign.

15.40. New railongs: Mike Taylor was really pleased to remove the rotten boards. (Author)

15.41. We had a spare bit of zig-zag to get castings made, bought some finials, posts and rail. Alan Griffiths helped Mike Taylor install it. (Author)

15.5.5 New staircase

In January, we spent three days making and putting in a new staircase in the bottom station. The old one was getting unsafe to use.

15.5.6 Replacement of the George White sign

From February to March, Mike Taylor made the intricate ironwork that used to be over the top gate. Rosettes, and big round railheads, had to be cast at the foundry. The curly brackets were made by Dorothea Restoration, the sign box and sign by Gould signs but the rest was Mike's own work. The ironwork was sent away for powder coating. This has been admired by everyone and has truly put Clifton Rocks Railway on the map. All the volunteers who helped us put it up have been with us since 2005, except new boy Jon Picken who is a very welcome addition. Look at the

Top: 15.42. James Tonkin looking at the old windows. The frames were absolutely rotten. Note the condensation on the concrete ceiling which stays here November to March. (Author)

Above: 15.43. Adam Whiting adding the finishing touches to James's windows. (Author)

Top: 15.44. Mike Taylor telling the story with the Made in Bristol television crew. (Author)

Above: 15.45. Peter Davey with Michael Portillo and his trusty 1863 Bradshaw timetable (which clearly holds no information on Clifton Rocks Railway). (Author)

15.46. Drinking fountain by Ham Baker of Westminster. (Author)

5.48. Replica Newnes sign with his initials on top. (Author)

15.47. Original Newnes sign at the junction of Sion Hill and Princes Lane (see chapter 4). It was removed by Sir George White. (Author collection)

15.49. Replica Newnes sign with hot air balloons in the distance. Note the views over Dundry. A truly new feature for Clifton. I did not notice it also said Clifton Rocks until the balloon fiesta. Very fitting. (Author)

289

15.50. Missing railheads replaced using new castings. An original was used as a pattern. (Author)

15.51. George White's original top gate decoration. This image was used as a pattern. (Peter Davey collection)

15.52. The assembly shows how many components had to be individually made. (Author)

right-hand side of the sign and you will see it is dedicated to me, 'Leader of the Heavy Gang'. This was added when I was not looking. The *Bristol Post* published an article about it.

Peter and I appeared on Made in Bristol TV. We also participated in the Radio Bristol navigation game 'Clueless' on April open day, describing what we were and giving a clue to the next location.

In June, we decided to spend one day a month doing maintenance and projects. I resurrected my work list from 2007 and we sorted out priorities. In July, maintenance was weeding, sweeping and painting just for a change, but at least the weeds were smaller. In September, we cleared out the trees and ivy from the top station reservoir. We also replaced the railing signboards which we had put up in April 2005 and had gone rotten. Gould Signs made a good job considering the spaces were trapezoidal. We also put a drawing based on the Loxton sketch (plate 4.9) on the bottom station wall showing a car going up, to help people understand what they would have seen pre-war.

For Doors Open Day, we handed out visitor feedback forms produced by the organisers from the Architecture Centre. These made good reading. People liked the enthusiastic tour guides, the artefacts, the history lesson etc. We had lots of applause again. We also had the great-granddaughter of Philip Munro the architect of the railway and Pump Room all the way from Beaconsfield. We were pleased to welcome her and show her the Munro medallions and clock.

In October, we spent a day helping to take out an old cable. On 11 June 1941, a high-voltage cable was installed down the north side of the tunnel to supply power from Clifton to the street lights on the Hotwell Rd, and to the BBC section. The same cable was removed from the tunnel by Western Power on 26 October 2016. They still had the original work sheets, which they gave to us. It still had power running through it until they disconnected the supply the day before. They had originally intended to carry it all out

15.53 Installing the new Newnes sign. Dominic Hewitt and Ed Scammell on the top row, Dave Strawford, Jon Picken, Mike Taylor and Tom Scammell on bottom row. The stone plinths are level. (Author)

Below right: 15.54 The other side is just as heavy to lift up. (Author)

by hand via the top station after cutting it up in to approximately 13ft pieces. After the first couple of trips carrying lengths back up the tunnel, they had a change of mind, and decided to take it all out of the bottom station as the segments were so heavy – still not an easy task but a wise move. We kept samples of the cable. Some months we seemed to spend as much time in the tunnel as at home.

In October, I was unexpectedly diagnosed with pancreatic cancer, and in November started six months of chemotherapy. I was determined to carry on with the Railway trips and write this book – it kept me fit and gave me focus.

In December, Mike and I got married, and had our reception in the Hotel, followed by trips down the tunnel – where else?

2017 We had 2,190 visitors for four open days. In January, Peter and I did a programme for BBC Radio 4 Countryfile. I thought I would sound puffed because I was describing the barrage balloon section, and had just climbed over 300 steps. I still had my chemotherapy bottle attached, but it was fine. We also helped with a BBC programme about the Bristol Blitz, but we have yet to see that. Amanda Ruggeri of the BBC also visited to create a web page about the railway.[16] A Spanish group made a short film for a movie called '7 limbos'. It was mainly shot in Spain but some scenes were filmed in Bristol, London and Porto. It was about some musicians that were located in different countries who had some relation with the idea of being lost. They wanted to shoot a Spanish musician (Miguel Prado), who lived in Bristol, singing in the tunnel because they thought there were some special sound qualities in there, and *because an abounded environment connects with this idea of lost*. The Hotel gave us permis-

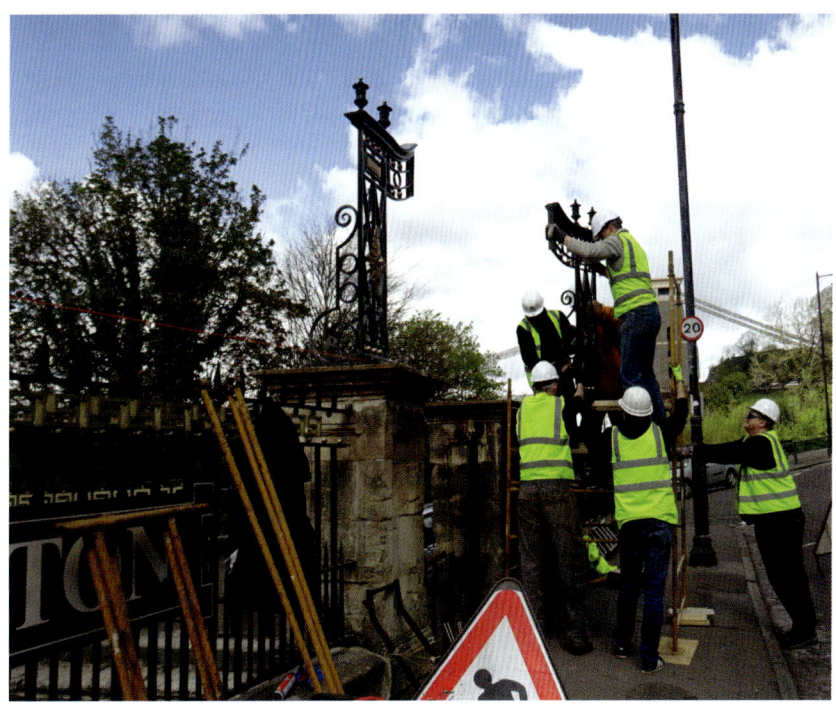

15.55. Left to right: Dave Strawford, Jon Picken, Mike Taylor, Dominic Hewlett, Tom Scammell and Ed Scammell looking very pleased with themselves. The sign is perfectly level. (Author)

15.56. Baroness Janke of Clifton came to 'open' the sign at our May open day. Left to right: Dave Strawford, Peter Davey, Mike Taylor, Baroness Janke, Ed Scammell, the author, Dave Hewgill, Pauline Barnes, Jon Picken and Sue Stops. (Author)

15.57. Gareth Mccarthy, who is a keen fan of Victorian Bristol and the Railway has a really good image of the bottom station tattooed on his arm.

sion, but we prefer to do films that have historical content.

For work days I got my gallant band of volunteers to help me with the book doing things like a rock plan, and checking the dimensions of the artefacts. I worked out where the BBC transmitting masts and aerials had been, and put a piece of scaffolding pole to mark the spot of one of them. I learnt so much doing the whole exercise, and my team was keen to learn with me.

In February, I was told my cancer was stable, and still is in September. I was told this was because of my positive attitude, being so fit and being focussed on writing this book. So I am still here much to everyone's surprise, mine especially.

We were blessed with good weather for our open days in April and May, and the usual exhibition in the Hotel went well. The numbers for September Doors Open Day were up despite sharp rain showers, and there were massive queues. (612 visitors on Saturday, 758 on Sunday.)

15.6 Number of visitors

We have now seen over 31,312 visitors on 53 open days and taken 7975 on 761 trips down the tunnel, about 39,000 in total. Demand is still strong.

Hours Working

In 2005 we spent 1,275 hours working since there was so much to do and we worked virtually every weekend. In 2016, we did 258 hours which was a massive increase. Work hours were reduced between 2008 and 2014 since we were only able to clean up for open days, sweep steps, check light bulbs and record artefacts.

Revenue

We get more revenue from trips than open day. Open days are a good focus point and good for publicity. Trips can be booked on open day and on the web site.

People from all over the world have now heard of us. On any day, one can see people looking over the railings and telling the story to their friends. This is very gratifying.

What is so wonderful is the atmosphere, and the visitors who keep coming back year after year to see progress. We get laughter, applause, lots of people turning up to tell us about their memories, so it is just like a big happy family. Visitors from all over the world as well as from all over the country all love to hear the history of the site and area, the latest research, and we love telling them. It is a very special place, and we have a lovely supportive team. Peter Davey our chairman also does sterling work in giving many talks about Clifton Rocks Railway (and Bristol trams) as well as taking groups down the tunnel, so is a great ambassa-

Table 15.2. Visitors 2006-2016

	2005	2006	2007	2008	2009	2010	2011	2012	2013	2014	2015	2016
Open day visitors April						374	377	402	431 (2)	294	288	385
Open day visitors May	3500 (2)	1132 (2)	1088 (2)	380	233	502	447 (2)	219	171	241	365	269
Open day visitors September (2 day)	3000	1665	2124	1247	1184	1372	1195	1444	1012	1056	1513	1122
Total	6500	2797	3212	1627	1417	2248	2019	2065	1614	1590	2166	1776
Group trip visitors	117	843	935	406	582	659	500	672	658	728	843	747
Number of trips	9	41	48	46	57	70	56	71	71	78	91	86
Hours spent on trips	30	225	229	207	264	309	265	304	331	362	435	378
Work hours	1275	697	377	36	56	63	92	71	78	109	83	258
Number of volunteers helping	62	45	43	32	28	24	22	22	24	29	28	29

dor.

Who would have thought that painting the railings in 2005 would have led us on such a voyage of discovery, and experiences, and that Clifton Rocks Railway would be a top site for Doors Open Day for the last 13 years. I am proud to have been so involved. What an amazing achievement by a small band of dedicated volunteers. I hope that my own contribution has been a catalyst to produce this remarkable result.

sixteen

Conservation *vs* Restoration
the Mary Celeste Experience

16.1 Introduction to the layers of history

Opened in 1893, Clifton Rocks Railway is a now defunct funicular railway constructed inside the cliffs of the Avon Gorge, and originally funded by George Newnes as part of a spa complex. It linked Clifton to pleasure steamers at Hotwells (the landing stage is now derelict), a rail link to Avonmouth (now long gone) and a tram terminus to Bristol centre (there is no room now even for a bus stop). It is the only cliff railway with four sets of tracks in a tunnel in the world. It is a superb example of great Victorian engineering. Because it is in a tunnel, there are no scenic views.

The Railway went bankrupt in 1908, George White bought it in 1912 and changed the façades and the turnstiles, of both the top and bottom station. The Railway's use declined again after the opening of the Portway in 1922, and it was finally decommissioned in 1934. It is a huge, impressive Marie Celeste-like structure with original railway features such as a sophisticated gas light, turnstiles, barriers, valves, pipes, cables, cable wheels, cable rollers and railway lines. Many memories have been gathered from this period.

The Railway then went on to find use as an air-raid shelter during bombing raids in WWII (see chapter 12), offices and workshops for the Barrage Balloon squadron (see chapter 11), as well as a clandestine transmission site for the BBC (see chapter 13). This entailed building many walls, ledges and steps over the rails (see chapter 10). Many artefacts have been found left by those who sheltered there, such as bottles, cups, childrens toys and domestic artefacts. Many memories have been gathered from this period too. Winston Churchill praised the complex with the words

> Let them make a last stand in their Bristol Tunnel – it's the best place for a last stand that I know of.

It comprises two distinct layers of history, plus two distinct façades (for George Newnes and George White) at both the top and bottom when the Railway was running. The bottom façade was changed again by the BBC during the war. The bottom façade was then spoilt in 1958 when massive external concrete supports were added. The interior of the bottom station has hardly been touched since before the war, still containing its original wooden panelling, but the top station was turned into a first-aid post during the war. In the current state, Clifton Rocks Railway presents a question to society: should it be conserved in its wartime form or restored to a working funicular railway? and to which railway façade?

16.2 Heritage Value

The Railway was locally listed in 2016. The listing panel did not think it possible to consider the lower station without considering the integral tunnel and upper station of the Railway. As a whole, the complex, with its long history of alteration and reuse, was considered to score highly on architectural, historic, artistic and community value. The panel was unanimous in recommendation for Local List entry.

In 2005, work started on the tunnel by volunteers to make it accessible to the public, having been left dormant for 45 years. The newly-formed Clifton Rocks Railway group, consisting totally of volunteers, became a charitable trust in 2007. This has meant that the group has been able to do a huge amount of research, and to speak to many experts about what the future should be, and what was feasible.

16.3 How to determine future use

When will it be open? and when will it get running again? are the most frequent questions I get asked, usually by people who have never been inside the tunnel.

Clifton Rocks Railway is a good example for understanding the wider effects and issues that relate to conservation and restoration. This chapter aims to investigate the question of how to safeguard this historic structure: 'should we continue to favour conservation (the norm) over restoration?' Should the goal of the project be to restore all or part of the environment back to its primary historical period? If so, it may be necessary to make some cars, restore some pieces, preserve some pieces, and conserve others.

We need to understand what these terms mean.[1] We also need to understand the importance of each period, and the problems of trying to get back to the original period in difference sections of the tunnel. One has to look at each section of the tunnel and work out which history is the most important – wartime or railway.

Bristol City Council Conservation Department have also told us that they want to preserve the top station as the first-aid post rather than putting it back to looking like the Victorian top station – an idea that was muted at one stage.

16.3.1 What does Restoration mean?

Restoration[3] is defined as 'a bringing back to a former position or condition'. In restoring an art object, a piece of furnishing, or architecture, the most important requirement is the final appearance. The restorer determines the most desirable period of an object's life; and does whatever is necessary to return the object's appearance to that period. The boards on the Railway top station railings are a good example. The railings and the sign boards were from George White's era. The boards had been removed, but we had photographs to show what was there (plates 7.61 and 4.31). We have put boards with raised letters back, and we had cast bits of railings so we could replace missing bits. Similarly we had photographs of the ironwork entrance signs from both George White's time (flamboyant and placed for all to see on Sion Hill plate 15.51), and the simpler one from George Newnes' time (plate 15.47) which had been placed discretely in Princes Lane. We fabricated both signs and put a GN on the Newnes sign to identify which era it was from. George White had removed the Newnes sign, but we wanted to show the signs from the both periods to help to tell the story. They do complement each other, and provide a draw to the site for visitors approaching from all directions.

16.3.2 What does Preservation mean?

Preservation involves keeping an object from 'decay, damage or destruction and seeing to it that the object is not irredeemably altered or changed'.[2] The word preservation is most commonly used in relation to architecture and built environments and means to maintain the current condition. Preserving an object places additional layers of requirements on the decisions regarding materials and methodology. In preservation, the final appearance is no longer the prime factor, but rather retaining the maximum amount of building fabric. It focuses on maintaining the object in the same quality as it is now. Museums often preserve artefacts and archives. The courts consider that preservation to be interpreted as 'preserve from harm' – that is harm to its significance, not simply its fabric. Preservation is generally associated with the protection of buildings, objects, and landscapes.

16.3.3 What does Conservation mean?

Conservation is generally associated with the protection of natural resources and means to use a resource wisely. However, in many cases the terms conservation and preservation are used interchangeably. I have stuck to Historic England's definitions,[3] and used the term conservation in preference to preservation because their booklets tend to have conservation in the title rather than preservation, and we talk about conservation areas rather than preservation areas. In conservation, the maximum amount of the original material, in as unaltered a condition as possible, is preserved. Any repairs or additions must not remove, alter or permanently bond/cross-link to any original material. All repairs or additions must be reversible and removable without affecting the condition of the original material now, and in the future. Conservation is the process of maintaining and managing change to a heritage asset in a way that sustains, and where appropriate, enhances its significance. Conservation (or preservation, when given its proper meaning) of the most sensitive and important buildings or sites may come close to absolute physical preservation, but those instances will be very rare.

To separate the terms of conservation and restoration, BS 7913: 2013 (*Guide to the Conservation of Historic Buildings*) more clearly states that conservation constitutes:

> a conservative approach of minimal intervention and disturbance to the fabric of an historic building in which there is a presumption against restoration is fundamental to good conservation.

For example, conservation of a vintage car can mean the upholstery is left in original condition since it tells more of its story and retains its character. Paintwork too would be left if possible, but only if the car will not deteriorate further. Clearly tyres and sundries have to be replaced, else the car will not be roadworthy. Restoration can mean the car looks better than when it came out of the factory, but it is always the original cars that people want to see and admire.

The ethos of conservative repair is 'to preserve the genuine and original, the different layers and transformations of history, as well as the patina of age' (Pendlebury, 2009).[4] Ruskin, alongside William Morris, was one of the founding fathers of the Society for the Protection of Ancient Buildings (SPAB). Morris then went on to write SPAB's famous 1877 manifesto.[5] SPAB exists to educate and advise on the treatment of historic buildings, the mantra of repairs over restoration is intrinsic to its existence. SPAB believe that misguided work to historic buildings can prove extremely destructive, they promote the virtues of Conservative Repair which, when working with historic fabric, results in minimal loss of fabric; retaining a building's romance and authenticity.

In the Railway bottom station, we propped up the wooden ceiling (plate 4.35) when it looked like it could have fallen down, in preference to replacing the wood and losing the colours of 1934. We have however, replaced the wooden stairs for safety reasons (but kept the original handrail) and allow visitors to view the water chutes, and the bottom reservoir, which are in original condition. The inside ironwork has deliberately been left unpainted to see the original colours of green and gold (plate 7.31). The outside ironwork has been repainted though, to show that someone is now caring for the structure. The chamber where the turnstile, valve and interior railings were retrieved from is still full of rubble (and two more turnstiles and other accessories) to help tell the story of where the turnstile had come from, why it was there, and that there are still artefacts still in there. Removing the rubble loses that story. If we put a car in there, we would remove the rubble. The current story of that section is about what happened to the railway artefacts during the war, and you can see two turnstiles sitting tidily on their barriers on top of rubble, whereas three heavy valves appear to have rolled down the 45-degree slope.

Concerns on the extent of restoration were raised by Girouard[6] who when referring to the restoration of Georgian terraces in Spitalfields, commented that returning the buildings to their original condition, 'robs them of the very quality for which they are prized – oldness and that "pleasing decay".' With quick visual inspection of the Railway, one would be hard-pressed not to see: its state of apparent neglect (though it is weeded and swept frequently, and much time has been spent in reducing water leaks in the top station), its substantial wartime modifications, and the perils of water ingress, which have all left their mark. There is no doubt that Clifton Rocks Railway's current form holds charm in its state of neglect, as a structure that is seemingly forgotten in time. The Mary Celeste look is preferred to the Disneyland experience. In the same context Matt Somerville of Feilden Clegg Bradley Studios suggests 'if we restore the theatre to a pristine condition we will destroy the very quality of the space that makes it so intriguing and unique', when referring to work being completed at Alexandra Palace.[7] Given the importance of heritage within tourism, a completely modern installation would strip the Railway of its perceived historic charm. There are many funicular railways running, but only one that has two very different and separate stories, with artefacts from both periods.

We have established its cultural significance (confirmed by being on the Local List). When contemplating the choice between conservation and restoration, primary consideration must be given to the approaches' impact on a place's cultural significance, as it is these values that we as people engage with and respond to; and what ultimately provides the building with its sense of place (Historic England, 2008).[8] The adaptation of the Railway to a working funicular railway would involve significant removal of its wartime construction, which is prevalent in its blast walls (which may also help support the tunnel roof), and concrete terracing used for seating. To make it water powered again would mean that all the BBC section be removed, since one needs access to the bottom water reservoir. This would be restoration at its worst. It is important to retain elements of its wartime function, minimizing loss of significance and highlight its layered history.

16.4 Statement of significance

The Clifton Grand Spa Complex is a multi-faceted site with a diverse and complex history of national importance. Valued by the local population as not only a last attempt to revive the Spa culture of Clifton using the most extravagant, elaborate, high Victorian architecture, but also the vital role it played in the defence of Bristol during WWII.

A long-term project such as this, needs a statement of significance. It is important that an assessment be made of the degree to which resources should be expended on worthy projects, and that priorities be established for conservation. Cultural significance can be taken as meaning importance of significance for past, present or future generations. Age, historic, cultural and social worthiness can be assessed by usage of the site.

It needs to address:
- Why the structure is significant, historically, socially and technically?
- What is its worth to the community?

The statement can be used:
- To develop policies to retain and enhance that cultural significance.
- To design Conservation strategies to achieve the long term viability.
- For investigation of the historical context, fabric and research potential by understanding the importance of the tunnel life to the people who built, ran and used the railway before the war, and those who used the tunnel during the war.
- For documentary evidence analysis, to determine the cultural, social and historical significance of the tunnel and its various uses.
- As a guide to conservators, curators, funicular enthusiasts, historians, volunteers, museum trustees, and potential funding donors in understanding the sometimes conflicting issues associated with the repair and conservation of the site, or operation of the cars.
- To apply for funding.
- To apply for a possible listing.

16.4.1 Public significance

It is so important to preserve the history because so many older Bristol residents have remembered it with such fondness, both before the war when it was a special treat to ride on the railway, and during the war when they felt safe in there. Their memories from all periods can be found in this book (chapters 8, 11, 12, 13, 14).

The interest shown has been enormous, the groups of people who have been shown down the tunnel have been amazed at its size and all the historic features. To make this structure in the heart of Clifton more accessible would be a testament to its original construction. Its size and mixed history means that several solutions can be considered, and there can be a phased progression. Individual stories in the history of the Railway are of significance, but together they make the site unique.

Between May 2006 and 2011, the Clifton Rocks Railway Trust did a survey (see section 16.5), whereby visitors completed questionnaires relating to the Railway. This is the only research conducted relating to the public's value of the complex. This book needed to be written to tell the whole story starting from the development of the spa industry in Hotwells to explain what happened to the spa industry in Clifton. It has taken me twelve years to try to understand aspects of civil engineering, the different façades, the operation and maintenance of the Railway, the BBC usage, the Barrage Balloon operation, the role of the Merchant Venturers, the artefacts, the surrounding transport systems, to gather up the social history by the use of oral history. This then enabled me to join up all the stories so that the public should be able to look at something and understand why it is important, and what role it had. I have been lucky to have so many historic photographs to help.

16.4.2 Historical significance

Clifton Grand Spa Complex is historically significant on the local, regional, national, and international stage. The complex is also important because it was the last attempt to revive the spa culture in Bristol.

When it was first constructed, Clifton Rocks Railway was the widest tunnel in the world. Other inclined railways were constructed in England, such as the Cliff Railway in Lynton, Devon; however, the Clifton Rocks Railway is the only inclined railway with four tracks in that is enclosed within a tunnel. It was one of the first places in this country where air powered rock drills were used.

Locally, Clifton Rocks Railway acted not only as a refuge for those escaping the bombs of the Luftwaffe, but also as a site for repairing the barrage balloons that were protecting Bristol during air raids. There are fragments of barrage balloon material in the barrage balloon section.

Regionally, the Complex is significant for the vital role the Pump Room played in coordinating the delivery of aircraft by the Air Transport Auxiliaries during the WWII.

The tunnel also played a nationally vital role as the emergency headquarters of the BBC during the WWII in case of invasion, so is hugely important. From here, the BBC intended to be able to continue broadcasting across the world.

The combination of the Imperial Airways, refuge areas and BBC usage during the war is unique. The Complex has a unique history of national significance.

16.4.3 Assessment of Cultural Significance
Uniqueness or rarity
Railway 16 cliff railways are still operating. This is the only four-track ever built. Only two were built in a tunnel (the other is at Hastings and still operating).
Imperial Airways Few relics of balloon history remain. There is a hangar at Pawlett in Somerset, two hangars at Pucklechurch, and a balloon and tender in the Imperial War Musem, Duxford.
Refuge Area Some communal shelters still survive. Some of the stations in the London Underground were never reopened after the war. The shelters under Bristol Temple Meads are open on special occasions. The tunnel near Bridge Valley Road was also used as a shelter but this has been sealed off. Others are Nottingham; Ramsgate Railway tunnel; London Bridge Underground station; Coldharbour Mill Museum at Uffcombe Devon; Chislehurst Caves.
BBC transmitting Some examples still survive and are accessible to the public as important museums. It is remarkable that the original ozoneator is still there to supply fresh air in a confined space.

Degree to which the section is illustrative of a type of activity that merits preservation
Railway The bottom station is relatively untouched from 1934 (except for extra blast walls, turnstile and crossley engine removal and staircase moved). Water chutes and 2 clack valves still present. The exposed rail section at top shows rails, cables, fittings and cable wheels. Original turnstiles, railings, water valves and gas light in the top station. Water reservoirs visible. Original floors still exist. The ornate railings at top station are made by Gardiners (there are also Gardiners lamp posts and inspection covers in the area). There are no other railings with a supplier's plate in the area. The original Victorian Railway gas light with its original bracket is on display.
Imperial Airways Holes in the walls, and fans demonstrate the need to ventilate the fumes from the glues used to repair the balloons. Balloon fabric is lying around. Wooden struts are attached to the tunnel roof. No excavation has been carried out until a full photographic record has been taken.
Refuge Area Shelters intact with canvas, drainage troughs, linoleum insulation on ledges, many artefacts being found showing what people took with them when staying there all night.
BBC transmitting Ingenious use of the space. The rooms are in a very bad state and equipment removed but pictures and photographs exist to show what they looked like. Manuals exist to describe operation. There are only fuse boxes, wiring and vents visible at the moment. No excavation has been carried out until a photographic record has been taken. The recording studio has an insulated (fibreglass) floor and there are rubber treads on the steps to deaden the sound of footsteps. Equipment was removed in 1960s. Location of aerials have been identified.

Extent to which the area represents an important technical or operational aspect
Railway National technical significance, since Railway is in an inclined skewed tunnel (the largest of its kind in the world when constructed). There are four tracks (normally there are two), and the water is recycled.
Imperial Airways National operational significance. Barrage balloon repair and offices. The adjacent Grand Spa Hotel was the administrative headquarters of BOAC and Barrage Balloon Headquarters
Refuge Area Local operational significance. Wartime shelter for residents at night, BOAC during the day.
BBC transmitting National operational significance due to transmission and recording techniques. Bristol was the backup to London if communications were lost there. The control room (known as control room no. 2), was manned continuously until the end of the war and acted as a replacement for the one at Whiteladies Road Bristol, handling all the programme lines (about 80) in and out of the West Country. It would certainly have fed the main transmitting stations. The Control Room handled all the home and overseas programmes of the BBC during the war, and for the Third Programme after the war.

Extent to which the area represents the social impact
Railway High since it was a big treat to travel on the railway to get to Clifton, the Zoo and the Downs. Memories have been recorded.
Imperial Airways High since barrage balloons are remembered for their size, shape, and the role they played in defending Bristol. Memories have been recorded.
Refuge Area High since refugees remember it fondly for its safeness. The artefacts show the social history of the time. Beverages local since hard to transport during war. Memories have been recorded. Bristol was the fifth most bombed city in UK.
BBC transmitting Transmissions of international importance, keeping everyone informed. Memories of staff have been recorded.

Extent to which the section forms part of an established collection
Railway Bristol Records Office, Kew National Archive.
Imperial Airways None.
Refuge Area None.
BBC transmitting Connections with Washford Radio Museum, Caversham BBC archives.

Extent to which the section represents an important stage in technical development
Railway Building an inclined, wide span, skewed tunnel is a challenge especially when there are faults in rock, and it is hard oolithic limestone. One of first tunnels to be constructed using air drills.
BBC transmitting Group H transmitters to prevent enemy aircraft using the signals from BBC transmitters for direction finding. Philips-Miller recording equipment very rare. There is still an example of the equipment in the Philips museum in Holland.

Involvement in some significant event
Imperial Airways WWII
Refuge Area WWII
BBC transmitting WWII

Associations with an important person
Railway Baron Marks of Woolwich, Sir George Newnes, Sir George White
Imperial Airways Viscount Knollys, Gerard d'Erlanger
BBC transmitting Engineer in chief Gerald Daly. Visitors such as Queen Mary in 1942

Associations with an organisation
Railway Merchant Venturers', Bristol Tramways
Imperial Airways Imperial Airways, BOAC. Air Transport Auxiliary, Bristol Balloon squadrons
Refuge Area Air Raid Precautions (ARP) wardens
BBC transmitting BBC

Local, regional or national importance
Railway The unique Clifton Suspension Bridge and the Avon Gorge are known throughout the world. The location of the Railway adjacent to the Bridge and under the Gorge, its vastness and its multiple history makes it a very secret, intriguing site in a high profile conservation area. Many engineering features. Many older local people recall travelling on the Railway as a special treat.
Imperial Airways Regional importance in helping protect Bristol Docks and Filton from bombing raids.
Refuge Area Culturally significant. Local social history has been identified by a survey filled out by 300 people.
BBC transmitting The BBC (and ITV) has promoted its regional allegiances as a matter of policy. The wartime activities strengthened the region.

16.5 Survey

The number of completed surveys was 313 (157 were completed by people who had been down the tunnel, the remainder who had visited top station on open days) between May 2006 and November 2011. The survey was modified in May 2007 to find out more details about the visitor, how they had heard that the railway was open to view, and what sections they found most interesting. It was gratifying that many people put a lot of thought into their answers. It is also interesting to see views altering as time has gone on with more people appreciating the WWII history as more artefacts were found and memories researched. Historical interest was the most common reason for coming, and a place to bring friends and family.

The average age was 47. 84 female (average age 44) and 106 male (average age 51), and 14 children.

68% were from Bristol (12% Clifton, 3% Hotwells), 17% Somerset, 4% Gloucestershire and others from as far apart as Lancashire, London, New Zealand etc.

38% passed the bottom station at least once a week (an average of 59 times per year), 66% at least once a month. 76% travelled by car, 16% by foot. 11% never went past the bottom.

Only three used the ferry in the Floating Harbour more than 50 times a year. 86% never used it, the average being once per year. However, 53% thought that a canal should be built to link to the ferry (despite the problems and cost of building one on a river with the second highest tide range in the world, some distance away from the docks, and the difficulty of crossing the road).

60% said they would use the railway, 4% as a commuter (three Clifton, one Hotwells, one Ashton, one Kingswood, one Knowle, one Southville, two Weston). There is only a short terrace of houses at the bottom, so the potential for commuters is small. 19% would visit on weekdays, the rest only on weekends and holidays and on average of 21 times per year (boosted by 6 people who said they would use it more than 100 times per year).

15% would take their bicycle. 26% said they would visit/shop in Clifton more.

The zig-zag was only used on average four times a year, 74% never using it.

If the tunnel stayed as a museum then the average

number of visits would drop to just over two per year. This would increase if events and exhibitions changed regularly. The main reason for coming was to bring friends and visitors. 17 cited seeing progress as main reason. Only 3 said to actually use the railway would be their main reason to come, but 17 would come more often if it was running. 15 people wanted a restaurant or café. There was a huge variety of things that people were interested in, and this suggests themes for exhibitions. Virtually everyone saw it as a viable treasure and very appreciative of our efforts.

Most people wanted cinema/ theatrical use and saw its educational use.

After the first 112 questionnaires, people were asked to give a score (1-4) for which section they found most interesting. Out of 244 with scoring, 128 had visited the tunnel, 116 had not. Score 1 is highest, so the funicular history they found most interesting, balloon section least.

thought it should be left as a museum rather than a railway running again, whereas of the people who had not visited, fewer thought it should be left as a museum and more thought it should run. This shows that people who visited the tunnel then realised the problems of reinstating it, and that it was not feasible. They were also impressed with the sheer size and atmosphere of the structures. Some wanted a small section of railway, but having both a full railway and museum is not practical, having a small section would be very expensive.

Many took the opportunity to tell us to keep up the good work, good luck, brave effort, brilliant trip, give thanks to volunteers, admire those working to preserve it, amazing progress so far, well done, very interesting etc. They just loved the whole history.

score	Funicular count	Shelter count	BBC count	Barrage count
1	183(85,98)	73(42,31)	31(12,19)	21(4,17)
2	32(23,9)	65(30,35)	73(47,26)	9(2,7)
3	13(9,4)	54(34,20)	70(38,32)	23(14,9)
4	5(4,1)	7(2,5)	18(9,9)	129(81,48)
Avg	1.313(1.438,1.179)	1.975(1.963,2.011)	2.391(2.415,2.36)	3.429(3.703,3.086)
Visit change	21%	-2.4%	2.33%	19.99%
Total count	233(121,112)	199(108.183)	192(108,91)	182(106,86)
Total score	306(171,132)	393(212,183)	459(256,203)	624(374,250)

Table 16.1, survey summary.

score	percent for funicular	shelter	BBC	balloon
1	75%(66%,84%)	30%(33%,27%)	12%(9%,16%)	9%(3%,15%)
2	13%(18%,8%)	27%(23%,17%)	30%(37%,22%)	4%(2%,6%)
3	5%(7%,3%)	22%(27%,17%)	29%(30%,28%)	9%(11%,8%)
4	2%(3%,1%)	3%(2%,4%)	7%(7%,8%)	53%(63%,41%)

Table 16.2

Survey summary

This shows that people gave better scores to the shelter areas, having seen them, and worse to the railway, the balloon sections (this area is sealed off), and BBC. This shows the people appreciated the shelter area more having seen it; the balloon section less since there was little to see and one could not enter (we want to put an exhibition space in here); the BBC section was rotting away and one could not enter, so they were appreciated less; the funicular less since people appreciated the other areas more.

	visited tunnel 137	not visited 107
museum	72 (52.6%)	25 (23.4%)
railway	47 (34.3%)	67 (62.6%)
both	16 (11.7%)	15 (14.0%)

Of those who had visited the tunnel, many more

16.6 Problems of getting a railway running again

In March 1983, the Junior Chamber of Commerce did a feasibility study (see chapter 14), and concluded that the limited returns would be insufficient to justify examining the matter further, so it had little development potential. Just removing the wartime structures would cost £300,000 to £500,000 (which would have to done manually). It would be hard to get funding since it was not listed, and hard to comply with the British Standard Specification for lifts (as funicular railways are officially called).

In February 1996, they drew up a business plan, having consulted many experts, and concluded that the cost of conversion far outweighed possible returns, and described it as an eccentric, exciting white elephant. The engineering viability report was particularly interesting, describing the problems that would be encountered when removing the blast walls, and getting rid of the rubble.

There is also the difficulty of the traffic problem at lower station; delivery of construction materials, rolling stock etc in safety; delivery of passengers to and from the bottom station when there is no room for a bus stop. The roadway is too narrow to build a footbridge over it. In 1999, they were quoted £300,000-500,000 to remove the wartime walls, and repoint the tunnel and prepare for a two or four car railway. It would cost £600,000 for two modern cars with a winch drive at the top. This did not include stations, rails, transformers etc. This would serve 12 trips per hour in each direction.

The Clifton Rocks Railway went bankrupt on two occasions in its 40 years of operation, finally closing in 1934, and there has been no appetite to reopen it since. It has hardly got a good transport record. It is difficult to imagine that getting the railway running again could ever be successful or viable. The hurdles of finance, building and running a train through a dark tunnel with no views would, surely, be insurmountable. And, it would destroy the very important and historic features of its unique three separate roles in WWII, which cannot be replaced. Most people are interested in both stories and it gives a bigger story to tell. If one gets the railway running again and it goes bankrupt again, then one has destroyed history needlessly.

Glossary

Adamant
An unbreakable or extremely hard substance. The tiles are fired to a higher temperature than normal quarry tiles and used for the paving of railway stations.

Accumulator cylinder
A hydraulic accumulator is a storage reservoir in which a non-compressible hydraulic fluid is held under pressure that is applied by an external source. The external source can be a spring, raised weight, or a compressed gas. An accumulator enables a hydraulic system to cope with extremes of demand using a less powerful pump, to respond more quickly to a temporary demand, and to smooth out pulsations. It is a type of energy storage device.

Accumulator-ram
A hydraulic system by which a non-compressible fluid held in an accumulator forces a movement under pressure.

Acro Prop
A telescopic tubular steel prop.

Aerial tuning hut/Aerial transmitter hut (ATH)
A connection is made between the mast and the earth using an earth mat, which is a set of copper wires buried in the ground and arranged radially from the foot of the mast. There is a hut where the feeder from the transmitter is connected to both the earth mat and the mast. A feeder is simply a cable of carefully controlled dimensions that joins a transmitter to an antenna. Large components in the ATH also match the transmitter to the antenna, which is a process very roughly analogous to a battery charger that converts mains voltage to the correct voltage for charging a battery. (See image 13.23, which also shows the location of the earth mat.)

Since the transmitter applies high voltages to the steelwork of the mast itself, insulators are also placed in the stays which support the mast and it is obviously necessary to switch the transmitter off before ascending the mast.

Air-raid warning
Sent by telephone to the appropriate Warning District from a central warning authority.

Air-raid warning Red
Raiding aircraft are heading towards certain districts that may be attacked within from five to ten minutes. Sound the Air Raid warning.

Air-raid warning Yellow
Preliminary Caution. Raiding aircraft are approaching the UK (22 minutes' notice.)

Air-raid warning Purple
warning to extinguish all lighting.

Air-raid warning Green/white
Raiding aircraft have left districts warned or no longer appear to threaten those districts. Sound the all-clear.

Alternating current
AC is easier to transform between voltage levels, which makes high-voltage transmission more feasible. DC, on the other hand, is found in almost all electronics. The two do not mix very well, and it is best to transform AC to DC to plug in most electronics into a wall outlet. DC provides a constant voltage or current.

Anchor-block
A massive block of concrete built to withstand a thrust or pull. A mass of concrete or similar material appropriately placed around a pipe to prevent movement when the pipe is carrying water.

Anderson air-raid shelter
Half-buried air-raid shelter made from corrugated iron for up to six people. The shelters were 6 ft high, 4.5 ft wide, and 6.5 ft long. They were buried 4 ft deep in the soil and then covered with a minimum of 15 inches of soil above the roof.

Architraves
Mouldings framing the top of a door or window. Architectural term.

Bed-plate
A metal plate forming the base of a machine.

Billet
Living quarters in a civilian's house to which an employee/evacuee/soldier is temporarily assigned to sleep.

Boring
In machining, boring is the process of enlarging a hole that has already been drilled (or cast) by means of a single-point cutting tool (or of a boring head containing several such tools), such as in boring a gun barrel or an engine cylinder.

Brake cylinder
A cylinder that contains brake fluid that is compressed by a piston.

Brake horsepower (bhp)
The power measured at the crankshaft just outside the engine, before the losses of power caused by the gearbox and drive train.

Brakesman
Conductor/driver of the car responsible for a train's brakes or for other duties such as those of a guard.

Brake windlass
Hand-controlled wheel. An apparatus for moving heavy weights. The action of the falling weight in this case is to keep the water pressure in the brake cylinder constant. Typically, a windlass consists of a horizontal cylinder (barrel), which is rotated by the turn of a crank or belt. A winch is affixed to one or both ends, and a cable or rope is wound around the winch, pulling a weight attached to the opposite end.

Bristol City Council
Formerly known as Bristol Corporation, and is the local government authority governing the city of Bristol. Bristol City Council is a unitary authority and ceremonial county in England. Originally formed on 1 April 1974 as a non-metropolitan district as a result of the Local Government Act 1972.

Bristol Corporation
Established in the nineteenth century. The office of Lord Mayor was created in 1888. Bristol was the first provincial town to be given the status of city and county in 1373. The Municipal Corporations Act 1835 established Bristol Corporation, which consisted of 48 councillors and 18 aldermen. The term Corporation of Bristol or Bristol Corporation, encompassing the mayor and common council, had been in use since the eighteenth century at least. Bristol became a county borough in 1888 and the boundaries were extended into Gloucestershire and Somerset. There were no council elections during WWII. The Corporation had local city authorities, such as the City Engineer, City Estates, Town Clerk (the secretary and chief administrative officer)

Bristol Junior Chamber
Formed in 1948, it has been at the forefront of many major projects and enterprises including starting the Bristol Balloon Fiesta. It has a firmly established reputation as one of the leading organisations for young professionals within the city. It establishes new social and business networks for members and a desire to give back to the Bristol community through volunteering.

Bristol Supply System change in 1947
The UK changed from the three round pin electrical plug to the three rectangular pin plug with base insulators and individual fuses (the BS 1363 standard) in 1947. It replaced BS 546, which has round pins. It is worth noting that these are of a completely different size to the standards now used in Europe. It is important to realise that, pre-war, differently rated plugs were actually different sizes – a 5A plug was physically different to a 15A plug, and they could not plug into each other's sockets. Basically, the change was to do with electrical safety and the way houses are wired. Prior to WWII, there were central fuse boxes. An individual set of wires ran from the fuse box to each socket in the house – with an individual fuse at the centre for each socket. A 5A fuse for a small socket, a 15A fuse for a bigger one. During WWII, Lord Reith (Minister of Works and Planning) commissioned a set of experts to design a safer system. A design aim was that a single form factor should be able to cope with 5A and 13A plugs, so any plug could fit into any socket. Partly, this was done through the introduction of a ring main (which has a ring of cables, and thus handles higher currents better). The ring could be as high as 32A (the previous maximum was 15A), so it was necessary to introduce a new plug so old devices could not be plugged into the new sockets by mistake. It was also necessary for each plug to have an embedded fuse, rather than rely on the central fuses.

Buttress
A structure of stone or brick built against a wall to strengthen or support it.

Cabot Quilt
Blanket insulation consisting of eel-grass matting between fire-resistant paper. Used for sound deadening.

Car
Carriage for passengers on cliff railway.

Caryatides
Stone carving of a draped female figure, used as a pillar to support the entablature of a Greek-style building.

Clack valve
A hinged valve that permits fluids to flow in only one direction and clacks when the valve closes.

Cipollino marble
Originally quarried in Greece, the marble has a white-green base, with thick wavy green ribs, held onto the path by stratas of mica. The marble was principally used for column shafts, including large and mainly smooth ones, by the Romans and Greeks up to 5AD. A green-greyish banded marble from Saillon in Wallis, Switzerland, which showed great similarity to the ancient Roman Cipollino, was discovered around 1873, was soon successfully marketed as Grand Cipolin antique across the world.

Colonnade
A row of shops (not completed) for visitors to the Hot Wells in 1786 (otherwise known as 414-420 Hotwells Road). It can also be spelt Colonade. Grade II* listed.

Conservation area
An area of special architectural or historic interest, the character or appearance of which it is desirable to preserve or enhance. Conservation areas were first designated in England in 1967.

Copenhagen plan
Since 1925 a conference has been held every few years by a union of experts representing most European governments to formulate plans allocating wavelengths to countries. This was in order to avoid stations interfering with each other and to prevent chaotic broadcasting conditions. Originally called the International Broadcasting Union, it was renamed after the war as the European Broadcasting Union (E.B.U.). Although little change had occurred in the number of wavelengths available, the number of broadcasting stations had greatly increased since the Lucerne Plan of 1930. So a meeting of the E.B.U. was called and took place in Copenhagen in 1948 where a plan was internationally agreed to re-allocate wavelengths and maximum transmitted powers to the countries of Europe, to be put into operation in 1950. With the exception of Droitwich, long-wave Light Programme (which was allowed an increase in power), all the BBC medium-wave Home, Light and Third Programmes were given new wavelengths and also allowed increased powers. Work started at Droitwich to convert H.P.M.W. (high power medium wave) to 1500 metres (200 kilohertz) in order to increase the aerial power of the Light Programme from 150 kilowatts to 400 kilowatts. To facilitate this the Third Programme was transferred to the ex-Forces 20 kilowatt transmitter as a temporary measure until the Plan was implemented, when a new transmitter at Daventry would eventually take over that service. On 15 March 1950 the

Copenhagen Plan came into force. At Droitwich the Light Programme transmitter, now known as H.P.L.W. (high power long wave), came into service at 400 kilowatts and 5GB radiated the Home Service on the same power of 50 kilowatts, but wave changed from 296 metres to 276 metres. Work started immediately to convert from 1500 metres to 276 metres and on June 25th it became the main Midland Home Service transmitter on 150 kilowatts, with 5GB now acting as a reserve.

Dead-man's handle

Safety switch that only allows operation while depressed by the operator.

Doors Open Day

Doors Open Day is a once-a-year chance to look behind closed doors and discover a city's hidden treasures. One can explore fascinating buildings, join guided tours and enjoy a range of events and activities – all free for the day. A fantastic celebration of history, architecture and culture. Heritage Open Days were established in 1994 as England's contribution to European Heritage Days, in which 49 countries now participate. Bristol has organised them since 1994, the first event having 28 sites to visit. Clifton Rocks Railway has participated since 2005, in the year that volunteers first started work, and was inundated with visitors.

Drill carriage

A platform or frame on which several rock drills are mounted and which moves along a track, for heavy drilling in large tunnels.

Entablature

The upper part of a classical building supported by columns or a colonnade, comprising the architrave, frieze, and cornice.

Fishplate

A good quality metal bar that is bolted to the ends of two rails to join them together in a track. The top and bottom edges are tapered inwards so the device wedges itself between the top and bottom of the rail when it is bolted into place. It has a simple structure and can largely reduce the impact of wheels to the connecting sections of the rail tracks and increase the continuity and steadiness of the train when passing through the connecting sections.

Fishplate bolts

Must be good quality and used to bolt fishplates to the sides of two rails or beams. Track bolts come with heavy square nuts and oval neck and diamond neck to prevent from turning when torqued.

Fly

A horse-drawn public coach or delivery wagon, especially one let out for hire. In Britain, the term also referred to a light covered vehicle, such as a single-horse pleasure carriage.

Funicular railway

A steep cable railway operating in such a way that the ascending and descending cars are counterbalanced. A cable attached to a pair of tram-like vehicles on rails moves them up and down a steep slope. Known variously as a lift railway or cliff railway or a cable lift or cable railway.

Gargoyle

A grotesque carved human or animal face or figure projecting from the gutter of a building, typically acting as a spout to carry water clear of a wall. Architectural term.

Grand Spa Hotel

The original name for the Avon Gorge Hotel, which was opened in 1898. Consisted of nos. 13 and 14 Princes Buildings. The name was changed in the early 1980s.

Green Goddess

Colloquial name for the Bedford RLHZ Self Propelled Pump, a fire engine used originally by the Auxiliary Fire Service (AFS), and latterly held in reserve by the Home Office until 2004.

Gripper brake

Hydraulic brakes which clamp each side of the crown of the rail making it impossible to move.

Gutta-percha

Yellowish or brownish leathery material derived from the latex of certain trees in Malaysia. On heating, gutta-percha becomes plastic and is very resistant to water. It has been widely used as insulation for underwater electrical equipment and cables.

Guttae

Cones protruding from triglyphs. Architectural term.

Horse gin

Winding engine driven by horses walking round in a circle to lift the water.

Horsepower (hp)

The term was invented by the engineer James Watt. Watt lived from 1736 to 1819 and is most famous for his work on improving the performance of steam engines. He found that, on average, a mine pony could do 22,000 foot-pounds of work in a minute lifting coal. The term was used to compare the output of steam engines with the power of draft horses. Two common definitions used today are the mechanical horsepower (or imperial horsepower), which is approximately 746 watts, and the metric horsepower, which is approximately 735.5 watts.

Hotwells

Hotwells takes its name from the hot springs which bubble up through the rocks of the Avon Gorge underneath the Clifton Suspension Bridge. The springs were documented in 1480 by William Worcester, the 15th century chronicler and antiquary. Just to confuse everyone, the road going through Hotwells is Hotwell Road.

Hydro (hydropathic spa)

Formerly called hydropathy and also called water cure, is a part of medicine and alternative medicine, that involves the use of water, both externally and internally, for pain relief and treatment. The Grand Spa Hotel became a hydropathic spa once the Turkish Baths were constructed in 1898.

Jute

A long, soft, shiny vegetable fibre that can be spun into coarse, strong threads.

Leaf spandrel

The space between a curved figure and a rectangular boundary – such as the space between the curve of an arch and a bounding moulding, or the wallspace bounded by adjacent arches. Architectural term.

Local list of valued buildings

Bristol City Council list of buildings, structures and sites without listed status but worth preserving because of their quality, style or historical importance.

Marl

Mudstone, a fine-grained sedimentary rock consisting of clay.

Medium wave (MW)

MW is the part of the medium frequency (MF) radio band used mainly for AM radio broadcasting. For Europe the MW band ranges from 526.5 kHz to 1606.5 kHz, using channels spaced every 9 kHz.

Merchant Venturers

The Society of Merchant Venturers is a charitable organisation in the English city of Bristol. For centuries, it was almost synonymous with the government of Bristol, especially its port. In recent times, the society's activities have centred on charitable agendas.

Otto engine

The Otto engine was a large stationary single-cylinder internal combustion four-stroke engine designed by Nikolaus Otto. It was a low-rpm (revolutions per minute) machine, and only fired every other stroke due to the Otto cycle, also designed by Otto. Otto found a way to layer the fuel mixture into the cylinder to cause the fuel to burn in a progressive, as opposed to explosive fashion. This resulted in controlled combustion and a longer push of the piston in the cylinder rather than the explosion which destroyed all the engines attempted previously. The engines were initially used for stationary installations, as Otto had no interest in transportation. Other makers such as Daimler perfected the Otto engine for transportation use. Crossley engines were used in the bottom station. The best-known builder of gas engines in the UK was Crossley of Manchester, who in 1869 acquired the UK and world (except German) rights to the patents of Otto and Langden for the new gas fuelled atmospheric engine. In 1876 Crossley acquired the rights to the more efficient Otto four stroke cycle engine and referred to their engines as Crossley Otto gas engines.

Ozoneator

Creates ozone and diffuses it into the air or water. The chemical reaction, which is produced when ozone does this particular phenomena expels pesticides, microorganisms, smells, and inorganic and natural mixes, from the encompassing environment. Ozone acts as a powerful antiseptic in contact with diseased mucous surfaces, consequently its beneficial action is quickly apparent in the treatment of bronchial and laryngeal affections, catarrh, hay fever and all diseases of the respiratory organs.

Patera

Oval features on frieze. Architectural term.

Pilasters

Ornamental supporting columns. Architectural term.

Piston-ram

In this context used for the braking system.

Plunger valve

A piston-ram is a solid disk or cylinder that fits inside a hollow cylinder, and moves under pressure (as in an engine) or displaces fluid (as in a pump). The pressure pumps are always at work to keep the hydraulic brakes gripping the rails unless the attendant is holding the wheel.

Pounds, shillings and pence

The currency before decimalization of British Pound Sterling in 1971. 20 shillings (s) in £1, 12 old pence(d) in 1 shilling, 240 old pence in £1. There are several websites to calculate purchasing power on a particular date, and often give different values.

£1 in 1751 is approximately equivalent to £200 in 2017

£1 in 1888 is approximately equivalent to £120 in 2017

£1 in 1893 is approximately equivalent to £90 in 2017; 1s to £4.50; 1d to 37.5p

£1 in 1927 is approximately equivalent to £44 in 2017; 2d to 37p

Comparing the value of a £30,000 Project (cost of constructing CRR) in 1893 there are three choices . In 2017 the relative:

historic opportunity cost of the project is £3,183,000.00

labour cost of the project is £12,990,000.00

economic cost of the project is £41,270,000.00

Portico

A row of columns supporting a roof at the entrance of a building. Architectural term.

Pump room

A room at a spa or hydro where medicinal water is dispensed.

Quoines

Masonry blocks at the corner of a wall. Architectural term.

RCA Corporation

Founded as the Radio Corporation of America, it was an American electronics company in existence from 1919 to 1986. RCA's greatest communications receiver creation was the AR-88, a receiver that achieved its renown by providing top performance and high reliability in service as a surveillance and intercept receiver during WWII. Most of the early AR-88 production was sent to Great Britain or Russia. An RCA Group H transmitter was used in Clifton Rocks Railway and other transmitting stations.

Rope-wheel

Cable wheel. The two cars are connected by wire rope or cable. The grooved wheel enables one car to go down as the other goes up.

Russian Bath

A room where a large amount of hot steam is created with the help of water and hot air. A classic Russian bath is heated with firewood. It is more like a sauna. Inside there are wide wooden benches along the walls upon which one sat or lay down. The higher up the bench the hotter the air. Once someone had warmed up well enough, they left the steam room and dipped into a pool of cold water. One could also pour water over oneself from a tub. A Russian bath has the same levels of humidity as the air we breathe every day: about 60%. The temperatures usually do not exceed 80C/180F.

Safety-gripper
Brakes that grip on the vertical part of the railway track.

Safety wedges
Tapered piece of metal/wood. Ropes are anchored independently to safety wedges, carried on the underframe so that the breakage or stretching of either rope would cause the wedges to close on the rails and stop the car.

Shot hole
A drilled hole in which an explosive charge is placed before detonation.

Simultaneous broadcasting (SB)
Broadcast by a number of transmitters within one broadcasting system of the same programme at the same time. Needs manual switchboard in control room. To broadcast a programme simultaneously from two or more locations or on two or more distribution channels. Pass from the studio amplifier to another mixing arrangement whereby several studios may be mixed or faded in and out at will. From microphone to control room to distribution by Post Office lines. The programme is then transmitted using high frequency wireless waves to carry low frequency (AF) current.

Skewed tunnel construction
The Clifton Rocks Railway tunnel was created 60 degrees to the Hotwell Road at the bottom, rather than 90 degrees. This makes it more difficult to construct since it is not at right angles.

Spa
A mineral spring considered to have health-giving properties.

Spring
A place where water or oil, wells up from an underground source.

Station inspector
Based at depots, liaises with railway staff.

Station manager
Takes responsibility for two or more stations on a local branch line, management of the other employees.

Station master
Takes responsibility for one station on a local branch line, management of the other employees.

Tie-plate
A plate set between the base of a rail and a crosstie to distribute the rail load over a greater area of the tie and thus reduce wear and damage to it.

Tonite
Patented on February 1877 and ceased use in 1906. It is an explosive that needs to be fired by a detonator. Its name was taken from the Latin verb *tonat* = it thunders.

Tram/tramway
Vehicle which runs on tracks along public urban streets. Can be horse drawn, powered by steam, cable hauled, or by electricity using overhead pantographs.

Trigliph
Consists of three square projections to represent the end of beams. Architectural term.

Turkish bath
A method of cleansing and relaxation that became popular during the Victorian era. The process involved in taking a Turkish bath is similar to that of a sauna, but is more closely related to ancient Greek and ancient Roman bathing practices. Consists of a set of rooms at different temperatures through which the patient passes. The treatment is usually accompanied by massage and a spray. It is a hot air bath. Neither hot steam nor hot water is applied. The Turkish bath temperature varies between 35-45C/95-113F according to the season.

Transmitter
In telecommunications, a radio transmitter is an electronic device which, with the aid of an antenna, produces radio waves. The transmitter itself generates a radio frequency alternating current, which is applied to the antenna. When excited by this alternating current, the antenna radiates radio waves. The term transmitter is usually limited to equipment that generates radio waves for communication purposes; or radiolocation, such as radar and navigational transmitters. Generators of radio waves for heating or industrial purposes, are not usually called transmitters even though they often have similar circuits. The term is popularly used more specifically to refer to a broadcast transmitter, a transmitter used in broadcasting, as in FM radio transmitter or television transmitter. This usage usually includes both the transmitter proper, the antenna, and often the building it is housed in. The term transmitter is often abbreviated XMTR or TX in technical documents. The purpose of most transmitters is radio communication of information over a distance.

Vapour bath
Consists of a cabinet in which the patient stayed for ten to thirty minutes at a temperature of about 71C/160F. The patient's head protrudes and a cold compress was placed around the head.

Zig-zag
A steep and winding footpath down the side of the Avon Gorge from the bottom of Sion Hill to river level at Hotwells offers a link between Clifton and the Spa. Originally, it was little more than a track and was approached at the back of the Colonnade by a long, steep, treacherous flight of steps and was impracticable in wet weather. This route was improved with the construction of the Zig-Zag Walk in 1828/9 – shown on Ashmead's map of 1828) by the Merchants' Society at about

the time that the Suspension Bridge project was first mooted. It was improved in 1849.

Abbreviations

AA gun	Anti-aircraft gun
AC	Alternating current
Ally-Pally	Alexandra Palace (the major production centre for BBC television 1936-1950s except during the war)
ARP	Air-raid Precaution
ATA	British Air Transport Auxiliary
ATC	Army Training Corps
AFS	Auxiliary Fire Service
ATH	Aerial Transmitter Hut
BBC	British Broadcasting Company
BEP	*Bristol Evening Post*
BDP	*The Birmingham Daily Post*
bhp	brake horsepower
BM	*Bristol Mercury*
BMAG	Bristol Museums and Art Galleries
BOAC	British Overseas Airways Corporation (from 1 April 1940 – previously Imperial Airways)
BTCC	Bristol Tramways and Carriage Company
BTM	Bristol Times and Mirror
CDC	Civil Defence Corps
CRR	Clifton Rocks Railway
DC	Direct current
diam	diameter
EBU	European Broadcasting Union
EiC	Engineer-in-Charge (BBC)
EP	*Evening Post*
EWO	Essential Works Order
F/L	Flight Lieutenant
GCM	George Croydon Marks
GN	George Newnes
HE	high explosive bomb
hp	horsepower
IB	incendiary bomb
lb	pound (weight)
LDV.	Local Defence Volunteer (Changed August 1940 to Home Guard)
LF	Low-frequency equipment
MIME	Member, Institution of Mechanical Engineers
MSA	Member of the Society of Architects
MVR	Merchant Venturers Report
PM	Philip Munro
PBP	*Pro Bono Publico* (for the public good, without payment)
PBX	Private Branch Exchange (private telephone network used within a company)
POW	Prisoner-of-war
RAFVR	Royal Air Force Voluntary Reserve
RCA	RCA Corporation. See glossary.
SB	simultaneous broadcasting. See glossary.
S/L	Squadron Leader
SPAB	Society for the Protection of Ancient Buildings
TA	Technical Assistant (BBC)
TAF	Technical Assistant Female
TAM	Technical Assistant Male
T77	low-power Air Ministry transmitter
TD7	type of gramophone
UXB	Unexploded bomb
VE	Victory in Europe (end of WWII)
WDP	*Western Daily Press*
WRVS	Women's Royal Voluntary Service, founded 1938
WWI	World War One
WWII	World War Two
YIT	Youth in Training
£sd	Pounds, shillings and pence, currency before decimalisation of British Pound Sterling in 1971

Endnotes

chapter 1
1 Clifton and Hotwells Character Appraisal Bristol City Council 2010. www.bristol.gov.uk
2 *Annals of Bristol in the seventeenth century*, Latimer
3 *The Merchant Venturers of Bristol*, p.191-2, McGrath, P. (1975)
4 'Concerning Clifton Green', Armytage, 1922
5 'St Vincent's Spring, Avon Gorge. An interpretation of what survives from use of a thermal spring'. Wood and Shapland, *Bristol Industrial Archaeological Society Journal* 48, pp 4-14
6 www.locallearning.org.uk/Westbury/Downs%20History%20Final%20Report%20Feb%202006.pdf
7 *Topographical Dictionary of England in 4 volumes*, Samuel Lewis, 1840
8 *The Bristol Times*, 25 October 1851
9 *Bristol Mercury*, 7 August 1858, *Western Daily Press*, 8 July 1859
10 *Bristol Mercury*, 11 February 1884
11 *Chilcott's Descriptive History of Bristol*, 1840
12 *Annals of Bristol in the Nineteenth Century*, Latimer
13 *Bristol Mirror* 11 June 1814
14 *Bristol Times and Mirror* 22 January 1853
15 *Chilcott's New Guide to Bristol, Clifton and the Hotwells*, etc., 1853
16 'Springs and Wells of Gloucestershire', Richardson, 1925
17 *Western Daily Press* 3 January 1861
18 *Western Daily Press* 6 February 1885
19 *Western Daily Press* and *Bristol Mercury* 9 April 1885
20 'The People's Carriage, 1874-1974', The History of Bristol Tramways Co. Ltd.
21 Calendar of Records of the Merchant Venturers Bristol Records Office SMV/10/6/1/6

chapter 2
1 *Life of Sir George Newnes*, Hulda Friederichs, 1911
2 *Cambridge Independent Press* 19 August 1892
3 *Baron Marks of Woolwich*, Michael Lane 1986
4 'Recent Developments in Gas Engines', Dugald Clerk, *Minutes of the Proceedings of the Institute of Civil Engineers, vol. 124*, 1896
5 *Cliff Railways of the British Isles*, Keith Turner 2002
6 *Bristol in 1889-99, Contemporary Biographies vols.1 & 2*, Pike
7 'Album Guide to Aberystwyth containing a Description of the Cliff Railway'
8 'Progress Commerce 1893: The Ports of the Bristol Channel; Wales and the West'
9 *Sir George White of Bristol 1854-1916* Charles Harvey and Jon Press, 1989
10 'Tramlines to the Stars George White of Bristol', George White 1995
11 'The People's Carriage 1874-1974', The History of Bristol Tramways Co. Ltd

chapter 3
1 'Explaining tunnel construction by joint mapping', S.J. Knight, *Transport 160* Issue TR2
2 'A century of tunnelling and where we go now', John King. The 2000 Harding Lecture, British Tunnelling Society
3 *Modern Mining Practice vol. 3*, G.M. Bailes, Bennett and Company 1906?
4 'Concrete Filler-joist floors and the development of the Lancashire Cotton Spinning Mills', Roger Holden, *Industrial Archaeology Review vol. 34*, 2012
5 'Bristol Railway Stations 1840-2005', Mike Oatley; 'Bristol Railway Panorama', Colin Maggs
6 'Marks on Cliff Railways' – *Selected papers, Institute of Civil Engineering vol. 66* pp.318-326, 1894
7 *The Severn Tunnel*, Thomas A. Walker (1888), facing page 86
8 *Practical Masonry and Bricklaying*, Nicholson, 1847
9 'Practical Tunnelling: Explaining in detail the setting out of the Works as exemplified by particulars of Blechingley and Saltwood Tunnels', Simms, 1860
10 'Report on the Stability of the cliffs of the Avon Gorge in the Vicinity of the Clifton Rocks Railway for the Bristol City Engineer, Surveyor and Planning Officer', Skempton and Henkel, Imperial College, London 1959. Skempton was one of the founding fathers of soil mechanics, starting his research in 1937 at the Building Research Station. He established the soil mechanics course at Imperial College in 1950. He was Britain's leading authority on soil mechanics for half a century. He was knighted in 2000 for services to engineering.
11 *Clifton Illustrated* 'Official Description of the Clifton Rocks Railway', 1893

chapter 4
1. Clifton Illustrated *The Official Description of the Clifton Rocks Railway*, 1894.
2 *Operation and Pay in Great Britain 1906-79*, Guy Routh, pp.99-107
3 Bristol Junior Chamber: 'A feasibility study on Clifton Rocks Railway', chairman M Williams, March 1983
4 www.cliftonrocksrailway.org.uk/history_06.htm
5 'Marks on Cliff Railways', *Selected papers Institute of Civil Engineering volume 66*, pp.318-326, 1894
6 *The Engineers Handbook*, 1929
7 'Recent Developments in Gas Engines', Dugald Clerk. *Minutes of the Proceedings of the Institute of Civil Engineers, vol.124*, 1896
9 'Engineering 1893-1901', p.625
10 Some lists of shareholders between 1894 and 1907 are held in Kew Archives BT 31/5845/41016
11 Financial records between July 1908 and 1913 are held in Kew Archives BT 31/5845/41016
12 *Mostly Clifton*, Cedric Barker, 1997, Redcliffe Press Ltd.
13 Bristol Records Office reference 39735/8 (Bristol Tramways and Carriage Company: Correspondence file regarding Clifton Rocks Railway)

chapter 5
1 'The People's Carriage. Bristol 1874-1974' Bristol Omnibus Company Limited
2 1897 Bristol Tramways and Carriage Company Ltd timetable for Joint Station, Tramways Centre and Hotwells Line. Horse Trams
3 'Bristol's Tramways', Peter Davey, Tramway Classics, 1995, Middleton Press
4 'Sir George White of Bristol 1854-1916', Charles Harvey and Jon Press, Bristol Branch of the Historical Association p25, *Western Daily Press* 9, 25, 29 April 1914

chapter 6
1 'Work of an engineer in 1900: Alfred Woolley', Maggie Shapland, *Bristol Industrial Archaeological Society Journal Vol.44*, 2011, pp 4-24
2 Statement of receipts and expenses Bristol Records Office reference 39735/8 (Bristol Tramways and Carriage Company: Correspondence file regarding Clifton Rocks Railway)

chapter 7
1 'The Gas Petrol, and Oil Engine in Practice', Dugald Clerk 1919, *Minutes of Proceedings of Institute of Civil Engineers vol, 124* 1896; 'Recent Developments in Gas-engines', Dugald Clerk
2 *Chambers' Twentieth Century Dictionary*, 1908
3 'Played in Manchester: The Architectural Heritage of a City at Play' (*Played in Britain*), 2004, Simon Inglis
4 www.acmewhistles.co.uk/ and correspondence with the company
5 *Practical Gas-Fitting, including gas manufacture. With numerous engravings and diagrams*, ed. P. N. Hasluck, 1900, p109

chapter 8
1 'Work of an engineer in 1900: Alfred Woolley', Maggie Shapland. *Bristol Industrial Archaeological Society Journal vol.44*, 2011, pp. 4-24

chapter 9
1 *Western Daily Press* 8 February 1893
2 *Western Daily Press* and *Bristol Mercury* 2 August 1894. Two and a half columns including speeches
3 Royal Album of Bristol and Clifton
4 www.victorianturkishbath.org/6directory/AtoZEstab/England/BathGeorge/BathGeorgeSF.htm
5 *Derby Mercury* 22 January 1896
6 'The Great Harmonia: The Physician Andrew J. Davis' *The Physician vol.1* 1851
7 Victorian Turkish Baths: their origin, development, and gradual decline, Malcolm Shifrin
8 Sales catalogue
9 A very detailed account of the facilities is to be found in the *Bristol Mercury* of 1 April 1898. Bristol Records Office Building plan/Volume 28/70b, Building plan/Volume 28/70c, Building plan/Volume 28/70e Princes Buildings – Pump Room and Cinema – Munro & Son

chapter 10
1 *Bristol Blitz, The Untold Story* Helen Reid, 1988
2 'Bristol's Civil Defence during World War Two', John Penny, 1998
3 'Bristol Bombed. Pictures of Streets and Buildings damaged in the raids of 1940-41'. F.G. Warne Bristol Development Board 1943
4 Bristol Record Society. vol 65. pp.394-407 (Clifton College)
5 *Bristol Blitz, The Untold Story*
6 'Buildings in Bristol of Architectural or Historic Interest damaged or destroyed by enemy action 1940-42', Transactions for the year 1944
7 Gerald Daly's Reminisces in a letter to Patrick Handscome 1974. File reference R35/207 BBC Written Archives Centre, Caversham Park, Reading RG4 8TZ. Bristol Records Office reference 39735/8
8 'CLIFTON ROCKS TUNNEL The story of a BBC Wartime Fortress', Frank Gillard. In 1936 he became a part-time broadcaster and in 1941 joined the BBC full-time becoming a war correspondent. 1945 to 1963 Gillard worked in the BBC's western region, becoming its director in 1955

9 'Miniature BBC in Avon Rocks' 20 March 1946 *Bristol Evening World*
10 Work 28/162 Kew Archives WORK 28 - Office of Works and successors: Civil Defence and Prison Building: Registered Files and Papers CENTRAL REGISTER OF UNDERGROUND ACCOMMODATION WORK 28/162 - Underground accommodation: Cliff Rock Railway Tunnel, Clifton Rocks Railway, Bristol

chapter 11
1 http://www.bbrclub.org Barrage Balloon Reunion Club
2 Balloons and bombs dominated the skies *Bristol Evening Post, Bristol Times* 6 January 2004 p27
3 http://www.fishponds.org.uk Fishponds Local History Society
4 'Up, Up And Away! An Account Of Ballooning In And Around Bristol and Bath 1784 To 1999', John Penny, 1999, Bristol Branch of the Historical Association
5 *Balloons at War, Gasbags, Flying Bombs and Cold War Secrets*, John Christopher, 2004, Tempus
6 'Barrage Balloons and Imperial Airways in Clifton and Bristol', Maggie Shapland with David Wintle, John Penny and John Christopher, CRR booklet
7 https://en.wikipedia.org/wiki/Air_Transport_Auxiliary
8 Author's garage at 97 Princess Victoria Street, Clifton, which was owned by the Hotel between 1905 and 1978
9 Gerald Daly's Reminisces written in a letter to Patrick Handsome 1974.
10 Memories of Doreen Whitfield
11 Jean Gillingham memories

chapter 12
1 ARP Training Manual No 1. Basic Training in Air Raid Precautions, HMSO, 1940
2 *Bristol at War*, C.M. MacInnes, 1962, Museum Press
3 'Wartime Artefacts found in Clifton Rocks Railway', Maggie Shapland: *Bristol Industrial Archaeological Society journal 40* 2007 p.4-18
4 http://www.gracesguide.co.uk/File:Im1951BGIDir-p055.jpg
5 'The Brooke and Prudencio Family of Companies', *Antique Bottle Collector UK*, 25, 2005, pp.32-33
6 *Pints West* Spring 2007
7 *Pints West* Summer 2010
8 'Smoking in British Popular Culture 1800-2000': Perfect Pleasures', Matthew Hilton
9 *Bristol Blitz*, Helen Reid
10 'Bristol Bombed. Pictures of Streets and buildings damaged in the Raids of 1940-41, F.G. Warne with Bristol Development Board

chapter 13
1 'A History of the BBC in the Clifton Rocks Tunnel' C Neil Wilson, *BBC History* 8 September 2006. Various papers and documents from the archives of the Washford Radio Museum (being moved to 5 Anchor Street, Watchet, TA23 0AZ)
2 Gerald H Daly, letter to Patrick Handcombe, 23 April 1974
3 Frank Gillard, 'Clifton Rocks Tunnel', *BBC Yearbook,* 1946
4 *The History of Broadcasting in the United Kingdom*, Asa Briggs, Oxford University Press. Max Barnes
5 'Miniature BBC in Avon Rocks',

Bristol Evening World, 20 March 1946
6 'Sound Recording and Reproduction', J.W. Godfrey and S.W. Amos. (BBC Engineering Training Manual). J.W.Godfrey, 'The History of BBC Sound Recording', *BSRA Journal, Vol.6, no. 1*, 1959
7 B*BC Engineering 1922-1972*, BBC Publications, Edward Pawley, 1972. *BBC Sound Broadcasting, Its Engineering Development*, BBC, 1962
8 'The War-time Activities of the Engineering Division of the BBC', H. Bishop. *Journal IEE, vol. 94*. Part IIIA, no. 11, 1947, p169
9 Richard Hope-Hawkins http://www.cliftonrocksrailway.org.uk/history_06.htm
10 BBC *Bristol 50 years*, 1984, BBC Bristol
11 'The Early Years Of The Telephone Service In Bristol 1879-1931'; M.J. Hall Ellis states that the West exchange was in Clifton, resurrected as an automatic exchange in 1931, although there had been an exchange there previously. The West and Central exchanges were geographically separate. It is hard to pinpoint where the West exchange was, because it was listed in Exchange Directories as a sub-exchange to Central, so the location is not listed
12 Training book. Maida Vale. Lecture 14 Transmitting section
13 http://frequencyfinder.org.uk/trans_hist1.html

chapter 14
1 'The Stability of the cliffs of the Avon Gorge in the Vicinity of the Clifton Rocks Railway', Skempton and Henkel, Imperial College, 15 April 1959
2 62/00775/P_U
3 *The Fight for Bristol. Planning and the growth of public protest*. Gordon Priest and Pamela Cobb, The Civic Society and Redcliffe Press 1980
4 *Unbuilt Bristol: The city that might have been 1750-2050*, Eugene Byrne, 2013 Redcliffe Press Ltd
5 'An investigation of the stability of a proposed hotel in the Avon Gorge'. Dr. W.J. Larnach and R. Bradshaw. *Quarterly Journal of Engineering Geology Vol 7*, no 1 1974
6 Bristol Junior Chamber: A feasibility study on Clifton Rocks Railway. Chairman M Williams March 1983
7 Yakometties: Dance Music (Video) https://youtu.be/Rd4kpT0HRTM

chapter 15
1 http://www.cliftonrocksrailway.org.uk and http://www.b-i-a-s.org.uk/rocks_railway_refurbishment.html
2 'It's time to rock 'n' rail'. *Bristol Evening Post* 19 April 2005
3 Clifton Rocks Railway 'A Ha'penny Down, A Penny up' narrated by Peter Davey
4 'Clifton Rocks Railway Refurbishment', Maggie Shapland, *Bristol Industrial Archaeological Society Journal 38* 2005
5 WARNING . . . THIS MAY BE A ROCKY RIDE 24 October 2006 *Bristol Evening Post*
6 'Cliff Lifts of Great Britain', DVD, Martin Easdown and W. Watters. 86 minutes
7 RAIL TRACK WORK HITS THE ROCKS 8 November 2007 *Bristol Evening Post*
8 'Wartime Artefacts found in Clifton Rocks Railway', Maggie Shapland, *Bristol Industrial Archaeological Society Journal 40* 2007

9 Groups Safe to Make Tracks 23 May 2008 *Bristol Evening Post*
10 *Britain's Weirdest Railways*, Robin Williams Mortons Media Group
11 http://www.bristol.ac.uk/news/2013/9462.html
12 *The Avon Gorge. Bristol's spectacular route to the sea*. Gordon Young 2014
13 *Inside Bristol. 20 years of Doors Open Day,* Penny Mellor, 2013 Redcliffe Press
14 'Book Storage Problem Solved', *Bristol Evening Post* 1 April 2014
15 *Auntie's War: The BBC During the Second World War*, Edward Stourton 2017, Random House
16 http://www.bbc.com/future/story/20170718-the-underground-railway-that-became-a-secret-wartime-base

chapter 16
1 https://historicengland.org.uk/advice/hpg/generalintro/heritage-conservation-defined/
2 *Webster's New Collegiate Dictionary* (1975)
3 https://historicengland.org.uk/advice/hpg/generalintro/heritage-conservation-defined/
4 *Conservation in the Age of Consensus*, John Pendlebury, 2008, Routledge
5 SPAB's Purpose Available at www.spab.org.uk.
6 'The Saving of Spitalfields', Mark et al. Girouard, 1989. Spitalfields Historic Buildings Trust p.164
7 FCB Studios, *Suspended animation*, Issue 16, 2015. Available at: http://fcbstudios.com/explore/view/16
8 'Conservation Principles Polices & Guidance', English Heritage, 2008

Contributors of memories

in the order they appear

Chapter 8 Railway Memories
from page 116
Eileen Amesbury
Mr Bartlett
Ann Beaver
John Bidgood
David Bickerton
Bill Bonner
Colin Boyce
Brian Bradley
Betty Britton
William Broadhurst
John Buckley
RG Burgoyne
Keith Cornish
Muriel Cox
Don Cullen
Trevor Dean
Mona Duguid
Phyllis Farmer
Molly Foss
Ralph Fryer
Judith Gauchi
Jean Gillingham
Norman Goldsworthy
Christine Hamm
Mary Hingston
Harold George
Ray George
Shyll Slade
Ruby Hengrove
R Howell
Reginald Incledon
Lilian Jenkins
Lorna Leach
Cecilia McCarthy
Heather Mcomie
Doreen Mallett
Jon Miell
Pam Millard
Iris Mitchell
Ann Osbourne
Brian Owens
David Pearce
Vera Price
Mr Roach
Joan Shapland
Joyce Smith
Stan Snook
James Steeds
Elise Symonds
Herbert George Wall
Peter Webley
D Williams
Robert Willis
Robert Woolley

Chapter 11.10 Barrage Balloons Memories
from page 181
David Wintle
Paul Adams
Stephen Alexander
P.F.H. Clarke
Mona Duguid
Mike Farr
Tony Fowles
Peter Garwood
Donald Lear
Brian Long
Iris Mitchell
Marina Rich
Tony Riddell
Barbara Salter
Stan Snook

Chapter 11.11 BOAC Memories
from page 184
Jean Gillingham
Patricia Jean Morris
Doreen Whitfield

Chapter 12.4 Shelter Memories
from page 199
Mike Farr
Tony Riddell
Barbara Salter
Raymond Wade
Paul Adams
Valerie Barrett
Trevor Beacham
John Bidgood's brother
Colin Britton
Maureen Budd
Margaret Bygrave
Rosemary Clinch
M Crosier
Jenny Evans
Mike Farr
Mrs Ford
B Fouracres
Tony Fowles
Donald Lear
Marina Rich (née Cohen)
Tony Riddell
Barbara Salter
Norman Frank Thomas
Raymond Wade
Hilary Thyer (née Wood

Chapter 12.6 Portway Tunnel Memories,
from page 211
Colin Boyce
Maureen Budd
Ralph Smith
Mrs Gregory
Keith House

Chapter 12.7 General Wartime Memories,
from page 212
Stephen Alexander
Mr Bartlett
John Humphries
Ruth Trapnell
Mrs Deverson
Phyllis Farmer
Iris Mitchell
David Summers

Chapter 13.19 BBC Memories
from page 236
Arthur Bradley
Molly Foss
David Pearce
Frank Shepherd
Elizabeth Taylor
Ernie Fletcher
Alan Urwick
Cliff Voice
Tudor Gwilliam-Rees

Image sources

The following list of images are those provided by public archives.

Chapter 1
1.2 Old Hotwell House Fine Art (Bristol Museums and Art Gallery M923)
1.4 The Avon, with the New and Old Hotwell Houses, by Moonlight Fine Art (Bristol Museums and Art Gallery M969)
1.9 Map of Bristol, G.C. Ashmead showing pipes of Sion Spring Water Works, 1811, and Richmond Spring Water Works, 1815, also proposed line by Merchant Venturers bringing water from Avon Gorge 1833 40662/1
1.13 Bus for High Level Station and Clifton Rocks Railway (43207/9/35/40 Bristol Records Office)

Chapter 3
Clifton Rocks Railway – Plan And Sections Of Tunnel – Part 1 Of 3 (Bristol Records Office 42054.G.Drawer 4/18831)
3.13 Clifton Rocks Railway – Plan And Sections Of Tunnel – Part 2 Of 3 (Bristol Records Office 42054.G.Drawer 4/18832)
3.7, 3.9 Clifton Rocks Railway – Plan And Sections Of Tunnel – Part 3 Of 3 (Bristol Records Office 42054.G.Drawer 4/18833)

Chapter 4
4.9 Clifton Rocks Railway Loxton (Bristol Central Library)
4.29, 4.30 Company No: 41016; Clifton Rocks Railway Ltd. Incorporated in 1894. Dissolved before 1916 1894-[1916] (National Archives Kew BT 31/5845/41016)
4.42 Bristol Tramways and Carriage Company: Correspondence file regarding Clifton Rocks Railway– referred to in operation and in maintenance (Bristol Records Office 39735/8)

Chapter 6
6.12 Clifton Rocks Railway – Plan And Sections Of Tunnel – Part 2 of 3 (Bristol Records Office 42054.G.Drawer 4/18832)
6.2, 6.13 Bristol Tramways and Carriage Company: Correspondence file regarding Clifton Rocks Railway– (Bristol Records Office 39735/8)

Chapter 9 (images 9.14 and 9.22) Princes Buildings – Pump Room and Cinema – Munro & Son 1892-1893 (Bristol Records Office Building plan/Volume 28/70b, Sion Hill – Pump Room and Cinema Theatre – Munro for Newnes 1892-1893 (Bristol Records Office Building plan/

Volume 28/70c
Sion Hill – Pump Room and Cinema Theatre – C.A. Hayes & Son 1892-1893 (Bristol Records Office Building plan/Volume 28/70e

Chapter 10
10.2, 10.7 Clifton Rocks Railway (Bristol Records Office 42054.G.Drawer 2/08208)
10.9, 10.13, 10,15, 10.16, 10.17 Clifton Rocks Railway (Bristol Records Office 42054.G.Drawer 2/08153)
10.27 Clifton Rocks Railway (BBC Archives Caversham R35/207)

Chapter 13
13.22 Clifton Rocks Railway (BBC Archives Caversham R35/207)

Chapter 14
14.3 Clifton Rocks Railway: reinforcement work (Bristol Records Office 37167/97)

Index

pictures are in **bold**

Aberystwyth Cliff Railway 26, 31, **32**, 120
Air Ministry 152, 153, 171, 172, 177, 190, 220
Air-raids 167, 177, 186, 220
    Artefacts 191-198, **191-199**
    Bombs/bombing 150-155, 159-161, 171, 173, 176, 179, 183, 184, 186, 199, 201-218, 229, 241-249, 29, 298-305
    memories 199-210
    precautions 151-152, 189-190
    regulations 154
    shelters 7, 38, 150, 153, 154, 161, 165, 170, 177, 189-216, 251, **254**
    wardens 151, 189-191, **189**, 197, 200-13, 249, 300
Air Transport Auxiliary 173, 175, 184, 298, 300
Alexandra Palace, 'Ally Pally' 186, 244, 245, 297, 309
Anti-aircraft guns 176, 183, 208, 212, 309
Army Training Corps 239, 309
Auxiliary Fire Service (AFS) 181, 305, 309
Avon Gorge **10**, 11, 14, 36, 48, 88, 101, 120, 128, 181, 253, **254**, 280, 295, 300, 305, 308
Avon Gorge Hotel – *see also* Grand Spa Hotel 9, **10**, 11, **126**, 142, 187, **256**, 264-266, 286, 305

Barrage balloons 8, 150, 153, 159, 173-188, **174, 175, 178-180, 183,** 202, 205-207, 212, 280, 291, 295, 298, 299
    artefacts 180
    cable 173, 174, 178, 179, **180**, 181, 182, 183
    command 149, 174-179
    memories 181-188
    operation 173-174, 298
Barlow, Douglas **224**, 236, 237, 240, 241, 245, **245**
BBC 8, 97, 116, 152, 160-172, 177, 184, 185, 188, 200, 202, 206, 207, 209, 211, 215, 217-252, 264-266, 268, 274, 280, 282-284, 286, 290, 292, 298, 300, 301, 309
    aerials 228-229, **231**
    artefacts 197, 235, **235**
    canteen 223
    cable, electric and telephone 164, **165**, 172, **221**, **227**, 236, 240, 250, 286, 290, 291, 303, 304, 305
    control room **224**, 225, 226, 227, 232, 233, 234, 235, 236, 237, 238, 239, 240, 241, 242, 243, 244, 245, 246, 247, 248, 249, 250, 299, 307
    conversion 163-165
    Copenhagen Plan 304, 305
    crystal drive 225, **225**, 248
    generator *see* Lister diesel generator
    Home Service 150, 186, 210
    move to Bristol 153-156, 217
    memories 236-252
    ozoneator 164, 223, **223**, 299, 306

piano 168, 226, 227, 241, 245, 247, 251, 282
recording equipment 226, 227
    Phillips-Miller 226, 300
search for a secure site 217-219
studios 227-228, **228**
TD7 gramophone **228**, 237, 242, 247, 309
telephone system 163, 164, 168, 170, 171, 172, 220, **221**, 222, **224**, 225, 226, 227, 228, 240, 246, 248, 249, 286, 309
Third Programme 169, 170, 219, 220, 232, 233, 299
transmitting 165-166, 219-220, 224-229, 293, 299, 300, 304
transmitters
    Group H transmitter 165, 166, 168, 169, 171, 172, 219, 220, 225, **229**, 232, 233, 300, 307
    Harvey-McNamara transmitter 220, 228
    RCA 165, 220, **220**, 228, 229, **229**, 241, 244, 247, 307

Bell, David 266, 278
British Overseas Airways Corporation (BOAC) 152, **156**, 177, 189, 194, 205, 208, 309
    creation of 174
    operating from the Grand Spa Hotel 175, 299
    plans for wartime use of railway 161, 162, 164, 167, 170
    staff memories of wartime 175, 184-188
Bolton, James 13, 14
Boult, Sir Adrian 153, 218, 238, 241, 245
Bradley, Arthur 238-240
Bridgnorth railway 26, 27, 38, 120, 264
Bristol blitz 152, 160, 181, 186, 201, 204, 205, 211, 214, 215, 278, 280-282, 291
Bristol Corporation 14, 33, 80, 83, 87, 143, 150, 157, 162, 164, 166-169, 171, 211, 218, 232, 159, 304
    Town Clerk 161, 162, 164-167, 170, 304
Bristol Docks 14, 32-34, 49, 55, 81, 84, 88, 91, 174, 176, 200, 201, 204, 206, 207, 214-216, 242, 300
Bristol Junior Chamber 260, 262, 263, 304
    Plans for railway 260, 301
    Business plan 262, 263, 301
Bristol Supply System 169, 304
Bristol Tramways and Carriage Company (BTCC) 7, 34, **34**, 74-77, 79, 91, 96-98, 153, 154, 159, 161, 162, 164-167, **278**
Bristol Water Works Company 18
Bristol Zoo 57, 85, 116, 121, 201, 202, 212, 213, 216, 245, 299
British Air Transport Auxiliary 173, 175, 184, 298, 300
Brooke & Prudencio 192, **193, 194**

BTCC *see* Bristol Tramways and Carriage Company
Bull, Chris **274**
Cabot Quilt 304
Campbell's Steamers 88, 117, 118, 121
Cars 7, 8, 19-21, 38, 47, 54, 60-73,**62, 64,** 78, 79, 81, 82, 85, 87, 88, 91, 92, 94, 98, 114, 116-124, 130, 153, 154, 156, **163**, 164, 165, 200, 206-208, 214
    boarding of 61, 72, 121
    braking 66
        brake cylinder 67, 303
        brake windlass 303
        brakesman 24, 61, 67, **76**, 119, 122
        dead man's handle 24, **64**, 66, 67
        patents 24, 26
        gripper 66, 67, **68**, 72, 92, 95, 305
    capacity of 54, 61, 62, 72
    cables and wheels 65, **65,** 66
    chassis 62
    dismantling of 165, 168, 206
    operation of 62
    safety 66-68, 307
    scale models of
    speed governor 66, 67
    water balance principle 47, 54, 63
    water reservoir/tank 62, 63
Caledonia Place 16, 17, 19, 46, 47, 122, 124-126, 130, 175, 176, 179, 182, 189, 190, 203, 204, 207
Cinemas 82, 96, 145-147, 149, 185, 187, 246, 301
Civil Defence Corps 170, 189, 190
CDC *see* Civil Defence Corps
Clerk, Dugald 25, 65, 101
Cliff Railways 19, 21, 23, 25-28, 31, 36, 38, 49, 66, 69, 149, 264, 278, 295, 298, 299
Clifton College 152, 184, 202, 215
Clifton Hydro *see* Grand Spa Hotel
Clifton Rocks garage 8, 143, 147, 149, 175, 286
Clifton Rocks Railway Limited
    Creation 73
    Dissolution 75
Clifton Rocks Railway Trust 262, 278, 285, 286
    Creation 278, 295
    Objectives 267
    public survey 198, 298, 300, 301
    tours 260, 262, 274, 282, 290
Clifton Spa Pump Room and Rocks Railway Group 260, 262
Clifton Suspension Bridge 7, 11, 19-21, 27, 32, 34, 38, 47, 57, 82, 84, **84**, 85, 116, 120, 128, 133, 136, 152, 161, **176**, 183, 184, 185, 202, 205-207, 212 214, 218, 261, 266, 268, 286, 300, 305, 308
Colonnade, also Colonade **10, 12, 13,** 11, 12, 15, 26, 54, 60, 81, 90, 91, 134, 136, 137-140, 146, 204, 241, 247, 304, 308

314

Conservation area 11, 296, 300, 304
Cradock, George & Co 65, 96
Crossley engines 25, 36, 38, 58, 63, 65, 66, 91, 92, 101, 123, 134, 157, 284, 299, 306
Daly, Gerald 153, 160-163, 166, 168, 171, 175, 217, **218**, **220**, 225, 227, 228, 233, 234, 237, 239, 242, 244, 247, 300
Davey, Peter 266, **267**, 272, 285, **288,** 293
Doors Open Day, (see also Open Days) 31, 276, 277, 280, 282-285, 290, 294, 305
Downs, Clifton 46, 121, 186

Essential Works Order 185
European Broadcasting Union 304
EWO *see* Essential Works Order

Fighter Command 174, 219, 220
Foss, Molly 119, 223, 226, 233, 236, 240, 247, 250

Gabbitass , P. **49**
George, George 31, 32, **31**, 36, 38, 39, **50**, 95, 119, 120, 283, 284
George, Harold 119, 120
Gillard, Frank 164, 217, 225, 227, 229, 232, 240, 242, 248, 249
Grand opening 133-135
Grand Spa Hotel (Clifton Hydro), *see also* St Vincent Rocks Hotel and Pump Room 11, 16, **125, 129**, 142, 148, 149, 173, 175, 185, 187, 188, 190, 204, 208, 259, 299, 305
    changes of ownership 11, 148
    cinema 145
    concerts 79
        Hydro **142**, 306
        development 124, 140
        opening 125, 142, 143
        Pump Room (aka Ballroom) 8, 14, 17-19, 22, 27, 31, 47, **58**, 72, 76, 79, 124-150, **125, 129, 131, 133, 148,** 260, 262, 264, 290, 306
            closure 140
            concerts and entertainments 19, 134-137, 140, 148, 149, 179
            opening 125, 128, 133, 134, 137, 179
            construction 125
            conversion to cinema 82, 145, 146
            demolition proposed and campaign to prevent 260, 261, 266
            licensing difficulties 143, 144, 147
            listing of 264
            marble 19, 29, 126, 127, 132, **132,** 146, 260
            uses 135, 136
        Spa 141-149, **141**
            competition to 139
            creation 137
            gardens 140, 157
            health treatments 136, 137, 142
            opening 137, 140
            wartime use 173 et seq
            requisition 175

    return to civilian use 179
Gutta-percha 305
Hall, Johny **277**
Hayes, Christopher Albert 32, 33, **33,** 46-52, 54, 68, 70, 96, 127
Hewitt, Dominic 285, **291**, 292
Hewgill, Dave **274, 275**, 277
Hayward, Roy 246
Howard, Roger 266, 267, **267**, **269**, 285
Home Guard (LDV) 200, 213, 214, 218, 219, 240, 241, 250, 309
horse buses **20**, 34, 85, 87
Horse gin 14, 305
Hotwell springs 11, **12, 13**, 14
Hotwell Road 11, 13, 15, 19, 21, 34, 37, 39-41, 46, 47, 50, 53, 54, 55, 58, 60, 68, 69, 71, 80, 84, 95, 97, 116, 118, 121, 134, 153, 165, 167, 168, 177, 202, 205, 239, 253, 254, 260, 263, 276, 305, 307

Imperial Airways *see* BOAC
Landmark achievements of tunnelling and concrete 37, 38
Lease of the Clifton sites 80-82, 125, 135, 137, 144, 146, 147, 149
    to BBC 160-172, 232
    to Bristol Tramways 82, 150, 161, 162
    to Corporation 80, 150, 161-167
    to Clifton Rocks Railway 80, 82, 135, 147
    to Ministry of Works for Imperial Airways 161-167
    to Newnes 80, 137
    to the Hotel and Spa 140, 144, 149

LDV *see also* Local Defence Volunteer 213, 214, 309
Lister diesel generator 168, 241, 247
Local Defence Volunteer *see* Home Guard 182, 309
Local List 295, 297
Luckhurst, Pete **274**, 282
Lynton and Lynmouth funicular railway 21, 23-28, 38, 47, 49, 57, 62, 81, 104, 124, 264, 298
Marks, Sir George Croydon 21, 23-29, **25,** 36, 37, **37**, 47, 48, 50-52, 60, 62, 65, 68, 69, **70, 97,** 120, 127, 134, 135, 137, 138, 140, 284
    early years 24
    election as MP 26
    and cliff railways 25, 26
    and Clifton Rocks Railway 26, 38, 73
        difficulties 47, 68, 70
        first journey 55
    'Marks on Cliff Railways' 66
    and Matlock railway 27
    as patent lawyer 25-27, 69

Mary Celeste effect 295-297, 306
Matlock tramway 27, 47
Munro, Philip 18, 21, **23**, 28-32, **29**, 46-49, 52, 62, 66, 68, 70, 72, 80, 125-127, 134, 272, 286, 290, 309

MVR *see* Society of Merchant Venturers

Nelson, Kathleen 29, 30, **30**, 31
Newnes, Sir George 7, 18, 19, 21-28, **24**, 36, 46-50, 54, 55, 57, 58, **59**, 69, 70, 72-74, 81, 88, 91, 106, 124-128, 130, **130**, 286, **289, 290**, 296, 300, 309
    failure and death 24, 74, 144, 149
    as financier of railway 23, 295
    proposal to build railway 21, 22, 124
    as publisher 21, 124
    as shareholder 73
    and spa 21, 22, 126-138, **140**, 142, 149, 295

Observatory 19, 116, 117, 120, 176, 177, **177**, 182, 183, 184, 187, 201, 202, 203, 212, 213, 283, 285
Old Hotwell House 12, **12**, 14, **14**, 15, **126**
Open Days (see also Doors Open Day) 9, 35, 124, 267, 268, 271, 272, 276-278, 280-287, 290, 291-294, **292**
Otto engine 101, 306

Paperweight 49, **49**, 271, 272
Parker, E.T. 162
Passmore, Mark 282
Pearce, David 121, 236, 237, 241, 242, 250
Pearce, Walter 26, 60, 91, 92, 286, 314
Peel, Robert 263, 264, 266, **267**
Pembroke Road 152, 165, 212, 215, 238
Phillips-Miller recording equipment 165, 226, 227, 237, 243, 250, 300
Perkin, John **279**
Picken, John 287, **291,** 292
Point Villa 31, 32, **32,** 119
Port and Pier drinking fountain 286, **289**
Port and Pier Railway 20, 32, 34, 39, **43**, 80, 84, **84**, 124, 157, 160, 211
    construction 84
    demise 80, 85, 88
Portway, The 11, 34, 38, 87, 116, 118, 121, 207, 245, 282, 286
    construction of 32, 34, 38, 85
    Port and Pier Railway, impact on 80, 88, 295
    trams and buses on 87
Portway Tunnel 153, 160, 161, **209**, 211, 217, 245
    construction 38
    wartime use 160, 161, 170, 189, 211
    wartime memories 152, 203, 211-216
Pottery 110, 194
Preservation vs conservation: *see* chapter 16
Price, Mr F.J. 145
Princes Lane **37**, 41, 42, 47, 54, 57, **57**, **58**, 76, 96, 99, 124, 127, 135, **129**, 136, 142, 145, 155, 204, 209, 229, 253, **261**, 271, **272**, 274, 285, **289**, 296
Prince's Buildings 18, 19, 21, 27, 48, 57, 60, 124, 127, 135, 137-146, 204, 305

Queen Mary, HM 161, 211, 237, 241, 251,

315

RAFVR *see* Royal Air Force Voluntary Reserve
Railway
    artefacts 99-115
    buffer spring **93, 273**, 284
    cable 57, 64, 65, 66, 81, 93, 102, **103, 159, 270, 271,** 295, 299, 303, 305, 306, 307
    clack valve 63, **63**, 104, **104, 105**, 280, 299, 304
    construction 24, 27, 36-56
        accidents 50, 51
        blasting 39, 46, 49, 50, 51, 54
        compensation 52, 54
        completion 46
        contractors 47, 50, 54
        costs 36
        design and plans 37, **40, 41, 42, 43, 44., 45, 46**, 52, 53
        delays 36
        drills and drilling 36-39, **37**, 52-54,
        shot holes 37, 39, 53, **53**
        Tonite 39, 48, 307
    criminal damage to 82, 97, 98
    closure 74, 82
    colours 61
    demolition, proposed and
        campaign to prevent 260-262
    electricity 87, 165
    electrical fittings 114
    façades 57, 76
    fares 69, 71, 72, 78-80
    finance 72-78
    debentures and shares 73, 74
    operating profit 73, 76
    operating losses 74, 76
    future use 295-296
    gradient 17, 25, 27, 38, 39, 41, 61, 68, 69, 71, 163, 191, **221**
    ironwork 11, 108, 109, 11, 112, 108, 109
    journey times 55, 71
    lighting 61, **102**, 112, **113**
    Listing, failure of 264
    Listing local 295
    maintenance 91-98
    medallions **23, 29,** 30, 69, **70**
    memories 116-123
    models 277, **279,** 280
    neglect 82, 256-259
    opening day 68-72
    operating hours 80
    ownership 7, 47, 72, 73, 74, 97, 150, 151, 162, 167, 260, 261, 262, 263, 266, 268, 278, 280, 281, 286
    passenger numbers 78-80
    rails 20, 21, 25, 38, 47, 54, 56, **56**, 62, 66-68, 72, 76, 81, 96, 109, 118, 271, 299
    railings 57, **76**, 99, **102, 108**, 111, 154, 204, 265
    receivership 74, **74**, 75, 95
    restoration/conservation 112, 265-302

bottom station staircase 286
    railings 11, **112,** 267, 268, **268,** 276, 286, **287,** 294, 296
    signage 286
    track 262, **270,** 272, **274, 275,** 276
    turnstiles 297
    restoration clause 150, 162
    sale of railway 82
        of site 82
scissor gate **76,** 109
signage 57, **59, 77,** 197
signalling and communication 69
spiral staircase 131
staff 57, 60, **61,** 69,
    conditions and wages 60
    posts 60
stations
    upper/top **42, 51, 58, 59, 61,** 62, 63, **65, 76,** 99, 106, **109, 112,** 116, 117, 118, 119, 120, 125, 135, **154, 155,** 172, 177, 202, 207, 209, 202, 207, 209, 247, 260, 262, **267,** 282, **279,** 286, 295
    lower/bottom **13,** 15, 25, 29, 34, 38, 40, **41,** 44, 45, 57, **59,** 60, 63, **63,** 66, 77, 88, **97,** 105, 108, 117-121, 123, 134, 156, **157,** 162, **163,** 165, 171, 172, 200, 204, 207, **222, 255, 256,** 262, 263, 295,302
    Inspector 96, 307
    Manager 60, 91, 96, 307
    Master 211, 307
    tickets 49, 69, 79, **79,** 87, **107,** 13
    turnstiles 38, 55, 58, **59,** 70, 71, 76, **76, 77,** 99, 102, 106, 112, 119, 120, 121, 154, 295, 297, 299
    vending machines 59, 77
    water reservoir 41, 62, 105, 130
Rew, Tim 262, 263, 266, **267**
Rownham Ferry 88
Royal Clifton Spa **13,** 16
Russian baths 125, 136
St Vincent's Rocks Hotel see also Grand Spa Hotel 16, 18, 47, **124,** 143, 144
Scammell, Ed **273,** 274, 277, 280, 284, **291, 292**
Scammell, Tom **273,** 274, **274, 275,** 278
Secret Underground Bristol 199, 236, 250, 262, 276
Skempton, Prof. survey 116, 253, **255,** 284
Sion Spring 15-18, **15,** 127
Sion Spring House 15, 16, **16, 124**
Sion Hill Tunnel 16, 17, **17,** 110, 277
Society for the Protection of Ancient Buildings 297, 309
Society of Merchant Venturers 7, 11, 12, 14, 19-22, 34, 36, 46, 54, 73, 96, 97, 146, 148, 149, 153, 162, 164, 298, 306
    conditions imposed on construction work 22, 36, 46
    conditions imposed on development 22, 124, 125, 143
    dealings with George Newnes 21, 22, 46, 47, 50, 124-127, 135-137, 143

    development of Pump Room and mineral baths 124-127, 135-137, 140, 141, 143, 144
SPAB *see* Society for the Protection of Ancient Buildings
Statement of Significance 297, 298
Steadman, Bob **274**
Strawford, Dave **275, 291, 292**
Taylor, Mike 9, **269, 273, 274, 281,** 286, 287, **287, 288, 291, 292**
Tonkin, James 264, 266, **267, 271,** 280, 284, **288**
Trams/Tramway *see also Bristol Tramways* 19, 20, 27, 33, 34, 36, 47, 61, 70, 85, 117, 120, 307
    operating Times and Fares 79, 85, 87
    problems with 20, 27, 87
Transport links 57, 84-90, 150
Tuffleigh House **10,** 41, 42, 48, 54, 82, **124,** 137-139, 147, 149, 152, 175, 179, 204, 207, 259
Tunnel 16, **17,** 28, 31, 32, 36-56, **41, 42,** 58, 68, 70-73, 81, 82, **84,** 117, 119, 121, 122, 150-172, **159,** 184, 191, 192, 194, 199, 202-212, **221**
    bricks and brickwork 17, **40, 43, 45, 46**
    bonding 44, **44**
    skewed brick design and construction 40-42, **42,** 45, **45,** 46
    types of 44-46
    construction 17, 36, 46-55
    decision to line with brick 177
    dimensions 17, 40
    lighting **60,** 61, 71, 72, 74, 112, **113**
    profile 38-40, **43**
    reservoir 41, 47, 55, 57, **59,** 63, **63, 64,** 66, **105,** 112, 130, **259,** 277, 280, 290, 297, 299
    sections **37, 150,** 155, **178, 221**
    ventilation 36, 116, **178,** 183
    water chute 55, **63,** 297, 299
    water tank **64,** 47
    war-time conversion and partition of 8, 66, 104, 150-172
Tunnelling in Clifton area 38, 39
Turkish baths **10,** 17, 27, 29, 79, **125,** 127, 134, 136-138, 140, 142, **142,** 146, 149, 260, 306, 307

University of Bristol 8, 144, 280, 284, 285
Urwick, Alan 241, 250, 314

Vapour bath 13, 136, **142,** 308
Voice, Cliff 236, 241, 244, 247, 250, **251,** 314
volunteers and volunteering 7, 8, 262, 264, 266, 268, 272, 278, 280, 282, 283, 285, 287, 293, 294, 295, 298, 301, 304, 305

Washford radio station and museum 165, 217, 219, 235, 239-241, 246, 300
White, George Sir, MP 32-35, **33,** 57, 58, 82, 106, **114,** 124, 263, 296, 300
    and Bristol Aeroplane Company 34
    and Bristol Tramways Company 7,

    19, 32-34, 162
    death 87
    proposal to build a funicular railway in Clifton 7, 19
    purchase of Railway 27, 34, 74, 295
Whiting, Adam **288**
Wilson, Neil 217
Women's Royal Voluntary Service, WRVS 194, 209, 309
Woolley, Alfred 91-95, 101, 123, **123**
Woolley, Robert 91, 123
Woolley, James 123
Woolley, Robert 91, 123, 284, 314
Woolley Brothers Engineers 91-92, **91, 93**, 123
Yeatman, A.A. 73, 74, 91
Zig-zag, the **10**, 11, 12, **14**, 46, 55, 116, 117, 119, 120, 122, 124, **125**, 205-207, 214, 300, 308